Two-Fluid Model Stability, Simulation and Chaos

Martín López de Bertodano • William Fullmer
Alejandro Clausse • Victor H. Ransom

Two-Fluid Model Stability, Simulation and Chaos

Springer

Martín López de Bertodano
School of Nuclear Engineering
Purdue University
West Lafayette, IN, USA

William Fullmer
School of Nuclear Engineering
Purdue University
West Lafayette, IN, USA

Alejandro Clausse
School of Exact Sciences
University of Central Buenos Aires
& CONICET, National Atomic
Energy Commission
Buenos Aires, Argentina

Victor H. Ransom
School of Nuclear Engineering
Purdue University
West Lafayette, IN, USA

Videos to this book can be accessed at
http://link.springer.com/book/10.1007/978-3-319-44968-5_4

ISBN 978-3-319-44967-8 ISBN 978-3-319-44968-5 (eBook)
DOI 10.1007/978-3-319-44968-5

Library of Congress Control Number: 2016956260

Printed on acid-free paper

This Springer imprint is published by Springer Nature
The registered company is Springer International Publishing AG
The registered company address is: Gewerbestrasse 11, 6330 Cham, Switzerland

to our Parents

Foreword

Many important technologies of the late industrial revolution required an understanding of the underlying science and the ability to model difficult-to-describe materials. Out of this need evolved the engineering discipline of two-phase fluid dynamics. While two-phase flow and heat transfer technology was initially an empirical-based science, in the middle of the twentieth century research being done in Russia, Europe, and the USA (i.e., the development of drift-flux theory) and the publication of several significant books[1] led to a revolution in the development of the theory and solution methods for this emerging discipline. Two-phase flow models had the interesting feature that an overly simple model is ill-posed (see Chap. 3), possessing small-scale exponentially growing errors. This book provides a thorough exploration of the state of the art in, and the fundamentals of, two-fluid modeling (TFM).

It is significant that while many of the mathematical concepts discussed and explained in this book are familiar in the fields of engineering mechanics and applied mathematics, they are not necessarily familiar to two-phase flow researchers and practitioners. However, an understanding of these analytical procedures is essential if one is to develop physically based and reliable multidimensional two-fluid models. Indeed, numerous essential concepts are presented by the authors such as: what it means to have a well-posed mathematical model of two-phase flow, the dispersion relations and Eigenvalues (which are the high frequency limit of the roots of the dispersion relation) of various models and their relationship to linear stability, the implications of the numerical "fixes" (e.g., numerical viscosity, upwinding) that have been widely used in the past to compensate for incomplete physics in mathematical models of two-phase flow, some of the nonlinear instabilities that may occur (including supercritical Hopf bifurcations leading to limit cycles and potentially dangerous subcritical bifurcations which, for

[1] For example, G. Wallis, *One Dimensional Two-Phase Flow*; M. Ishii, *Thermo-Fluid Dynamic Theory of Two-phase Flow*.

large enough perturbations, can lead to divergent oscillations well before the linear stability threshold is reached, and period-doubling bifurcations leading to chaotic oscillations having the so-called strange attractor type of phase space topology), the relationship of these instabilities to Ledinegg-type excursive instabilities and density-wave oscillations (DWO), which are important instabilities that can be induced by the method of coupling of phase change devices, and, finally, the multidimensional lateral phase distribution phenomena which have been observed due to the omission of important nondrag forces (e.g., lift and dispersion) in multidimensional two-fluid models.

The authors present a specific approach to achieving well-posed two-fluid models through introducing more realistic physics into the models (e.g., interfacial pressure models and phasic collision forces). This type of approach is, of course, valuable and it appears that it could be further developed such that mechanistic flow regime transitions due to physically realistic instabilities of the void waves (i.e., the kinematic or continuity waves of the system of conservation equations) signaling the transition from one flow regime to another. If successful there will no longer be a need for the notoriously inaccurate flow regime maps which have been widely used in the past, and this would represent a welcome and long overdue advance in the field of phase change thermal hydraulics.

The authors also present an interesting summary of drift-flux formulations, how they are related to two-fluid models, and show how their algebraic closures lead to rather robust conservation equations which can be useful in many applications. In addition, they carefully differentiate between the instabilities inherent in the partial differential equations which describe two-phase flows and in the discrete numerical approximations of these equations. Indeed, there is a good review of the implications of the numerical "fixes" that have been made in the past for USNRC safety analysis codes such as RELAP5 and TRACE because of inadequacies in the physical models being evaluated. Suffice it to say that the numerics cannot compensate for situations when important physics is not adequately modeled. Moreover, it is potentially dangerous to attempt to do so.

While the book has not been specifically structured to address two-fluid model validation, it presents elements of the relevant approaches to do so. In particular, for a two-fluid model of two-phase flow to be valid, the mathematical formulation must satisfy the classical postulates of continuum mechanics, such as objectivity (i.e., invariant under coordinate system translation and rotation), and they must be well-posed. In addition, these models need physically based drag and nondrag forces which can be independently verified against separate-effect experiments (e.g., sonic velocity and critical flow tests to validate the virtual mass force formulation), and they must agree with the data from various experiments to verify the ability of the model to predict, for example, important phase separation and distribution phenomena.

Also, two-phase turbulence formulations are vital since they can influence phase distribution predictions. Much more research remains to be done in the area of two-phase flow turbulence modeling; however the structure presented in this book allows more advanced models to be incorporated and evaluated (e.g., the spectral two-phase turbulence models which are beginning to appear in the literature).

The authors are to be congratulated for compiling, discussing, and relating many disparate but important topics associated with the field of two-phase flow. We believe that this is a very useful book and that it will become an important reference for researchers and practitioners in the field of two-phase flow. We strongly encourage its adoption and use by our colleagues in this field.

Troy, NY
St. Augustine, FL
August 5, 2016

Donald A. Drew
Richard T. Lahey, Jr.

Contents

1 Introduction . . . 1
1.1 Summary . . . 1
1.2 Outline of Contents . . . 6
References . . . 8

Part I Horizontal and Near Horizontal Wavy Flow

2 Fixed-Flux Model . . . 11
2.1 Introduction . . . 11
2.2 Compressible Two-Fluid Model . . . 13
2.2.1 One-Dimensional Model Equations . . . 13
2.2.2 Characteristics . . . 14
2.3 Incompressible Two-Fluid Model . . . 17
2.3.1 One-Dimensional Model Equations . . . 17
2.3.2 Derivation of the Fixed-Flux Model . . . 19
2.4 Linear Stability . . . 23
2.4.1 Dispersion Relation for the Kelvin–Helmholtz Instability ($F = 0$) . . . 23
2.4.2 Dispersion Relation for the SWT Instability ($F \neq 0$) . . . 28
2.4.3 Sheltering Effect . . . 33
2.5 Numerical Stability . . . 35
2.5.1 Obtaining a Well-Posed Numerical Model . . . 35
2.5.2 First-Order Semi-Implicit Scheme (Inviscid) . . . 35
2.5.3 First-Order Semi-Implicit Scheme (with Viscous Terms) . . . 42
2.5.4 First-Order Fully Implicit Scheme (with Viscous Terms) . . . 44
2.5.5 Second-Order Semi-Implicit Scheme . . . 46

2.6 Verification . . . 49
2.6.1 Kreiss–Yström Equations . . . 49
2.6.2 Characteristic Analysis . . . 49
2.6.3 Dispersion Relation . . . 51
2.6.4 Method of Manufactured Solutions . . . 53
2.6.5 Water Faucet Problem . . . 57
2.7 Kelvin–Helmholtz Instability . . . 59
2.8 Summary and Discussion . . . 61
References . . . 62

3 Two-Fluid Model . . . 65
3.1 Introduction . . . 65
3.2 Incompressible Two-Fluid Model . . . 66
3.3 Linear Stability . . . 68
3.3.1 Characteristics . . . 68
3.3.2 Dispersion Analysis . . . 70
3.3.3 KH Instability . . . 71
3.4 Numerical Stability . . . 73
3.4.1 TFIT Two-Fluid Model . . . 74
3.4.2 Staggered Cell Structure . . . 74
3.4.3 First-Order Semi-Implicit Scheme . . . 77
3.4.4 Implicit Pressure Poisson Equation . . . 78
3.4.5 von Neumann Analysis . . . 79
3.4.6 Numerical Regularization . . . 82
3.4.7 Second-Order Semi-implicit Scheme . . . 84
3.5 Verification . . . 87
3.5.1 Sine Wave . . . 87
3.5.2 Water Faucet Problem . . . 88
3.5.3 Modified Water Faucet Problem . . . 91
3.5.4 Convergence . . . 92
3.6 Nonlinear Simulations . . . 93
3.6.1 Thorpe Experiment . . . 93
3.6.2 Viscous Stresses . . . 95
3.6.3 Wall Shear . . . 96
3.6.4 Interfacial Shear . . . 97
3.6.5 Single Nonlinear Wave . . . 97
3.6.6 Thorpe Experiment Validation . . . 98
3.6.7 Convergence . . . 100
3.7 Summary and Discussion . . . 103
References . . . 104

4 Fixed-Flux Model Chaos . . . 107
4.1 Introduction . . . 108
4.2 Chaos and the Kreiss and Yström Equations . . . 109
4.2.1 Nonlinear Simulations . . . 109
4.2.2 Sensitivity to Initial Conditions . . . 111

4.2.3 Lyapunov Exponent 112
4.2.4 Fractal Dimension 114
4.2.5 The Route to Chaos 116
4.2.6 Numerical Convergence 120
4.3 Fixed-Flux Model Chaos 122
4.3.1 Nonlinear Simulations with the FFM 122
4.3.2 Extension of Thorpe Experiment into Chaos 122
4.3.3 Fixed-Flux Model for Fully Developed Laminar Flow in a Pipe 124
4.3.4 Kelvin–Helmholtz Instability 128
4.3.5 Nonlinear Simulations 130
4.3.6 Lyapunov Exponent 132
4.3.7 Numerical Convergence 133
4.3.8 Fractal Dimension 134
4.4 Summary and Discussion 134
References 137

Part II Vertical Bubbly Flow

5 Fixed-Flux Model 141
5.1 Introduction 141
5.2 Compressible Two-Fluid Model 142
5.2.1 Compressible Model Equations 142
5.2.2 Virtual Mass Force 143
5.3 Incompressible Two-Fluid Model 144
5.3.1 Interfacial Pressure 145
5.3.2 Fixed-Flux Model Derivation 145
5.4 Linear Stability 148
5.4.1 Characteristic Analysis 148
5.4.2 Collision Force 149
5.4.3 Dispersion Relation: Kinematic Instability 151
5.4.4 Drag Force 152
5.5 Nonlinear Simulations 155
5.5.1 Stable Wave Evolution 155
5.5.2 Kinematically Unstable Waves in Guinness 158
5.6 Summary and Discussion 161
References 161

6 Drift-Flux Model 163
6.1 Introduction 163
6.2 Void Propagation Equation 165
6.3 Applications of Void Propagation Equation 167
6.3.1 Level Swell 167
6.3.2 Drainage 170
6.3.3 Propagation of Material Shocks 172

6.4 Dynamic Drift-Flux Model 173
6.4.1 Mixture Momentum Equation 173
6.4.2 Integral Momentum Equation 176
6.5 Delay Drift-Flux Model 179
6.6 Flow Excursion 184
6.6.1 Homogeneous Equilibrium Model 184
6.6.2 Drift-Flux Model 186
6.7 Density Wave Instability 187
6.7.1 Homogeneous Equilibrium Model 187
6.7.2 Transfer Function 189
6.7.3 Drift-Flux Model 192
6.8 Summary and Discussion 192
References 192

7 Drift-Flux Model Nonlinear Dynamics and Chaos 195
7.1 Introduction 195
7.2 Nonlinear Mapping of the Boiling Channel Dynamics 197
7.3 Model of a Boiling Channel with Moving Nodes 202
7.4 Dynamics of a Boiling Channel with an Adiabatic Riser 209
7.4.1 Summary of MNM Equations for the Channel-Riser System 211
7.4.2 Low Power Oscillations at Low *Fr* Numbers in a Heated Channel with Adiabatic Riser 213
7.4.3 Experimental Validation of Quasi-periodic Oscillations 218
7.5 Summary and Discussion 221
References 222

8 RELAP5 Two-Fluid Model 225
8.1 Introduction 225
8.2 Material Waves 226
8.2.1 RELAP5 Adiabatic Two-Fluid Model 226
8.2.2 Characteristics 228
8.2.3 Bernier's Experiment 230
8.3 Low Pass Filter Regularization of the TFM 233
8.3.1 Dispersion Analysis 234
8.3.2 Numerical Viscosity 236
8.3.3 Artificial Viscosity Model 240
8.3.4 Water Faucet Problem 242
8.4 Summary and Discussion 243
References 244

9 Two-Fluid Model CFD 247
9.1 Introduction 247
9.2 Incompressible Multidimensional TFM 248
9.2.1 Model Equations 248
9.2.2 Interfacial Momentum Transfer 249

9.2.3 Drag Force . . . 249
9.2.4 Lift Force . . . 250
9.2.5 Wall Force . . . 251
9.2.6 Laminar Pipe Flow . . . 252
9.3 RANS Two-Fluid Model . . . 253
9.3.1 Reynolds Stress Stabilization . . . 253
9.3.2 Single-Phase k–ε Model . . . 253
9.3.3 Two-Phase k–ε Model . . . 256
9.3.4 Decay of Grid Generated Turbulence . . . 257
9.3.5 Turbulent Pipe Flow . . . 261
9.3.6 Turbulent Diffusion Force . . . 265
9.3.7 Bubbly Jet . . . 267
9.4 Near-Wall Two-Fluid Model . . . 271
9.4.1 Wall Boundary Conditions . . . 271
9.4.2 Two-Phase Logarithmic Law of the Wall of Marie et al. (1997) . . . 271
9.4.3 Near-Wall Averaging . . . 273
9.4.4 Laminar Pipe Flow Revisited . . . 275
9.4.5 Turbulent Bubbly Boundary Layer . . . 276
9.4.6 Turbulent Pipe Flow Revisited . . . 278
9.5 URANS Two-Fluid Model . . . 280
9.5.1 Stability . . . 280
9.5.2 Constitutive Relations . . . 281
9.5.3 Plane Bubble Plume . . . 281
9.6 Summary and Discussion . . . 287
References . . . 288

Appendix A: One-Dimensional Two-Fluid Model . . . 293

Appendix B: Mathematical Background . . . 299

References . . . 347

Index . . . 351

Nomenclature

English

A	Cross-sectional area
$\boldsymbol{A}$	Matrix of coefficients of inertial terms
$\boldsymbol{B}$	Matrix of coefficients of convective terms
C	Coefficient of void gradient term in SWT and FFM
c	Wave speed, eigenvalue
C_0	Drift flux distribution parameter
C_{D}	Drag coefficient
C_{L}	Lift coefficient
Co	Courant–Friedrichs–Lewy number (Courant number)
C_{p}	Interfacial pressure coefficient
C_{S}	Smagorinsky model constant
C_{TD}	Turbulent dispersion coefficient
C_{VM}	Virtual mass coefficient
C_{wall}	Wall force coefficient
C_{μ}	k-ε model constant
D	Diameter
$\boldsymbol{D}$	Matrix of coefficients of diffusion terms
$\boldsymbol{E}$	Matrix of coefficients of stiffness terms
f	Friction factor
F	Algebraic source terms in momentum equation
Fr	Froude number
F_{u}	Derivative of algebraic source terms wrt u
g	Acceleration due to gravity
$\boldsymbol{G}$	Growth matrix
H	Channel height
$\boldsymbol{I}$	Identity matrix
I, i	Imaginary unit ($\sqrt{-1}$)

j	Volumetric flux
k	Turbulent kinetic energy, wave number
$\boldsymbol{M}$	Interfacial force per unit volume vector
N_{SUB}	Subcooling number
N_{pch}	Phase change number
p	Pressure
r_ρ	Density ratio
Re	Reynolds number
S	Source
t	Time
u_*	Friction velocity
$\boldsymbol{u}, u$	Velocity vector, x-velocity
$\boldsymbol{u}'$	Velocity fluctuation vector
u^+	Nondimensional velocity in wall units
V_{gj}	Drift velocity
v_w	Kinematic wave speed
x	x-Coordinate
y	y-Coordinate
y^+	Nondimensional distance in wall units

Greek

α	Void fraction
β	Two-phase friction velocity scaling factor
δ	Perturbation
$\varDelta$	Filter size
Δt	Finite difference time step
Δx	Finite difference space interval
ε	Turbulent eddy dissipation, void fraction diffusivity
θ	Angle of inclination of channel
k	von Karman constant, wave number
λ	Wave length, length scale
μ	Dynamic viscosity
ν	Kinematic viscosity
ρ	Density
σ	Surface tension
τ	Time constant
$\boldsymbol{\tau}$	Shear stress
ϕ	Dependent variable
ω	Growth rate
ω_L	Lyapunov exponent

Superscripts

coll	Collision
D	Drag
e	Exit
L	Lift
n	Time level
T	Turbulent component
TD	Turbulent diffusion
VM	Virtual mass
W	Wall
x	Two-phase
Δ	Finite difference scheme
*	Dimensionless

Subscripts

B	Bubble
BI	Bubble induced
i	Interfacial, cell center node, inlet
j	Junction node
k	Phase k
L	Left
LLE	Largest Lyapunov exponent
m	Mixture
n	Subcooled node
p	Pressure
R	Right, relative velocity, riser
r	Ratio, riser node
s	Subcooled
SI	Shear induced
VM	Virtual mass
w	Wave
x	x-Direction
y	y-Direction
0	Initial condition
1, l	Liquid phase
2, g	Gas phase
2ϕ	Two-phase
∞	Free stream

Abbreviations

1D	One-dimensional
2D	Two-dimensional

3D	Three-dimensional
CFD	Computational fluid dynamics
CFL	Courant–Friedrichs–Lewy criterion
DFM	Drift-flux model
DW	Density wave
FDE	Finite difference equation
FFM	Fixed-flux model
FFT	Fast Fourier transform
FOU	First order upwind
HEM	Homogeneous equilibrium model
KH	Kelvin–Helmholtz
KY	Kreiss–Ystrom
LES	Large Eddy simulation
MMS	Method of manufactured solutions
MNM	Moving nodes model
MUSCL	Monotonic upstream-centered Scheme for Conservation Laws
PDE	Partial differential equation
QUICK	Quadratic upwind interpolation for convective kinematics
RANS	Reynolds averaged Navier–Stokes
RDF	Radial distribution function
RELAP5	Two-Fluid Model nuclear reactor safety code
RHS	Right-hand side
SGS	Subgrid Scale
SMART	Sharp and monotonic algorithm for realistic transport
SWT	Shallow water theory
TFIT	Two-Fluid Model research code
TFM	Two-Fluid Model
TVD	Total-variation diminishing
URANS	Unstable Reynolds averaged Navier–Stokes

Chapter 1
Introduction

1.1 Summary

The stability of the Two-Fluid Model (TFM) is an outstanding problem since the inception more than 40 years ago. The difficulty stems from the combined challenges of turbulence in each fluid field and the dynamics of the interfacial structure, both of them nonlinear phenomena. This book addresses the linear and nonlinear two-phase fluid dynamic stability of the one-dimensional (1D) TFM material waves and the numerical methods used to solve it. The objective is to analyze two-phase linear and nonlinear physical stabilization and artificial regularization, separating two-phase structure from turbulence.

The TFM was first proposed by Landau in his Nobel theory of superfluidity of Helium-4 (1941). The first book on the TFM approach to two-phase flow analysis may be attributed to Wallis (1969), who also made ample use of the stable 1D Drift-Flux Model (DFM). The first book on the mathematically rigorous derivation of the TFM, starting from first principles, was Ishii's (1975), which also included the derivation of the dynamic DFM from the TFM. The dynamic 1D DFM had been previously applied by Ishii (1971) to the analysis of the density wave instability (DW). An early application of the 1D TFM was the RELAP5 nuclear reactor safety code of Ransom et al. (1982) for loss of coolant accident simulations. The first application of the DW analysis was performed by Lahey and Moody (1977) for the thermal-hydraulic stability of boiling water reactors. Finally, Kocamustafaogullari (1985) was the first to adopt a full 1D TFM formulation for the stability analysis of thin falling liquid films with interfacial shear and the results were very similar to those obtained with the Orr–Sommerfeld equation. The current book is an attempt to tie together these early and remarkable TFM advances with more recent research and to provide a consistent and complete picture of TFM fluid dynamic stability.

The TFM is used for several applications, both engineering and scientific. From an engineering standpoint it was originally promoted by the U.S. Nuclear Regulatory Commission for light water nuclear reactor safety design and in particular for

M.L. de Bertodano et al., *Two-Fluid Model Stability, Simulation and Chaos*,
DOI 10.1007/978-3-319-44968-5_1

the analysis of loss of coolant accidents. Since then it has been applied to boiling and condensation systems and two-phase flow transport systems in the power industry, the oil industry, the chemical industry, heat transfer, refrigeration equipment, etc. Furthermore, there is a flourishing particle transport community of scientists and engineers that have advanced the TFM theory and practice far beyond the early applications.

Rigorous derivations of the TFM, starting from first principles, have already been performed several times in the open literature and currently there exist three monographs that employ different averaging approaches starting with the Navier–Stokes equations: ensemble averaging by Drew and Passman (1999), time averaging by Ishii and Hibiki (2006), and volume averaging by Morel (2015). Furthermore, the particle physics community has developed their own TFMs starting from kinetic theory, see Garzó et al. (2012) for a recent example and review of previous works. Given these, there is no need to rederive the TFM and it is assumed that the reader is somewhat familiar with these TFM derivations. The present text book focuses instead on the stability of the TFM material waves and the approximations that best simplify the analyses. The approach is mechanistic rather than mathematical and the models are simplified as much as possible to clarify the complicated TFM stability behavior. Nevertheless, a significant amount of mathematics is inevitable in the linear stability analysis of the TFM partial differential equations, even when the greatest possible simplifications are applied, and the task has been facilitated by *Mathematica* (Wolfram Research, Inc., 2016). The nonlinear stability analyses are performed based on numerical simulation.

Starting with the 1D approximation the next simplification is incompressibility, of course, since it removes the acoustic characteristics and leaves only the material waves. While the global stability of the TFM material waves is well understood in terms of the DFM, the local stability has been more elusive. The difficulty arises from incomplete TFM averaging that removes physically stabilizing short wave mechanisms, e.g., surface tension and viscosity, and is most acute in the case of 1D models. In particular, the Kelvin–Helmholtz instability (KH) makes the material part of an Euler 1D TFM linearly ill-posed or elliptic (Gidaspow 1974), even though the acoustic part of the model remains well-posed or hyperbolic like the single-phase 1D Euler equations. The ill-posed problem is caused by the embedded KH instability of the material waves arising from the two distinct velocity fields of the TFM. In this book the KH instability is defined in the strict sense of Section 4 of Drazin and Reid (1981), as a transverse discontinuity in the velocity which only occurs at a vortex sheet or a vortex line for single-phase flow and is ill-posed for inviscid flow. We distinguish between it and the shear layer instability, which is generally referred to as the KH instability, where the velocity changes across a finite width and which is well-posed. A fundamental problem of the TFM is that the KH instability occurs throughout the whole domain unless the velocities of the two phases are equal. For single-phase flow, on the other hand, the KH instability is sustainable only in inviscid flow since viscosity turns a vortex sheet into a shear layer immediately. Of course for two-phase flow there is also an abrupt change in the density but this is not the source of the ill-posed condition. To deal with this problem artificial regularizations are often used to remove the KH instability, e.g., it is well known that the TFM may be hyperbolized by adding an artificial interfacial force that contains a void gradient term or parabolized with artificial viscosity. While regularization may be convenient, it is not strictly necessary. On the other hand, physical stabilization is not only desirable in principle but it is also feasible in practice.

The book focuses on two flow regimes: horizontal and near horizontal stratified wavy flow and vertical bubbly flow. Three instabilities are considered: the first two are the local Shallow Water Theory (SWT) and the KH instabilities, and the third is the global DW instability. The Ledinegg instability, which is also considered, may be viewed as a special case of the DW instability. These comprise a reduced number of all the possible 1D TFM instabilities, and an even smaller fraction of all possible two-phase instabilities, e.g., viscous, heat and mass transfer instabilities are not considered, neither are any nonequilibrium thermodynamics. Nevertheless, the chosen instabilities are sufficient to clarify the behavior of the TFM over a range of wavelengths that include local and global phenomena of engineering interest, and furthermore they address directly the ill-posed stability question. Four stability approaches are used: characteristics, dispersion analysis, the Laplace transform, and numerical nonlinear simulations. The von Neumann analysis, which is the numerical equivalent of the dispersion analysis, is employed for the numerical methods but we preserve the dispersion analysis nomenclature so we may compare the numerical stability to the differential model stability directly.

We first focus on the local stability of material waves in near horizontal stratified flow. In particular, we include viscous and surface tension terms to make the model linearly well-posed. But linear stability only prevents immediate blowup and an unstable well-posed linear model still exhibits exponential blowup. Therefore, the next step is to look into nonlinear stability where we apply the extrapolation to short wavelengths of Whitham (1974) for 1D SWT, even though the 1D TFM is strictly valid only for long wavelength phenomena. To support this assumption, it is demonstrated that the stable or kinematically unstable TFM for horizontal stratified flow may be reduced exactly to SWT using the fixed-flux condition, and that consequently the short wave nonlinear behavior of the TFM leads to the same material shocks and expansion waves as Whitham's SWT. We call this approximation the Fixed-Flux Model (FFM). This simplification of the TFM has been used in the past, but the complete stability analysis past KH and the mathematical association of the FFM and the DFM stability theories are new.

Although the FFM is analogous to SWT under certain operating conditions, the two-phase stability behavior differs because the FFM incorporates the KH instability, which introduces the unique TFM ill-posed condition. It will be shown that short wave physics remedy the situation and lead to a well-posed unstable model. Yet in practice the linear growth of the waves lasts a very short time before the evolution becomes nonlinear and the nonlinear behavior of the FFM past the KH instability remains almost unexplored. Remarkable exceptions are the pioneering mathematical studies of Kreiss and Yström (2002), Keyfitz et al. (2004), and more recently Fullmer et al. (2014). This book applies those studies to a case of near horizontal wavy flow. Our analyses show that a viscous 1D TFM that is KH unstable results in surface wave chaos; it is found that viscosity is involved in a key nonlinear mechanism that makes the FFM Lyapunov stable, i.e., that energy is transferred across wavelengths with a net nonlinear transport to short scales where it is dissipated, similarly to single-phase turbulence. The corresponding estimation of the Largest Lyapunov Exponent results in a growth rate which is one order of magnitude smaller than the linear counterpart, and eventually diverging trajectories become bounded. This typical nonlinear behavior has not been analyzed before for the TFM because the stability discussion has centered around the ill-posed condition. For example, the linear stability analysis of an unstable TFM with low viscosity performed by Drew and Passman (1999) shows that short wavelengths have fast exponential growth so the model is practically ill-posed. Even if surface tension and other interfacial forces slow down the growth, linear stability analysis is not enough because surface waves in nature stop growing when they either break or peak, and both of these phenomena are nonlinear. So the ultimate question is not whether the model is ill-posed but whether an unstable well-posed TFM is Lyapunov stable, i.e., whether there is a nonlinear mechanism to bound wave growth. Hence, it is desirable to move past the ill-posed issue in order to analyse the TFM using modern nonlinear stability theories.

The analyses developed for stratified wavy flow may be applied to vertical bubbly flow. The virtual mass force and the interfacial pressure are introduced to obtain a conditionally well-posed FFM. Then a collision induced pressure is considered and we adopt an interfacial collision force derived from the Enskog kinetic equation that removes the Kelvin-Helmholtz-type instability completely, resulting in an unconditionally well-posed model. It is demonstrated that these mechanisms are also associated with the acoustic and material wave speeds and so they have an effect on the fidelity of the model. Therefore, while the pursuit of a complete TFM may be impractical, it is at least possible to obtain a well-posed model with the correct wave speeds. Furthermore the model may become SWT unstable and exhibit nonlinear behavior similar to stratified wavy flow.

Second, we review the global stability of bubbly flow material waves in terms of the DFM. It is shown that the kinematic assumption applied to the TFM, instead of the fixed flux assumption, automatically removes the SWT and KH instabilities and results in the well-posed and hyperbolic DFM, which is therefore locally stable but globally unstable. Two cases of global instability are analyzed: flow excursion and density wave oscillations. It is shown that the DFM is the optimal TFM approximation to analyze global material wave instabilities in bubbly flows, precisely because it precludes the local instabilities specifically addressed by the FFM. Thus, the FFM and DFM models are counterparts that allow the analysis of two-phase local and global instabilities separately within a unifying framework. Furthermore, it is shown that applying the fixed flux and kinematic assumptions simultaneously results in the stable DFM of Wallis (1969) which is comprised of the nonlinear void propagation equation for bubbly flow. The stable DFM is of considerable engineering interest because of its simplicity and accuracy when there are no instabilities.

Third, we review the stability of some simple finite difference schemes used for the TFM because it is important to separate numerical stability from the stability of the differential model. Usually both get mixed up, in particular when numerical regularization is performed on an ill-posed TFM. We analyze the effect of numerical, i.e., artificial, viscosity because it not only regularizes the TFM, but it also provides artificial dissipation that saturates the growth of the nonlinear waves and usually dampens them altogether. However, there is a price to be paid: numerical convergence cannot be obtained beyond the KH instability, because numerical viscosity decreases with the mesh size so the wave growth rate increases as the mesh is refined. This leads to the problem of numerical convergence, which is of significance given today's emphasis on verification. We review the simplest finite difference first-order upwind methods and develop an original second-order finite-difference scheme suitable for the computations presented along the chapters. Higher order or more advanced numerical methods, like Lagrangian and spectral schemes, are not included here because they do not add significantly to the understanding of the stability of the differential TFM, which is the main focus. However, users and developers of more advanced methods should find the discussions in this book applicable to their pursuits.

The 1D-TFM assumption is used throughout the book because it leads to the simplest stability analyses. It also allows the separation of the two-phase instabilities of interest (SWT, KH, DW) from turbulence theory which does not apply to 1D. Furthermore, it has been shown by other researchers (Kocamustafaogullari 1985, Barmak et al. 2016) that, for specific cases of stratified two-phase flow, the 1D-TFM and the Orr–Sommerfeld stability analyses give similar results, i.e., they overlap, so the 1D stability analysis is not only appropriate but also fairly accurate. In addition, the analysis of TFM chaos would be computationally much more demanding with TFM CFD compared to 1D-TFM simulations. Also the 1D-TFM allows the mathematical definition and analyses of the ill-posed condition which is one of the driving motives of the book. In the end, 1D assumption is sufficient to advocate that the TFM need not be ill-posed, provided it includes enough short wave physics. Finally, the current 1D analyses demonstrate that a multidimensional approach does not per se render the TFM well-posed. In short, the 1D formulation has proved sufficient to probe the ill-posed question and the nonlinear behavior of the TFM in greater depth than has been done in the past.

Last, there is a concern when short wave phenomena are resolved by a 1D TFM, e.g., material shocks that go beyond the slope range of validity of the theory. After all both the 1D TFM and SWT are long wave theories which are valid for $L \gg D$. We approach this concern with the extrapolation of Whitham (1974) for 1D SWT: "Yet (wave) breaking certainly does occur and in some circumstances does not seem to be too far away from the description given by (the 1D SWT equations)...; moreover bores, breakers and hydraulic jumps are sometimes reasonably well described by (the 1D equations) ... But the shallow water theory goes too far: it predicts that *all* waves carrying an increase of elevation break. Observations have long since established that some waves do not break." This implies that the short wave extrapolation may be performed for some cases, and we take advantage of this

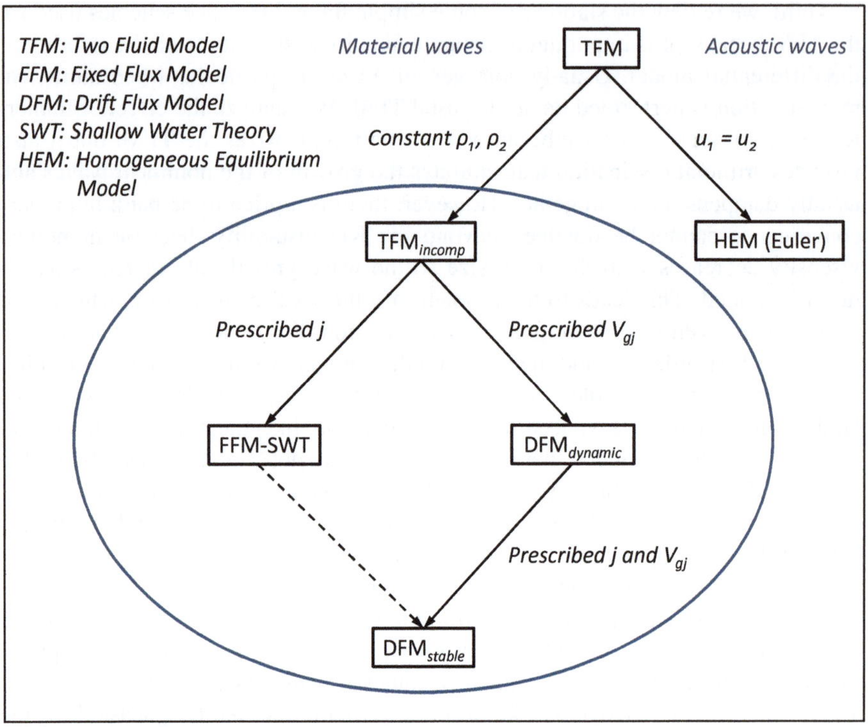

Fig. 1.1 Flow chart of simplified models derived from the TFM

for one case that is similar to Whitman's surface waves where it is possible to demonstrate that the nonlinear 1D TFM is Lyapunov stable and chaotic.

1.2 Outline of Contents

Figure 1.1 illustrates various 1D TFM simplifications addressed in this book. Two of them resemble the well-known single-phase SWT and Euler (or Navier–Stokes) equations, whereas the other two, the DFM and the FFM beyond the KH criterion, are distinct two-phase flow theories. The mathematical assumptions that isolate the different stability aspects of the TFM are indicated by the arrows. First, the homogeneous flow assumption leads directly to the Homogeneous Equilibrium Model (HEM), i.e., the TFM equivalent of the Euler equations of single-phase flow. While other assumptions may be made to analyze acoustic waves, e.g., the frozen flow assumption, the point is that the acoustic part of the 1D TFM is hyperbolic in general like the 1D Euler equations. On the other hand, the incompressible assumption removes the acoustic waves isolating the material waves which may not be strictly hyperbolic. As already mentioned there are two possible paths for material wave analysis. The kinematic assumption leads to the dynamic DFM which is hyperbolic. However, the fixed flux assumption leads to the FFM which is not hyperbolic beyond the KH instability but becomes Lyapunov stable if it is well-posed, i.e., hyperbolic-parabolic.

The book chapters follow approximately a counterclockwise trajectory around the ellipse of Fig. 1.1. Chapter 2 begins with the FFM derivation from the incompressible TFM, followed by the local linear stability analysis for horizontal and near horizontal wavy flows. In Chapter 3 there is a detour to the KH unstable incompressible TFM and then the trajectory is resumed toward the nonlinear analysis of the FFM which culminates in chaos in Chapter 4. Next the FFM methods are applied to bubbly vertical flows in Chapter 5. Then the DFM is derived and used to perform global linear and nonlinear stability analyses of vertical bubbly flows in Chapters 6 and 7. Finally the incompressible TFM stability analyses performed in the previous chapters are extended to two bubbly flow applications, the 1D RELAP5 code in Chapter 8 and some cases of TFM CFD in Chapter 9. Other than that the book is divided in two parts; horizontal and near horizontal wavy flows from Chapters 2 to 4 and vertical bubbly flows from Chapters 5 to 9. Chapters 1 through 7 are concerned with the analysis of the TFM and the FFM and DFM approximations. A broader spectrum of flow regimes may be found in Wallis (1969). Abstracts are given at the beginning of each chapter.

There are also two appendices. Appendix A is the formal derivation of the 1D TFM for near horizontal stratified flows starting from the TFM of Ishii and Hibiki (2006). Then, since this book is intended for an engineering fluid dynamics audience that may not be familiar with some of the theories, Appendix B presents a review of the stability of one equation wave models and SWT to introduce the analyses that are applied in various chapters. It starts with the one-way wave equation (Strang 2006) with the addition of viscosity and a source term. The basic linear methods of characteristics and dispersion stability analyses are introduced. Next the formation of nonlinear kinematic shocks is illustrated in terms of Burgers' equation, again a one-way wave equation model. The Drift Flux void propagation equation is introduced next and it is shown that it bears a strong resemblance to Burgers' equation. Moving to numerical methods, the Von Neumann analysis of the discretized one-way wave equation is presented and compared with the dispersion analysis for various numerical schemes to illustrate the relationship between the differential and the numerical stabilities. This is followed by the 1D SWT equations (Whitham 1974) dispersion analysis to identify the kinematic instability condition. This section also includes the Water Faucet problem of Ransom (1984), a traditional benchmark of the 1D TFM. Finally, there is a brief introduction to some specific mathematical tools of chaos theory that are used in the book.

Acknowledgments First of all we would like to thank our families, friends, and the many people, too many to mention, who provided support, read the manuscript, and offered comments.

We would like to express appreciation to Profs. Mamoru Ishii, Richard Lahey, and Donald Drew who contributed to the ideas in this book through instruction, mentoring, and many discussions over the course of the years. Former collaborators Prof. Min Chen and Dr. Deoras Prahbudarwadkar and former students Drs. Raj Krishnamurthy, Himanshu Pokharna, Brahma Nanda Reddy Vanga, and Raúl Marino also made significant contributions. Current students Trevor Kyle and Krishna Chetty helped with the figures and calculations. Above all, we owe a debt of gratitude to Dr. Avinash Vaidheeswaran who contributed heavily to the manuscript, particularly Chapters 5 and 9.

Special thanks go to Marta Moldvai, the Editor, who enabled us to publish this book. Finally, we are indebted to the sponsors and funding institutions that made the research possible: Drs. Stephen Beus and John Buchanan from Bettis Atomic Power Laboratory, the NEUP Program from the U.S. Department of Energy, Purdue University, and CNEA and CONICET from Argentina.

References

Barmak, I., Gelfgat, A., Ullmann, A., Brauner, N., & Vitoshkin, H. (2016). Stability of stratified two-phase flows in horizontal channels. *Physics of Fluids, 28*, 044101.

Drazin, P. G., & Reid, W. H. (1981). *Hydrodynamic stability*. Cambridge: Cambridge University Press.

Drew, D. A., & Passman, S. L. (1999). *Theory of multicomponent fluids* (Applied mathematical sciences). Berlin: Springer.

Fullmer, W. D., Lopez de Bertodano, M. A., & Clausse, A. (2014). Analysis of stability, verification and chaos with the Kreiss–Yström equations. *Applied Mathematics and Computation, 248*, 28–46.

Garzó, V., Tenneti, S., Subramaniam, S., & Hrenya, C. M. (2012). Enskog kinetic theory for monodisperse gas-solid flows. *Journal of Fluid Mechanics, 712*, 129–168.

Gidaspow, D. (1974). Round table discussion (RT-1-2): Modeling of two-phase flow. In *Proceedings of the 5th International Heat Transfer Conference*, Tokyo, Japan, September 3–7, 1974.

Ishii, M. (1971). *Thermally induced flow instabilities in two-phase thermal equilibrium*. Ph. D. Thesis, School of Mechanical Engineering, Georgia Institute of Technology.

Ishii, M. (1975). *Thermo-fluid dynamic theory of two-phase flow*. Paris: Eyrolles.

Ishii, M., & Hibiki, T. (2006). *Thermo-fluid dynamics of two-phase flow* (1st ed.). New York: Springer.

Keyfitz, B. L., Sever, M., & Zhang, F. (2004). Viscous singular shock structure for a non-hyperbolic two-fluid model. *Nonlinearity, 17*, 1731–1747.

Kocamustafaogullari, G. (1985). Two-fluid modeling in analyzing the interfacial stability of liquid film flows. *International Journal of Multiphase Flows, 11*, 63–89.

Kreiss, H.-O., & Yström, J. (2002). Parabolic problems which are ill-posed in the zero dissipation limit. *Mathematical and Computer Modelling, 35*, 1271–1295.

Lahey, R. T., Jr., & Moody, F. J. (1977). *The thermal-hydraulics of a boiling water nuclear reactor*. Chicago, IL: American Nuclear Society.

Landau, L. (1941). Theory of superfluidity of helium II. *Physical Review, 60*, 356–358.

Morel, C. (2015). *Mathematical modeling of disperse two-phase flows*. New York: Springer.

Ransom, V. H. (1984). Benchmark numerical tests. In G. F. Hewitt, J. M. Delhay, & N. Zuber (Eds.), *Multiphase science and technology*. Washington DC: Hemisphere.

Ransom, V. H., Wagner, R. J., Trapp, J. A., Carlson, K. E., & Kiser, D. M. (1982). *RELAP5/MOD1 code manual, NUREG/CR-1826-V1*. Washington, DC: Nuclear Regulatory Commission.

Strang, G. (2007). *Computational Science and Engineering*. Wellesley-Cambridge Press, Wellesley, Massachussetts.

Wallis, G. B. (1969). *One-dimensional two-phase Flow*. New York: McGraw-Hill.

Whitham, G. B. (1974). *Linear and nonlinear waves*. New York: Wiley.

Wolfram Research, Inc. (2016). *Mathematica*, Version 10.4, Champaign, IL.

Part I
Horizontal and Near Horizontal Wavy Flow

Chapter 2
Fixed-Flux Model

Abstract In this chapter, the compressible, one-dimensional Two-Fluid Model (TFM) for stratified two-phase flow is introduced first. For conditions of practical interest, a characteristic analysis demonstrates that the acoustic roots are always real and that the origin of ill-posedness (or well-posedness) of the model lies in the material roots. Therefore, compressibility is neglected and an incompressible model is used to derive a reduced two-equation model by applying the fixed-flux approximation, which is the key to study the local material waves in isolation, and is referred to as the Fixed-Flux Model (FFM). It is demonstrated that the FFM reduces to the well-known Shallow Water Theory (SWT) under some conditions. While it is not possible to cover all the possible local instabilities for stratified flow with the FFM, we are interested in two significant cases: SWT and Kelvin–Helmholtz (KH) instabilities, otherwise known as kinematic and dynamic instabilities.

Furthermore, the local linear material stability analysis allows us to address the question of the ill-posed TFM, which is caused by the KH instability which also precisely differentiates the FFM from SWT. The dispersion analysis of the FFM shows the well known results that the hydrostatic force makes the TFM stable up to the KH instability and surface tension makes the unstable model well-posed beyond it (Ramshaw and Trapp, Nuclear Science and Engineering, 66, 93–102, 1978). However, the well-posed FFM is still Lyapunov unstable and a bounding nonlinear viscous mechanism will be analyzed in Chap. 4, in terms of SWT material shocks (Whitham, Linear and Nonlinear Waves, Wiley, New York, 1974). Finally, the numerical stability and convergence of a few finite-difference schemes typically used to simulate TFM problems is also addressed with the von Neumann stability analysis of the FFM.

2.1 Introduction

Gidaspow (1974) was the first to perform a characteristic analysis of the incompressible potential flow Two-Fluid Model (TFM) and demonstrate that the problem is ill-posed. Furthermore, for compressible flow a characteristic analysis demonstrates that the acoustic roots are always real and that the origin of an ill-posed TFM lies in the material roots.

M.L. de Bertodano et al., *Two-Fluid Model Stability, Simulation and Chaos*,
DOI 10.1007/978-3-319-44968-5_2

First the compressible, one-dimensional characteristics for stratified two-phase flow are presented. Then an incompressible TFM is used to derive a reduced two-equation model by applying the fixed-flux approximation, which is the key to study the local material waves in isolation. Furthermore, the simplified Fixed-Flux Model (FFM) reduces to the well-known Shallow Water Theory (SWT) under some conditions.

Horizontal stratified wavy flows are among the simplest topologies of two-phase flows. Waves on oceans and rivers are some of the earliest phenomena analyzed and there is a wide body of literature. They are a good starting point to analyze the local stability of the TFM material waves. While we do not intend to cover all the possible local instabilities, encountered in stratified flows, which include several possible variations of the Orr–Sommerfeld equations (Barmak et al. 2016) as well as heat and mass transfer instabilities, we are interested in two significant fluid dynamics cases: SWT and Kelvin–Helmholtz (KH) instabilities, otherwise known as kinematic and dynamic instabilities. A good discussion on the fundamentals of kinematic waves (also referred to as continuity waves) and dynamic waves, and the associated instabilities, may be found in Wallis (1969).

The local material stability analysis allows us to address the question of the ill-posed TFM. This arises from averaging that removes several short wavelength physical mechanisms and is most acute in the case of one-dimensional models. The issue may be stated in its simplest form as follows: once the FFM goes beyond the KH instability the problem is no longer hyperbolic; it may be ill-posed elliptic or it may be well-posed in which case it is parabolic–hyperbolic. Sometimes an ill-posed TFM is regularized by adding significant numerical diffusion in which case the parabolic behavior damps the instability, others by adding an artificial interfacial force that returns the TFM to hyperbolic, in which case there is no instability. Therefore, the effect of regularization is to modify the local linear dynamics of the TFM.

A central objective of this chapter is to obtain an exact definition of an ill-posed TFM. First of all, it is important to remark that the ill-posed condition is a linear stability concept, and in the present case it is limited to the KH instability. However, the KH instability is not sufficient to make the TFM ill-posed. In fact, the problem arises only when a TFM is not sufficiently complete, i.e., it is possible to have a well-posed unstable TFM with a judicious choice of short wavelength physical mechanisms that balance the KH force. In the case of horizontal stratified flows, some of these mechanisms were identified early on: the hydrostatic force that makes the model stable up to the KH instability and surface tension (Ramshaw and Trapp 1978) or viscosity (Arai 1980), which make the unstable model well-posed beyond it. However, the unstable well-posed TFM remains Lyapunov unstable or practically ill-posed (Drew and Passman 1999, chap. 20), a problem that is left for Chap. 4.

First a two-equation incompressible FFM is derived from the incompressible isothermal four-equation TFM for horizontal stratified-wavy flow. It is shown that, below the KH instability limit, it reduces exactly to SWT for small density ratios. Well-known linear stability analyses may be applied that are relevant to stable and kinematically unstable cases, i.e., characteristics and the dispersion relation. The results are consistent with the TFM linear stability analysis of Barnea and

Taitel (1993) who identified the kinematic instability for stratified two-phase flows and Brauner and Maron (1993) who improved the analysis by including the wave sheltering effect. Eventually, these analyses may be extended with the nonlinear analysis of SWT (Whitham 1974) to material shocks and expansion waves for stable and kinematically unstable cases.

Secondly, the stability of a few finite-difference numerical schemes is analyzed. The development of a numerical scheme is complicated by the difficulty to separate physical, nonphysical, and/or numerical instabilities. It is well known that a numerical scheme must satisfy the von Neumann stability condition. In particular, we require that any numerical scheme be stable for the shortest wavelengths that can be represented numerically, i.e., $2\Delta x$, otherwise the discretization noise would be amplified until it dominates the solution. The specific objective is to develop a second-order finite-difference scheme for a well-posed model to perform nonlinear simulations that are reasonably accurate.

However, when the TFM is ill-posed, as in several industrial codes, it becomes necessary to use a numerical method that has sufficient damping at short wavelengths to suppress the unstable tendencies of the differential model. That is why we also review first-order upwind numerical schemes that have significant numerical, i.e., artificial, viscosity.

2.2 Compressible Two-Fluid Model

2.2.1 One-Dimensional Model Equations

A wavespeed analysis for compressible two-phase flow is performed with a TFM that is similar to Wallis (1969) separated flow model, to demonstrate that the acoustic characteristics of the TFM are always real. This is done to demonstrate that the ill-posed behavior of the TFM is only associated with the material waves, which is the concern of the rest of the book. For the characteristic analysis we only need to consider a simplified version of the 1D TFM for stratified flow, cf. Appendix A.

Generally, the density is given by the pressure and temperature through an equation of state. Here, we take the isentropic case so that the density material derivatives for compressible flow may be written as:

$$\frac{D_k\rho_k}{Dt} \equiv \frac{\partial}{\partial t}\rho_k + u_k\frac{\partial}{\partial x}\rho_k = \frac{1}{c_k^2}\frac{D_k p_{2\mathrm{i}}}{Dt}, \quad k = 1,2. \tag{2.1}$$

To include compressibility the speed of sound, given by $c_k = \sqrt{\partial p_k / \partial \rho_k}$, is needed to derive the material derivative of the density for each phase, where it has been assumed that $p_k = p_{2\mathrm{i}}$. Inserting this into the continuity equations yields

$$-\frac{D_1\alpha_1}{Dt} + \alpha_1 \frac{\partial u_1}{\partial x} + \frac{\alpha_1}{\rho_1 c_1^2} \frac{D_1 p}{Dt} = 0, \tag{2.2}$$

$$-\frac{D_2\alpha_1}{Dt} + \alpha_2 \frac{\partial u_2}{\partial x} + \frac{\alpha_2}{\rho_2 c_2^2} \frac{D_2 p}{Dt} = 0. \tag{2.3}$$

The momentum equations for this simple TFM for stratified flow only include the hydrostatic force, i.e., first-order derivatives only,

$$\rho_1 \frac{D_1 u_1}{Dt} = -\frac{\partial p}{\partial x} + \rho_1 g_y H \frac{\partial \alpha_1}{\partial x}, \tag{2.4}$$

$$\rho_2 \frac{D_2 u_2}{Dt} = -\frac{\partial p}{\partial x} + \rho_2 g_y H \frac{\partial \alpha_1}{\partial x}, \tag{2.5}$$

where α, ρ, u, p, and c are the time and space-averaged void fraction, density, velocity, pressure, and sound speed. The momentum equations remain the same as those in Appendix A because they are nonconservative so there are no derivatives of the densities. The void fractions are related by $\alpha_1 + \alpha_2 = 1$ and $g_y H$ is the product of the transverse gravitational constant and the channel height (Fig. 2.3).

Of course the analysis of the compressible 1D TFM is much more extensive and only the simplest possible case is considered in this section. For a thorough up-to-date analysis of the gas dynamics aspects of the TFM the reader is referred to Stadtke (2006).

2.2.2 *Characteristics*

The characteristics analysis follows those of Gidaspow (1974) and Stadtke (2006). Expanding the derivatives in Eqs. (2.2)–(2.5), the original system of partial differential equations can be cast into vector equation form as:

$$\boldsymbol{A}\frac{d}{dt}\underline{\phi} + \boldsymbol{B}\frac{d}{dx}\underline{\phi} = \boldsymbol{0}, \tag{2.6}$$

where $\underline{\phi} = [\alpha_1, u_1, u_2, p]^T$ is the vector of the independent variables. The coefficient matrices, obtained from Eqs. (2.2)–(2.5), are as follows:

$$\boldsymbol{A} = \begin{bmatrix} 1 & 0 & 0 & \frac{\alpha_2}{\rho_2 c_2^2} \\ -1 & 0 & 0 & \frac{\alpha_1}{\rho_1 c_1^2} \\ 0 & \rho_2 & 0 & 0 \\ 0 & 0 & \rho_1 & 0 \end{bmatrix}, \tag{2.7}$$

$$\boldsymbol{B}=\begin{bmatrix} u_2 & \alpha_2 & 0 & \dfrac{\alpha_2 u_2}{\rho_2 c_2^2} \\ -u_1 & 0 & \alpha_1 & \dfrac{\alpha_1 u_1}{\rho_1 c_1^2} \\ -\dfrac{1}{2}\rho_2 gH & \rho_2 u_2 & 0 & 1 \\ -\dfrac{1}{2}\rho_1 gH & 0 & \rho_1 u_1 & 1 \end{bmatrix}. \tag{2.8}$$

This system of equations must satisfy $\det[\boldsymbol{B}-c\boldsymbol{A}]=0$ for a nontrivial solution to exist. Neglecting the hydrostatic forces and making $u_1=u_2=0$, the solutions are $c=0,\ 0,\ +c_{2\phi},\ -c_{2\phi}$, where the first two characteristics are the void wave speeds and the last two characteristics are the (isentropic) acoustic wave speeds:

$$c_{2\phi}=\sqrt{\frac{\alpha_2\rho_1+\alpha_1\rho_2}{\dfrac{\alpha_2\rho_1}{c_1^2}+\dfrac{\alpha_1\rho_2}{c_2^2}}}. \tag{2.9}$$

This is the well-known sound speed of stratified flow with no relative velocity (Wallis 1969).

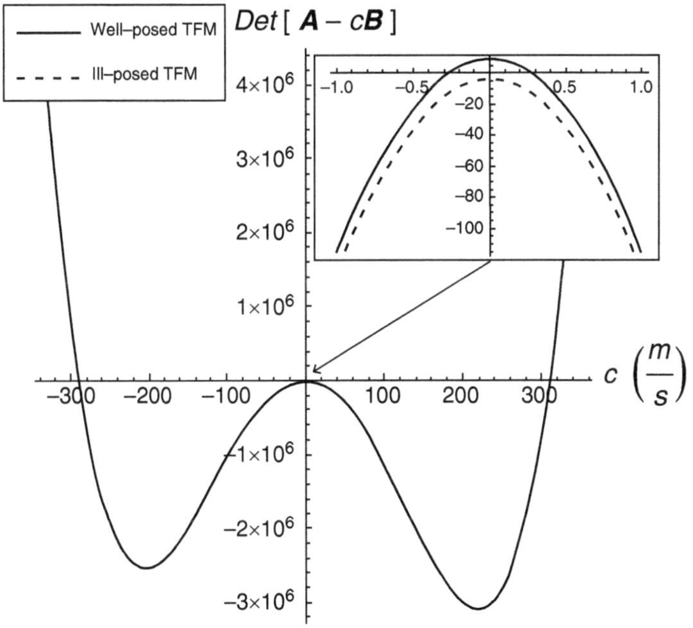

Fig. 2.1 Characteristic polynomial vs. wave speed

Of great significance to the present discussion is the characteristic polynomial including hydrostatic forces. Consider the full characteristic polynomial:

$$\det[\boldsymbol{A} - c\boldsymbol{B}] = \frac{1}{2}\alpha_1\alpha_2\left[\alpha_2\rho_1\left[\alpha_1 gH - 2(u_1 - c)^2\right]\right.$$
$$\left. + \frac{(u_2 - c)^2\left\{c_1^2\left[2\alpha_2\rho_1(u_1 - c)^2 - \alpha_1\rho_2 c_2^2 - \alpha_1\alpha_2\rho_1 gH\right] - 2\alpha_1\rho_2 c_2^2(u_1 - c)^2\right\}}{c_1^2 c_2^2}\right] \tag{2.10}$$

which is plotted in Fig. 2.1 for the case of stratified air–water flow at atmospheric conditions: $\rho_1 = 1000$ kg/m^3, $\rho_2 = 1$ kg/m^3, $c_1 = 1500$ m/s, $c_2 = 300$ m/s, $\alpha = 0.5$, $u_1 = u_2 = 0$, and $H = 0.03$ m. What is significant about this figure is that the characteristic equation, $\det[\boldsymbol{A} - c\boldsymbol{B}] = 0$, may have four or two real eigenvalues, depending on whether the system is well-posed or not. The two acoustic roots are approximately ± 300 m/s and the material wave speeds are approximately ± 0.3 m/s. Figure 2.1 also shows a close-up of the origin where it is seen that when the gas velocity is increased from $u_2 = 0$ to 10 m/s the polynomial no longer intersects the x-axis, so the material wave speeds are now imaginary (i.e., ill-posed) because the KH criterion has been exceeded. On the other hand, the acoustic wave speeds are always real. Therefore, when compressibility is included, the 1D TFM is always acoustically hyperbolic, as the 1D Euler equations. It is the TFM material waves that differ, and this occurs only beyond the KH instability. This aspect of the TFM is different from the Euler equations and it is better described by a model for surface waves, i.e., the FFM that is related to SWT and is developed in the next section.

In order to validate the model, the largest characteristic value is compared to the data of Henry et al. (1971) in Fig. 2.2. The phase velocities are set for a horizontal

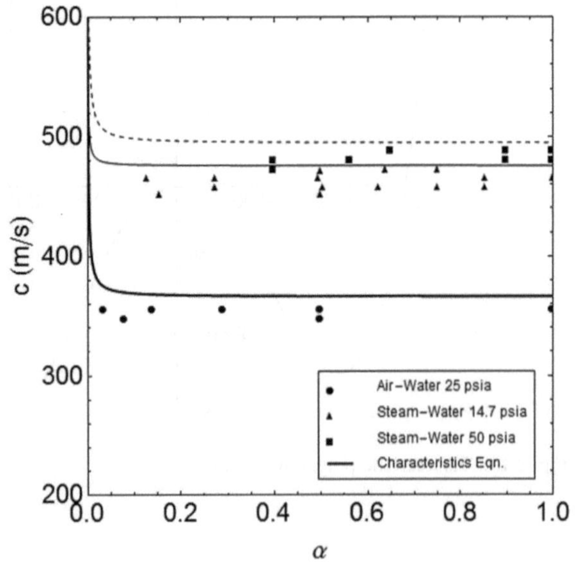

Fig. 2.2 Comparison of the sound speed model with the data of Henry et al. (1971)

pipe with $u_1 = 2$ m/s, $u_2 = 11$ m/s and $H = 0.05$ m. The graphic shows that the characteristic analysis compares well with the data. This comes as no surprise since the acoustic speed obtained with Eq. (2.6) matches the frozen separated sound speed model of Wallis (1969), which is known to work well for stratified flows.

Before moving on to an incompressible model it is worthy of note that industrial TFM codes with upwind first-order numerical schemes may predict acoustic waves quite accurately. The reason for this is that they have near second-order accuracy for acoustic propagation when the time step is sufficiently small so that the acoustic Courant number is significantly lower than unity (Tiselj 2000). This is necessary because these codes employ the pressure Poisson equation method which treats the pressure implicitly, cf. Sect. 3.4.3, and are prone to significant wave dispersion otherwise.

2.3 Incompressible Two-Fluid Model

2.3.1 One-Dimensional Model Equations

The analysis of the ill-posed TFM is performed with an incompressible model since it was previously established that the model acoustics are unconditionally well-posed and only the material waves render the system ill-posed. Therefore, the rest of the chapter, and the book, will be devoted to the analysis of the incompressible TFM so the analysis becomes simpler. The incompressible isothermal 1D TFM derived in Appendix A for horizontal stratified flow is a system of four partial differential equations (PDE). This model does not include compressibility but is otherwise more complete than Eqs. (2.2)–(2.5) because it includes additional terms for gravity, surface tension, and friction, some of which play a significant role in the stability of the model. Another simplification is that the velocity distribution parameters, i.e., shape factors, are assumed to be unity corresponding to uniform velocity profiles. This constraint will be removed in Chap. 4. Finally the viscous terms have been left out. The conservation of mass and momentum equations (A.22) to (A.25) applied to the flow between two parallel plates and excluding the viscous terms reduce to,

$$\boxed{\frac{D_1\alpha_1}{Dt} + \alpha_1\frac{\partial u_1}{\partial x} = 0,} \tag{2.11}$$

$$\boxed{-\frac{D_2\alpha_1}{Dt} + \alpha_2\frac{\partial u_2}{\partial x} = 0,} \tag{2.12}$$

$$\rho_1 \frac{D_1 u_1}{Dt} = -\frac{\partial p_{2i}}{\partial x} - \rho_1 g_y H \frac{\partial \alpha_1}{\partial x} + \sigma H \frac{\partial^3 \alpha_1}{\partial x^3} + \rho_1 g_x + \frac{1}{\alpha_1 H} \frac{f_i}{2} \rho_2 (u_2 - u_1)^2 - \frac{1}{\alpha_1 H} \frac{f_1}{2} \rho_1 |u_1| u_1, \tag{2.13}$$

$$\rho_2 \frac{D_2 u_2}{Dt} = -\frac{\partial p_{2i}}{\partial x} + \rho_2 g_y H \frac{\partial \alpha_2}{\partial x} + \rho_2 g_x - \frac{1}{\alpha_2 H} \frac{f_i}{2} \rho_2 (u_2 - u_1)^2 - \frac{1}{\alpha_2 H} \frac{f_2}{2} \rho_2 |u_2| u_2, \tag{2.14}$$

where α_k, ρ_k, u_k, and p_{2i} are the averaged void fractions, densities, velocities, and the lighter phase interfacial pressure, respectively. The subscripts $k = 1, 2$ indicate the heavier and lighter phases respectively, as shown in Fig. 2.3. The void fractions are related by the total void fraction condition, $\alpha_1 + \alpha_2 = 1$, and the remaining parameters, g, H, and σ, are gravity, channel height, and surface tension. The coefficients f_1,f_2, and f_i are the phasic wall and interfacial Fanning friction factors, respectively.

The first term on the RHS of the momentum equations is the pressure gradient. The next term is the hydrostatic force which is derived in Appendix A. In any industrial 1D TFM code the phasic pressures are assumed to be in equilibrium and the terms appearing in Eqs. 2.13 and 2.14 are the area averaged pressure whereas here it is assumed that the phases have distinct pressures that are related to a reference pressure with contributions from hydrostatic and surface tension forces. This is illustrated in Fig. 2.3. The effect of surface tension on the TFM was first analyzed in the open literature by Ramshaw and Trapp (1978), and it renders the model well-posed when KH is unstable.

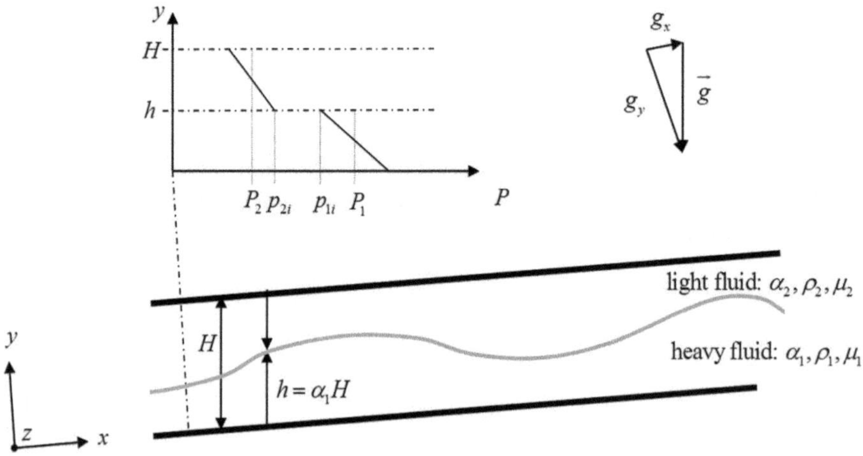

Fig. 2.3 Geometry and transverse pressure distribution for inclined channel flow, reprinted from Lopez de Bertodano et al. (2013), with permission from Begell House

2.3.2 Derivation of the Fixed-Flux Model

In spite of the simplifications made to derive the four PDEs of the previous section, they still present a formidable mathematical problem, even with the incompressible and isothermal assumptions. In this section, we make the additional fixed flux assumption and proceed to reduce the model to two PDEs following the approach of Holmås et al. (2008). Upon the further assumption of low density ratio the model may be simplified even more into the Shallow Water Theory (SWT) equations which have been the object of significant nonlinear stability research by Whitham (1974) among many others. The present derivation follows that by Lopez de Bertodano et al. (2013).

The first PDE is the sum of the two continuity Eqs. (2.11) and (2.12):

$$\frac{\partial}{\partial t}(\rho_1\alpha_1 + \rho_2\alpha_2) + \frac{\partial}{\partial x}(\rho_1\alpha_1 u_1 + \rho_2\alpha_2 u_2) = 0. \tag{2.15}$$

The second equation is the difference of the two momentum Eqs. (2.13) and (2.14), which eliminates the pressure:

$$\begin{aligned}\frac{\partial}{\partial t}(\rho_1 u_1 - \rho_2 u_2) + \frac{\partial}{\partial x}\left(\frac{1}{2}\rho_1 u_1^2 - \frac{1}{2}\rho_2 u_2^2 + (\rho_1 - \rho_2)g_y H\alpha_1\right) - \sigma H\frac{\partial^3\alpha_1}{\partial x^3} \\ = (\rho_1 - \rho_2)g_x - \frac{1}{\alpha_1 H}\frac{f_1}{2}\rho_1 u_1^2 + \frac{1}{\alpha_2 H}\frac{f_2}{2}\rho_2 u_2^2 + \left(\frac{1}{\alpha_1 H} + \frac{1}{\alpha_2 H}\right)\frac{f_i}{2}\rho_2(u_2 - u_1)^2.\end{aligned} \tag{2.16}$$

It is significant that the momentum equations are subtracted in this model because this links the local instabilities in some sense to the relative velocity. Two more equations are needed for closure. The first is the void fraction condition

$$\alpha_1 + \alpha_2 = 1 \tag{2.17}$$

and the second is the volumetric flux condition obtained from the time derivative of the void fraction condition combined with the continuity equations:

$$\frac{\partial}{\partial t}(\alpha_1 + \alpha_2) + \frac{\partial j}{\partial x} = 0, \tag{2.18}$$

where $j = \alpha_1 u_1 + \alpha_2 u_2$ is the total volume flux which, taking into account Eq. (2.17), ends up being a function only of time. We will further assume it to be constant in time so $j(x, t) =$ constant. This is the fundamental fixed flux assumption which greatly simplifies the TFM equations because it replaces the momentum PDE of the mixture, corresponding to the sum of the TFM momentum equations, with an algebraic equation. By doing so, it restricts the model from an engineering perspective. Furthermore, it removes the very long wave system instabilities, due to the dynamics of the total flux like flow excursions and density wave oscillations, but it preserves the local instabilities. The advantage for our purposes is that the single resulting momentum equation corresponds to the dynamics of the relative velocity which cause the TFM to be ill-posed.

We now follow the procedure of Holmås (2008) to recast this model in terms of the original variables $\underline{\phi} = [\alpha_1, u_1]^T$. First, Eqs. (2.15) and (2.16) are rewritten in matrix form, namely:

$$\frac{\partial}{\partial t}\underline{\psi} + \frac{\partial}{\partial x}\underline{\varphi} + \boldsymbol{E}\frac{\partial^3}{\partial x^3}\underline{\phi} = \underline{\varsigma}, \tag{2.19}$$

where

$$\underline{\psi} = \begin{bmatrix} \rho_1\alpha_1 + \rho_2\alpha_2 \\ \rho_1 u_1 - \rho_2 u_2 \end{bmatrix}, \underline{\varphi} = \begin{bmatrix} \rho_1\alpha_1 u_1 + \rho_2\alpha_2 u_2 \\ \frac{1}{2}\rho_1 u_1^2 - \frac{1}{2}\rho_2 u_2^2 + (\rho_1 - \rho_2) g_y H \alpha_1 \end{bmatrix}, \boldsymbol{E} = \begin{bmatrix} 0 & 0 \\ -\sigma H & 0 \end{bmatrix} \tag{2.20}$$

and the source terms are given by

$$\underline{\varsigma} = \begin{bmatrix} 0 \\ (\rho_1 - \rho_2) g_x - \frac{1}{\alpha_1 H}\frac{f_1}{2}\rho_1 u_1^2 + \frac{1}{\alpha_2 H}\frac{f_2}{2}\rho_2 u_2^2 + \left(\frac{1}{\alpha_1 H} + \frac{1}{\alpha_2 H}\right)\frac{f_i}{2}\rho_2 (u_2 - u_1)^2 \end{bmatrix}. \tag{2.21}$$

The next step is to convert this system of equations into primitive variables

$$\boldsymbol{A}\frac{\partial}{\partial t}\underline{\phi} + \boldsymbol{B}\frac{\partial}{\partial x}\underline{\phi} + \boldsymbol{E}'\frac{\partial^3}{\partial x^3}\underline{\phi} = \underline{F}, \tag{2.22}$$

where, applying the chain rule,

$$\boldsymbol{A} = \boldsymbol{I}, \boldsymbol{B} = \left[\frac{\partial \underline{\psi}}{\partial \underline{\phi}}\right]^{-1}\frac{\partial \underline{\varphi}}{\partial \underline{\phi}}, \quad \boldsymbol{E}' = \left[\frac{\partial \underline{\psi}}{\partial \underline{\phi}}\right]^{-1}\boldsymbol{E}, \ \underline{F} = \left[\frac{\partial \underline{\psi}}{\partial \underline{\phi}}\right]^{-1}\underline{\varsigma}.$$

The matrices $\frac{\partial \underline{\psi}}{\partial \underline{\phi}}$ and $\frac{\partial \underline{\varphi}}{\partial \underline{\phi}}$ of the chain rule derivatives are obtained using the void fraction condition, Eq. (2.17), and the volumetric flux condition, $j = \text{constant}$, such that:

$$\frac{\partial \underline{\psi}}{\partial \underline{\phi}} = \begin{bmatrix} \rho_1 - \rho_2 & 0 \\ -\frac{\rho_2 (j - u_1)}{\alpha^2} & \rho_1 + \frac{(1-\alpha)\rho_2}{\alpha} \end{bmatrix} \rightarrow \left[\frac{\partial \underline{\psi}}{\partial \underline{\phi}}\right]^{-1}$$
$$= \begin{bmatrix} \frac{1}{\rho_1 - \rho_2} & 0 \\ \frac{\rho_2 (j - u_1)}{\alpha(\rho_1 - \rho_2)(\alpha(\rho_1 - \rho_2) + \rho_2)} & \frac{\alpha}{\alpha(\rho_1 - \rho_2) + \rho_2} \end{bmatrix} \tag{2.23}$$

and

$$\frac{\partial \underline{\varphi}}{\partial \underline{\phi}} = \begin{bmatrix} u_1(\rho_1 - \rho_2) & (1-\alpha)(\rho_1 - \rho_2) \\ \dfrac{g_y H \alpha^3(\rho_1 - \rho_2) - \rho_2(j - u_1)(j - u_1(1-\alpha))}{\alpha^3} & \rho_1 u_1 + \dfrac{\rho_2(1-\alpha)(j - u_1(1-\alpha))}{\alpha^2} \end{bmatrix}. \tag{2.24}$$

From these equations it is possible to obtain the elements of matrix $\boldsymbol{B}$:

$$B_{11} = u_1, \tag{2.25}$$

$$B_{12} = \alpha_1, \tag{2.26}$$

$$B_{21} = \frac{(1 - r_\rho)(1-\alpha_1) g_y H - r_\rho (u_2 - u_1)^2}{1 - \alpha_1 + \alpha_1 r_\rho}, \tag{2.27}$$

$$B_{22} = \frac{(1-\alpha_1)u_1 + r_\rho \alpha_1 (2\, u_2 - u_1)}{1 - \alpha_1 + \alpha_1 r_\rho}. \tag{2.28}$$

For further simplification it is convenient to perform a Taylor series expansion of these elements in terms of the density ratio, $r_\rho = \rho_2/\rho_1$. Thus:

$$B_{21} = g_y H - \frac{(u_2 - u_1)^2}{1 - \alpha_1} r_\rho + O\left(r_\rho^2\right), \tag{2.29}$$

$$B_{22} = u_1 + \frac{\alpha_1}{(1-\alpha_1)}(2u_2 - u_1) r_\rho + O\left(r_\rho^2\right). \tag{2.30}$$

It is now assumed that $r_\rho \ll 1$, e.g., air water flows at atmospheric conditions, so the term of order r_ρ in B_{22} is neglected. In turn, the term proportional to r_ρ in B_{21} is preserved because we are interested in the Kelvin–Helmholtz instability, i.e., $r_\rho \frac{(u_2 - u_1)^2}{1-\alpha_1} > g_y H$. The low density ratio assumption will be removed in Chap. 4, where the FFM consisting of Eqs. (2.27) and (2.28) will be used, but for this introductory chapter it is convenient to work with the simplified matrix $\boldsymbol{B}$:

$$\boldsymbol{B} \cong \begin{bmatrix} u_1 & \alpha_1 \\ g_y H - \dfrac{r_\rho (u_2 - u_1)^2}{1 - \alpha_1} & u_1 \end{bmatrix}. \tag{2.31}$$

Finally, the surface-tension tensor is $\boldsymbol{E}' \cong \frac{1}{\rho_1}\boldsymbol{E}$, and the simplified source terms become

$$\underline{F} = \begin{bmatrix} 0 \\ F \end{bmatrix} \cong \begin{bmatrix} 0 \\ g_x - \dfrac{1}{\alpha_1 H}\dfrac{f_1}{2} u_1^2 + \dfrac{1}{\alpha_2 H}\dfrac{f_2}{2} r_\rho u_2^2 + \left(\dfrac{1}{\alpha_1 H} + \dfrac{1}{\alpha_2 H}\right)\dfrac{f_i}{2}\, r_\rho (u_2 - u_1)^2 \end{bmatrix}. \tag{2.32}$$

Therefore, the FFM may now be written in the familiar SWT form, including surface tension which is not usual but is needed for the following stability analysis,

$$\boxed{\frac{\partial \alpha_1}{\partial t} + u_1 \frac{\partial \alpha_1}{\partial x} + \alpha_1 \frac{\partial u_1}{\partial x} = 0,} \tag{2.33}$$

$$\boxed{\frac{\partial u_1}{\partial t} + u_1 \frac{\partial u_1}{\partial x} - C \frac{\partial \alpha_1}{\partial x} = \frac{\sigma H}{\rho_1} \frac{\partial^3 \alpha_1}{\partial x^3} + F,} \tag{2.34}$$

where

$$\boxed{C = r_\rho \frac{(u_2 - u_1)^2}{1 - \alpha_1} - g_y H.} \tag{2.35}$$

If C is negative and surface tension is neglected the two-equation model becomes the well-known 1D SWT equations. Furthermore, $C = 0$ is the approximate Kelvin–Helmholtz criterion, given by Eq. (2.147) in Ishii and Hibiki (2006), in the limit $r_\rho \to 0$:

$$(u_2 - u_1)^2 > \frac{1 - \alpha_1}{r_\rho} g_y H. \tag{2.36}$$

Therefore, if C is positive Eqs. (2.33) and (2.34) represent the KH unstable case which is no longer considered SWT, i.e., it is unique to the TFM. It is now possible to define the types of TFM waves and instabilities that will be analyzed in this chapter.

Following Wallis, kinematic waves are associated to the kinematic condition, i.e., $F(\alpha_1, u_1) = 0$, where the kinematic wave speed derived in Appendix B.5.1, is given by $v_w = -\alpha_1 \frac{\partial F / \partial \alpha_1}{\partial F / \partial u_1}$. If the KH condition is satisfied then dynamic waves occur with wave speed, derived also in Appendix B.5.1, given by $c = \sqrt{-\alpha_1 C}$. The respective instability conditions are $v_w > c$ and $C > 0$. The first, associated with kinematic waves, is referred to as the SWT instability in this book. The second, associated with dynamic KH waves, is the KH instability. These are the viscous and inviscid Kelvin–Helmholtz instabilities of Barnea and Taitel (1993). In section 2.4.2 it will be shown that the SWT instability occurs at lower flows than the KH instability. Because of the similarity between the FFM Eqs. (2.33) and (2.34) and SWT the linear and nonlinear behavior of the stable and kinematically unstable FFM (i.e., $C < 0$) may be understood in terms of SWT, e.g., Whitham (1974). If $C = 0$ and $F = 0$ the system becomes the Water Faucet Problem of Sect. 2.7 (Ransom 1984) which has been used for the verification of the TFM in nuclear reactor safety codes. The case $C > 0$ corresponds to the KH unstable TFM and is relevant for the analysis of two-phase flow in general and industrial codes in particular. However, the nonlinear behavior of the KH unstable TFM has not been explored beyond the pioneering mathematical analyses of Keyfitz et al. (2004) and Kreiss and Yström (2002). In this chapter, these works are extended to the physically relevant case of stratified two-phase flow.

It cannot be stressed too much that the present analysis for local instabilities is based on the fixed flux approximation. This approximation is removed in Chap. 6 where global instabilities will be analyzed with the Drift-Flux Model. It is remarkable that the kinematic assumption, used instead to derive the DFM, removes the local instabilities.

2.4 Linear Stability

2.4.1 *Dispersion Relation for the Kelvin–Helmholtz Instability* ($F = 0$)

Up to this point, we have reduced the TFM to a set of two equations for two material waves, which are remarkable similar to the equations of the SWT. The fact that the FFM resembles SWT is a major benefit because of the large body of existing work on the analysis of surface waves with SWT. We consider first the Kelvin–Helmholtz instability shown schematically in Fig. 2.4. The KH instability is defined in the strict sense of Section 4 of Drazin and Reid (1981) as a transverse discontinuity in the velocity which only occurs at a vortex sheet or a vortex line for single-phase flow and is ill-posed for inviscid flow. We should distinguish between the latter and the shear layer instability, which in classical fluid mechanics is usually referred to as the KH instability, where the velocity changes across a finite width and which is not ill-posed. This is the central issue, namely, the fundamental problem of the TFM is that the KH instability occurs throughout the whole flow field at all times unless the velocities of the two phases are equal. For single-phase flow, on the other hand, the KH instability is sustainable only in inviscid flow since viscosity turns the vortex sheet into a shear layer immediately. Of course for two-phase flow there is also an abrupt change in the density but that is not the source of the ill-posed condition.

Artificial and physical viscous terms are now added to Eqs. (2.33) and (2.34), respectively, in an ad hoc approximate manner. A more rigorous treatment is left for Chap. 4. While kinematic viscosity is a legitimate part of the TFM momentum equations, artificial viscosity in the continuity equation is not. However, it is

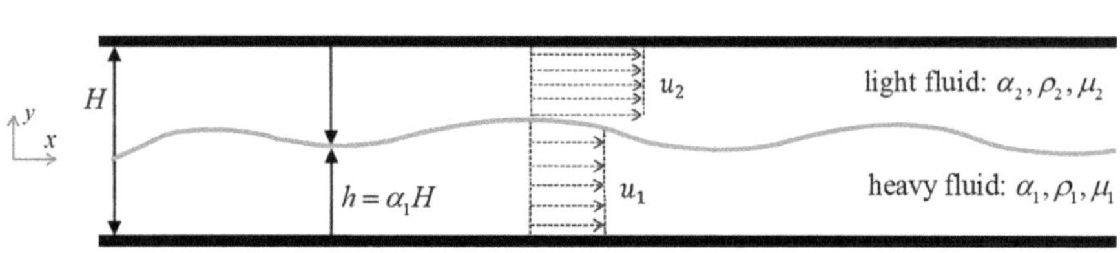

Fig. 2.4 Two-phase Kelvin–Helmholtz instability

included in this model because it is analogous to numerical viscosity often included in industrial codes to deal with the ill-posedness problem, and has a significant impact on the model stability. With these terms Eqs. (2.33) and (2.34) now become:

$$\frac{\partial \alpha_1}{\partial t} + u_1 \frac{\partial \alpha_1}{\partial x} + \alpha_1 \frac{\partial u_1}{\partial x} = \varepsilon \frac{\partial^2 \alpha_1}{\partial x^2}, \tag{2.37}$$

$$\frac{\partial u_1}{\partial t} + u_1 \frac{\partial u_1}{\partial x} - C \frac{\partial \alpha_1}{\partial x} = \nu \frac{\partial^2 u_1}{\partial x^2} + \frac{\sigma H}{\rho_1} \frac{\partial^3 \alpha_1}{\partial x^3} + F. \tag{2.38}$$

The viscous term in the momentum equation may include both physical and artificial viscosity.

We consider first the model with no friction terms, i.e., $F = 0$. The analysis follows that of Lopez de Bertodano et al. (2013). The two Eqs. (2.37) and (2.38) can be written as:

$$\boldsymbol{A} \frac{\partial}{\partial t} \underline{\phi} + \boldsymbol{B} \frac{\partial}{\partial x} \underline{\phi} + \boldsymbol{D} \frac{\partial^2}{\partial x^2} \underline{\phi} + \boldsymbol{E} \frac{\partial^3}{\partial x^3} \underline{\phi} = 0, \tag{2.39}$$

where $\underline{\phi} = [\alpha, u]^T$ is written dropping the subscripts such that $u = u_1$ and $\alpha = \alpha_1$. The matrix $\boldsymbol{D}$, which includes artificial and physical viscous terms in the continuity and momentum equations, is:

$$\boldsymbol{D} = \begin{bmatrix} -\varepsilon & 0 \\ 0 & -\nu \end{bmatrix}. \tag{2.40}$$

The characteristics, cf. Appendix B.2.2, given by the condition $\det[\boldsymbol{B} - c\boldsymbol{A}] = 0$ describe the behavior of the inviscid system such that:

$$C < 0 : c = u \pm \sqrt{|C|\alpha},$$
$$C = 0 : c = u,$$
$$C > 0 : c = u \pm i\sqrt{|C|\alpha}.$$

The first two cases are well-posed or hyperbolic and well understood. However, the last case is ill-posed elliptic, which leads to difficulties (Barnea and Taitel 1993) and will be analyzed next with a dispersion analysis.

The dispersion relation extends the results of the characteristic analysis at zero wavelength to the full spectrum of wavelengths as described in Appendix B.2.2. The analysis of Appendix B.5.1 is reproduced here, only now $F = 0$, but the effects of viscosity and surface tension are included. The first step is to linearize the two-equation system using $\underline{\phi} = \underline{\phi}_0 + \underline{\phi}'$ and keeping only first-order terms with respect to $\underline{\phi}'$. Then a Fourier solution is applied $\underline{\phi}' = \underline{\hat{\phi}}' e^{i(kx - \omega t)}$ to the linearized equations, where k and ω are the wavenumber and the angular frequency. The solution must satisfy:

$$\det\left[-i\omega \boldsymbol{A} + ik\boldsymbol{B} + (ik)^2\boldsymbol{D} + (ik)^3\boldsymbol{E}'\right] = 0 \tag{2.41}$$

in order to have a nontrivial solution. The dispersion relation is:

$$\omega = uk - \frac{i}{2}(\varepsilon + \nu)k^2 \pm k\sqrt{-C\alpha - \left[\frac{1}{4}(\varepsilon - \nu)^2 - \frac{\sigma}{\rho_1}\alpha H\right]k^2}. \tag{2.42}$$

The material wave speed is then given by $c_{\mathrm{w}} = \omega/k$:

$$c_w = u - \frac{i}{2}(\varepsilon + \nu)k \pm \sqrt{-C\alpha - \left[\frac{1}{4}(\varepsilon - \nu)^2 - \frac{\sigma}{\rho_1}\alpha H\right]k^2}. \tag{2.43}$$

For the inviscid case with no surface tension the wave speeds reduce to the characteristics, which are independent of wavelength. However, if there is viscosity or surface tension, different wavelengths travel at different wave speeds, i.e., the model becomes diffusive or dispersive. The critical point and the maximum growth rate (i.e., the most dangerous wave) may be obtained from $\omega_i = 0$ and $\dfrac{\partial \omega_i}{\partial k} = 0$, respectively. For the case where the viscosities are negligible the cut-off wavenumber reduces to $\sqrt{\dfrac{\rho_1 C}{\sigma H}}$ and the most dangerous wavenumber is

$$k_{\max} = \sqrt{\frac{\rho_1 C}{2\sigma H}} \tag{2.44}$$

with a corresponding wave growth rate,

$$\omega_{i,\ \max} = \sqrt{-\frac{\alpha\rho_1 C^2}{\sigma H}}. \tag{2.45}$$

It is worth mentioning that when the sign of g is reversed and $u_1 = u_2 = 0$ the 1D theory also predicts the critical Taylor wavenumber correctly and Eq. (2.44) predicts the most dangerous Taylor wavenumber within a factor of $\sqrt{2/3}$.

The KH stable case (i.e., $C \leq 0$) is valid up to gas speeds below the KH limit. The wave speeds in the limit of very small wavelengths (i.e., $k = 0$) are the characteristic speeds given by Eq. (2.43), which are real, so the stable model is hyperbolic. Hyperbolicity is an attribute of well-posed stable wave propagation models, cf. Appendix B.2.2. Stable waves propagate at the characteristic speeds but do not grow.

The inviscid Eqs. (2.37) and (2.38) without surface tension are the well-known SWT equations with linear stability characteristics that are very simple. However, the nonlinear behavior leads to material shocks equivalent to those appearing in the

Burgers equation of Appendix B.3.1. We shall address this nonlinear behavior throughout the book.

The results for finite wavelengths retain this characteristic behavior, but the effects of viscosity and surface tension make the system dissipative and dispersive, respectively. If the short wave diffusion and dispersion are neglected then

$$\frac{\omega}{k} = u \pm \sqrt{-C\alpha} \tag{2.46}$$

which is the equation for the characteristics.

The main difference between SWT and the FFM is that SWT does not account for the velocity of a second phase. Once the KH condition is exceeded this difference is fundamental. The dispersion analyses of the FFM including artificial viscosity, kinematic viscosity, and surface tension are shown in Fig. 2.5 for KH unstable conditions $H = 0.1\,\mathrm{m}$, $\alpha = 0.5$, $u_1 = 1\,\mathrm{m/s}, u_2 = 7\,\mathrm{m/s}$, $r_\rho = \dfrac{1}{50}$ and $\dfrac{\sigma H}{\rho_1} = 7 \times 10^{-6}\,\mathrm{m}^4/\mathrm{s}^2$. The basic, or Euler, 1D TFM is obtained from Eqs. (2.37) and (2.38) by setting $\varepsilon = \nu = 0$ and $\sigma = 0$. Beyond the KH limit the wave growth rates increase unboundedly as the wavelength shrinks to zero where there is a singularity. This singularity also occurs with the Euler equations but only for the case of the vortex sheet. The strict mathematical definition of ill-posedness is that at zero wavelength growth rate is infinite for any relative velocity other than zero. Furthermore, including wall and interfacial friction, even with unrealistically large coefficients values, the ill-posed nature of the dynamic instability at the zero wavelength does not change.

At this point of the analysis, it is relevant to compare the ill-posed TFM with the stability of the Euler and the Navier–Stokes equations of single-phase flow. First of all, both the 1D Euler equations and the 1D Navier–Stokes equations are not only

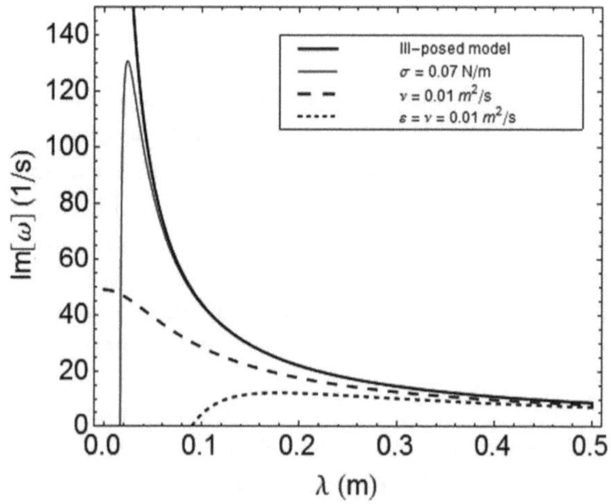

Fig. 2.5 Dispersion relation for FFM with $F = 0$, reprinted from Lopez de Bertodano et al. (2013), with permission from Begell House

well-posed but they are also stable. Furthermore, the multidimensional Euler vortex sheet is the only instance of an ill-posed single-phase flow problem (Drazin and Reid 1981), but it vanishes as soon as the shear-layer thickness becomes finite, i.e., the problem turns into an instability which is well-posed. Thus the Navier–Stokes equations are more physical in this sense because a vortex sheet becomes a shear layer immediately by the effect of viscosity, and so the problem is well-posed from the beginning. Therefore, the TFM has a unique condition compared to the Navier–Stokes equations because the averaging of the interfaces imprints the KH instability, i.e., the two distinct velocities that behave like an inherent and permanent vortex sheet. In this sense the ill-posed TFM question is relevant and the remedy will be pursued in this and further chapters adopting a variety of short wave physically stabilizing mechanisms that depend on the flow regime. Nevertheless, the linear growth rates of the well-posed models remain high. However, it is important to realize that the concern with the linear ill-posed issue of TFM stability goes too far sometimes, for it overshadows the more important issue of nonlinear stability. In the next two chapters it will be shown, with nonlinear simulations and analysis, that even when the initial growth rate of the waves is high a viscous mechanism stops the growth of the nonlinear waves.

The first such mechanism is surface tension. Figure 2.5 shows that it makes the model well-posed and introduces a cutoff wavelength. This cutoff wavelength is approximately 20 mm for this particular case of water–gas flow, i.e., $\sigma = 0.07\ \mathrm{N/m}$. Surface tension stabilization is the earliest published demonstration (Ramshaw and Trapp 1978) that the TFM may be rendered well-posed for unstable KH flow by including appropriate short wavelength physics. Nevertheless the maximum growth rate of the waves is high.

Furthermore Figure 2.5 shows that a very large kinematic viscosity of 0.01 m^2/s, makes the model well-posed too but with no cutoff, i.e., the growth rate is maximum and large at zero wavelength, $\omega_\mathrm{i} = \frac{\alpha_1 C}{\nu}$, which is unphysical and practically ill-posed.

Finally, industrial TFM codes are rendered well-posed adding numerical viscosity. Figure 2.5 shows the effect of a first-order upwind finite-difference discretization by arbitrarily setting a high artificial viscosity, $\varepsilon = \nu = 0.01\ \mathrm{m^2/s}$. The artificial viscosity values were selected to obtain a cut-off wavelength of a representative pipe diameter, i.e., $D = 0.1$ m. In addition to being well-posed, the model now exhibits and a cutoff wavelength for wave growth that sets a lower limit to the size of perturbations allowed to grow when the flow is unstable. The point of this exercise is that when the TFM is unstable, numerical viscosity makes it well posed. However, the cutoff wavelength may be arbitrarily adjusted changing the space and time discretizations. When the short wavelength components of the solution are not of interest, numerical regularization is often used, but it does not resolve the fundamental ill-posed behavior of the TFM in a physical sense and it precludes convergence beyond the KH criterion, as shall be demonstrated in Sect. 2.5.2.

The difference between an ill-posed TFM and an unstable one may now be addressed. It is clear from Fig. 2.5 that the TFM becomes ill-posed once the KH condition is exceeded. The physical reason in our case is that the averaging removes the interface with its shear layer, so it perpetuates the ill-posed KH condition, i.e., the vortex sheet, as opposed to the instability of the shear layer which is well-posed. Furthermore, it is well known that the TFM loses hyperbolicity in the process. However, the question remains: is the TFM unacceptable because it is not hyperbolic anymore? The answer depends on the choice of stabilization since the Fig. 2.5 also shows that the addition of short wave physics, in particular viscosity and surface tension for the purpose of this discussion, renders the model well-posed and, as we shall see when nonlinear stability is considered, well behaved. Furthermore, the physical mechanism that turns a vortex sheet into a shear layer in the Navier–Stokes equations is viscous diffusion, which is a parabolic process, and it is well known that the Navier–Stokes equations are hyperbolic–parabolic in addition to being well-posed. So a hyperbolic–parabolic viscous TFM should be a welcome alternative to an ill-posed model and preferable to a TFM that is artificially hyperbolic.

The next and more important question is whether a well-posed but unstable TFM is sufficient. Well-posedness, as defined in this section, is a linear stability property. Drew and Passman (1999) argued that a well-posed viscous TFM is still practically ill-posed because the wave growth rate remains high and the model has exponential blow up. Indeed the TFM would be unacceptable if the surface waves kept on growing unboundedly. But linear stability analysis turns out to be insufficient because unstable surface waves in their very nature either break or peak, and both these phenomena are nonlinear. So the ultimate question is whether a TFM is Lyapunov stable, i.e., whether there is a nonlinear mechanism to bound wave growth. It turns out that viscosity is involved in a key nonlinear mechanism that makes the largest Lyapunov exponent of wavy stratified flow one order of magnitude smaller than the maximum linear growth rate, as shall be demonstrated in Chap. 4, and bounds the growth of the waves after a certain point.

2.4.2 Dispersion Relation for the SWT Instability ($F \neq 0$)

We have considered the dynamic KH instability in the previous section. We now consider the kinematic instability caused by the wall and interfacial friction forces, i.e., the SWT instability, following the analysis of Lopez de Bertodano et al. (2013). Kocamustafaogullari (1985) was the first to adopt a full TFM formulation for the case of thin falling liquid films with interfacial shear. His analysis, which gives very similar results to Anshus and Goren (1966) using the Orr–Sommerfeld equation, includes the velocity shape factor term for the liquid film in addition to the effects of evaporation and condensation, which are not considered in this book. Later the TFM analysis for cocurrent adiabatic two-phase flow in a duct was performed by Barnea and Taitel (1993) who labeled this instability the Viscous Kelvin–Helmholtz instability. It is related to the kinematic waves described by Lighthill and Whitham (1955), also called void waves or continuity waves by Wallis (1969).

Before performing the stability analysis it is necessary to define a kinematic wave for the 1D FFM. To consider the mathematics that leads to a kinematic instability in the two-equation model we include the frictional and drag (i.e., viscous) forces, F, in the momentum equation and disregard viscosity and surface tension from Eqs. (2.37) and (2.38) to obtain the SWT equations, cf. Appendix B.5.1:

$$\frac{\partial \alpha}{\partial t} + u\frac{\partial \alpha}{\partial x} + \alpha\frac{\partial u}{\partial x} = 0, \tag{2.47}$$

$$\frac{\partial u}{\partial t} + u\frac{\partial u}{\partial x} - C\frac{\partial \alpha}{\partial x} = F, \tag{2.48}$$

where F is defined by Eq. (2.32). This results in the dispersion relation given by Eq. (B.56). We now apply the kinematic stability condition of Lighthill and Whitham (1955), i.e., $v_{\mathrm{w}} = c$. For the specific case of the frictional forces of the TFM we will use $F = F(\alpha_1, u_1, u_2)$ instead of $F = F(\alpha_1, u_1)$ to derive the kinematic wave speed. Then applying the definition of the kinematic wave speed (Wallis 1969) introduced in Appendix B.5,

$$v_{\mathrm{w}} = u + \alpha_1 \left(\frac{\partial u_1}{\partial \alpha_1}\right)_F \tag{2.49}$$

together with the fixed flux condition, it can be shown that

$$v_{\mathrm{w}} = u - \alpha_1 \left[\frac{(1-\alpha_1)\dfrac{\partial F}{\partial \alpha_1} + (u_2 - u_1)\dfrac{\partial F}{\partial u_2}}{(1-\alpha_1)\dfrac{\partial F}{\partial u_1} - \alpha_1\dfrac{\partial F}{\partial u_2}}\right]. \tag{2.50}$$

Combining Eqs. (2.49) and (2.50), an algebraic formulation for the kinematic wave speed may be obtained, which enables the kinematic instability condition, i.e., $c - v_{\mathrm{w}} = 0$, to be mathematically defined. But first it is convenient to obtain the dynamic instability condition in terms of the two independent variables, α_1, u_1. Starting with the kinematic condition, $F = 0$ with $g_x = 0$,

$$-(1-\alpha_1)\,\frac{f_1}{2}u_1^2 + \alpha_1\,\frac{f_2}{2}r_\rho u_2^2 + \frac{f_{\mathrm{i}}}{2}r_\rho(u_2 - u_1)^2 = 0 \tag{2.51}$$

yields u_2 in terms of $\alpha = \alpha_1$ and $u = u_1$, that is:

$$u_2 = \frac{f_{\mathrm{i}} r_\rho\, u + \sqrt{-r_\rho u^2 \left[f_2 f_{\mathrm{i}} r_\rho\, \alpha - f_1\,(f_{\mathrm{i}} + f_2\,\alpha)(1-\alpha)\right]}}{r_\rho(f_{\mathrm{i}} + f_2\,\alpha)}. \tag{2.52}$$

It can be seen that the term containing f_2 in Eq. (2.51) becomes negligible in the limit $\alpha \to 0$, which corresponds to low values of j_1. Neglecting the f_2 term greatly simplifies the analysis. The dynamic stability condition $C = 0$, with C defined by Eq. (2.35), results in the following critical liquid velocity for dynamic instability:

$$u_{\rm dyn} = \pm\sqrt{\frac{f_{\rm i}}{f_1} g_y H}. \tag{2.53}$$

The kinematic stability condition, $c = v_{\rm w}$, reduces into a quadratic equation for u:

$$f_{\rm i}\, gH\bar{f} - f_1\left[\bar{f} + f_{\rm i}\frac{9}{4}(1-\alpha)\alpha\right]u^2 = 0, \tag{2.54}$$

where

$$\bar{f} = \left(\sqrt{f_1(1-\alpha)^3} + \sqrt{f_{\rm i} r_\rho}\right)^2. \tag{2.55}$$

Finally, Eq. (2.54) may be solved to obtain the critical liquid velocity for the kinematic instability,

$$u_{\rm kin} = u_{\rm dyn}\sqrt{\frac{1}{1 + \frac{9}{4}\frac{f_{\rm i}}{\bar{f}}\alpha(1-\alpha)}}. \tag{2.56}$$

This simple analytic result leads to some straightforward insights for gas–liquid flows. First, $u_{\rm kin} \le u_{\rm dyn}$ unconditionally. Second, $u_{\rm kin} \to u_{\rm dyn}$ when $\frac{f_{\rm i}}{\bar{f}} \to 0$. For example, consider conditions $r_\rho = 0.001, f_1 = 0.005, f_{\rm i} = 0.014$ (Cohen and Hanratty 1965), and $H = 50$ mm, shown in Fig. 2.6. These values are similar to the TFM analysis of Barnea and Taitel (1993) for stratified flow shown in Fig. 2.7, the difference in the results is due primarily to the different geometries (i.e., parallel plates vs. a pipe). Furthermore, Barnea and Taitel (1994) performed an analysis with the FFM Eqs. (2.47) and (2.48) but with a slightly different form of Eq. (2.35), i.e., $C = r_\rho \frac{u_2^2}{1-\alpha_1} - (1 - r_\rho) g_y H$, for flow between parallel plates, where it is assumed that the gas velocity is much greater than the liquid velocity. For a round tube they got practically the same results as those in Fig. 2.7, which validates the use of the simplified FFM analysis given by Eqs. (2.47) and (2.48) instead of the TFM analysis.

Increasing the liquid viscosity by a large amount, from $\nu = 1 \times 10^{-6}$ m^2/s to $\nu = 500 \times 10^{-6}$ m^2/s, makes the kinematic and dynamic instabilities approach each other because $\frac{f_{\rm i}}{\bar{f}} \to 0$, see Eq. (2.56). Thus, the difference of the two stability

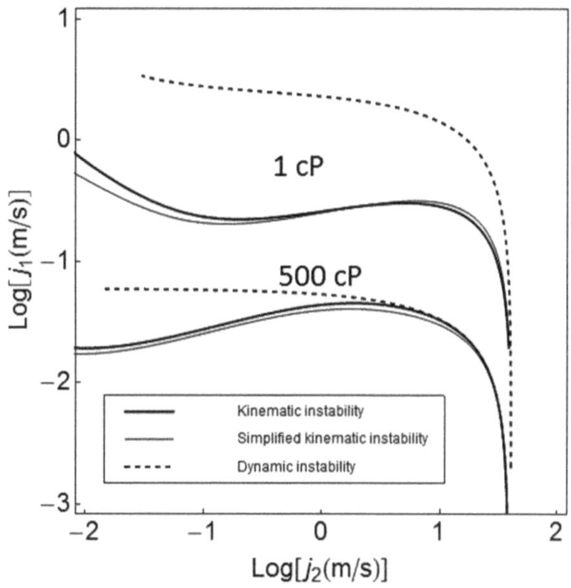

Fig. 2.6 Stability map for stratified flow between horizontal parallel plates, reprinted from Lopez de Bertodano et al. (2013), with permission from Begell House

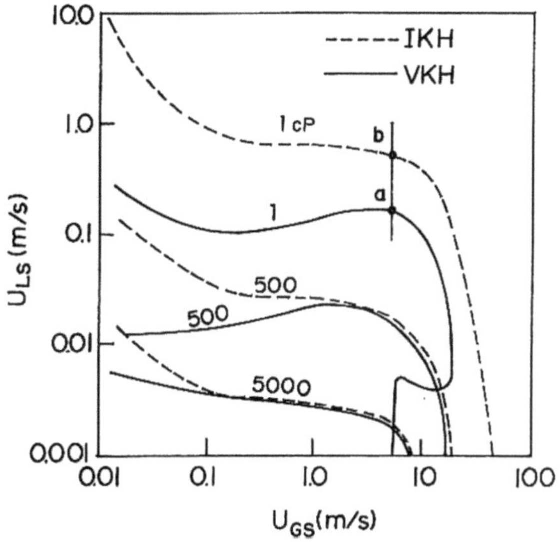

Fig. 2.7 Stability map for stratified flow in a horizontal pipe, reprinted from Barnea and Taitel (1993), with permission from Elsevier

boundaries is negligible for very viscous liquid flows, e.g., oil–gas flows vs. water–gas flows.

The stability result for the full Eq. (2.52), i.e., including f_2, way also obtained and a quadratic equation for u was derived, leading to an analytic solution considerably more complicated than Eq. (2.56), which will not be presented here. However, the results for $f_2 = 0.005$ are shown in Fig. 2.6 and they are very close to the simplified results with $f_2 = 0$.

Finally, it is desirable to demonstrate that the source term, F, may be neglected from the analysis when the flow becomes dynamically unstable, so that the dynamic stability analysis performed in Sect. 2.4.1 may be shown to stand on its own. The objective is to show that the kinematic instability becomes negligible once the KH condition is exceeded. This derivation follows the numerical results of Barnea and Taitel (1993) comparing the wave growth rates for the kinematic and dynamic instabilities for gas–liquid flows. It will be shown that the growth rate of the dynamic instability is at least one order of magnitude higher than the growth of the kinematic instability when these occur simultaneously, so the latter may be neglected.

The wave growth rate for the dynamic instability is given by Eq. (2.45),

$$\omega_{\mathrm{dyn,max}} = \sqrt{\frac{\alpha\rho_1}{\sigma H}\left[\frac{\rho_2}{\rho_1}\frac{(u_2-u_1)^2}{\alpha_2} - gH\right]^2} \sim \sqrt{\frac{\alpha\rho_1}{\sigma H}\left[g_y H\right]^2}. \tag{2.57}$$

The dispersion relation for the kinematic instability is given by Eq. (B.56),

$$\omega_{1,2} = uk + i\frac{F_{\mathrm{u}}}{2} \pm \sqrt{\left(i\frac{F_{\mathrm{u}}}{2} - ck\right)^2 + iF_{\mathrm{u}}(c - v_{\mathrm{w}})k} \tag{2.58}$$

which has the following approximate Taylor expansion in terms of $(c - v_{\mathrm{w}})$:

$$\omega_1 \cong \frac{F_{\mathrm{u}}}{2}\left[\mathrm{i}\frac{\dfrac{4c(c-v_{\mathrm{w}})k^2}{F_{\mathrm{u}}^2}}{\dfrac{8c^2k^2}{F_{\mathrm{u}}^2}+2} - \frac{2k(c-v_{\mathrm{w}})}{\dfrac{8c^2k^2}{F_{\mathrm{u}}^2}+2}\right] + (u+c)k. \tag{2.59}$$

The maximum wave growth rate then occurs in the limit as $k \to \infty$ which is given by

$$\omega_{\mathrm{kin,max}} = i\frac{F_{\mathrm{u}}}{4}\frac{(c - v_{\mathrm{w}})}{c} \sim -i\frac{F_{\mathrm{u}}}{4}. \tag{2.60}$$

Taking the derivative of Eq. (2.32), and assuming $f_2 = 0$, leads to

$$F_{\mathrm{u}} = -\left[\frac{1}{\alpha}\frac{f_1}{f_{\mathrm{i}}r_{\rho}} + \left(\frac{1}{1-\alpha} + \frac{1}{\alpha}\right)\sqrt{(1-\alpha)\frac{f_1}{f_{\mathrm{i}}r_{\rho}}}\right] f_{\mathrm{i}}r_{\rho}\frac{u}{H} \sim -\frac{f_1 u}{H\alpha} \tag{2.61}$$

because the second term in the brackets becomes negligible as $r_\rho \to 0$. Combining Eqs. (2.60) and (2.61) then yields an approximate kinematic wave growth rate,

$$\omega_{\mathrm{kin,max}} \sim \frac{1}{4}\frac{f_1 u}{H\alpha}. \tag{2.62}$$

The ratio of the dynamic and kinematic wave growth rates then is

$$\frac{\omega_{\mathrm{dyn,max}}}{\omega_{\mathrm{kin,max}}} \sim 4\alpha\frac{g_y H^2}{f_1 u}\sqrt{\alpha\frac{\rho_1}{\sigma H}}. \tag{2.63}$$

For water flow with $\alpha = 0.2$, $\nu = 1 \times 10^{-6}$ m^2/s and $j_1 = 0.1$ m/s, this ratio is approximately 1400, and for oil with $\nu = 500 \times 10^{-6}$ m^2/s the ratio is approximately 14. Therefore, for most cases of practical interest, once the dynamic instability is attained friction has little effect on the growth rate of the waves within the scope of the 1D TFM. This does not mean that friction terms are not important to the dynamics of the model, for they determine whether the flow reaches the KH instability, but they become irrelevant to the growth rate of the waves afterward and Eq. (2.42) suffices. Therefore, it is not necessary to develop a dispersion relation with the combined effects. In lieu of that, the dispersion relation is given by Eq. (2.58) when the liquid velocity reaches the kinematic instability velocity, i.e. Eq. (2.56), and by Eq. (2.42) when the liquid velocity is increased beyond the KH velocity, i.e. Eq. (2.52).

2.4.3 *Sheltering Effect*

While the FFM of the previous section is qualitatively correct for gas–liquid flows, it does not accurately predict the transition to wavy flow. To remedy the discrepancy Benjamin (1959) proposed a more complete picture of the interfacial force. There is a compression and expansion of the streamlines in the windward and leeward sides of the wave which leads to an increase and decrease in the interfacial shear respectively. Brauner and Maron (1993) incorporated this into the 1D TFM adding a void gradient in the interfacial shear:

$$\tau_{\mathrm{i}} = \frac{f_{\mathrm{i}}}{2}\rho_2(u_2 - u_1)^2 + C_{\mathrm{h}}\rho_2(u_2 - u_1)^2 H\frac{\partial \alpha_1}{\partial x}, \tag{2.64}$$

where the first term on the RHS is the original steady drag component and the second term represents the sheltering effect with the coefficient

$$C_h = C_{h0}\left(\frac{Re_1}{Fr_1^2}\right)^m, \quad m = \min[1,\ 1.565 - 0.072\ln(Re_1)]. \tag{2.65}$$

More recently, Kushnir et al. (2014) performed an Orr–Sommerfeld analysis that confirms this hypothesis, in particular for gas–liquid stratified flow. Furthermore, they obtained that the wall stresses also have a component that is in phase with the wave slope. This sheltering model is now incorporated into the FFM (Lopez de Bertodano et al. 2013).

The form of the coefficient, C_h, in Eq. (2.65) was determined by Brauner and Maron (1993) after an extensive comparison with experimental data. The leading coefficient, C_{h0}, was adjusted in the present model from 0.000245 up to 0.0004 to account for the difference in flow geometry, i.e., rectangular channel HAWAC data of Vallee (2010) vs. pipe flow data used by Brauner and Maron (1993).

Equations (2.64) and (2.65) were added to the FFM and the linear stability analysis of the previous section was repeated. The modified KH criterion, cf. Eq. (2.36),

$$(u_2 - u_1)^2 = \frac{1}{r_\rho}\frac{gH}{\frac{1}{1-\alpha_1} + C_h\left(\frac{1}{\alpha_1} + \frac{1}{\alpha_2}\right)} \tag{2.66}$$

was derived from the characteristic analysis.

The sheltering force has a considerable destabilizing effect. The SWT stability criteria with and without the sheltering effect are compared to the HAWAC data of Vallee (2010) in Fig. 2.8, where $j_k = \alpha_k u_k,\ k = 1, 2$. The stable region is in the lower left corner, the flow is SWT stable below the solid lines and KH unstable above the dashed line. It can be observed that all the wavy-stratified flow data,

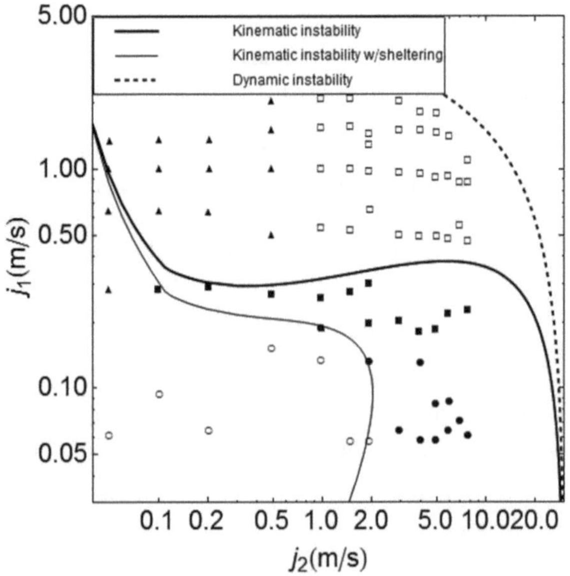

Fig. 2.8 Stability Map of the HAWAC experiment of Vallee et al. (2010). Symbols denote flow regimes: *open circle* stratified smooth, *filled circle* stratified wavy, *filled square* transition, *filled triangle* elongated bubble, *open square* slug, reprinted from Lopez de Bertodano et al. (2013), with permission from Begell House

which was previously in the stable region, is shifted to the unstable region. Therefore, including the effects of wave sheltering in the steady drag force marks a significant improvement to the prediction of unstable air–water stratified flow data.

2.5 Numerical Stability

2.5.1 Obtaining a Well-Posed Numerical Model

The ill-posed nature of an incomplete TFM has led to different approaches being used to obtain a well-posed numerical model. The first approach has been to seek added modeling detail to render the differential model well-posed. The general logic behind this approach is that since the physical problem and the Navier–Stokes equations are well-posed, something must have gone astray in the process of averaging the TFM to obtain a macroscopic description. Often the added differential model is artificial and its object is to hyperbolize the TFM. While this may not be physically correct it may be an acceptable compromise if supported by tuning with experimental data, at the cost of losing generality. An example will be given in Chap. 8.

The second approach has been to simply use numerical viscosity which damps the short wavelength components of the numerical solution. This approach parallels the methods which are used for modeling turbulent flow, in which the small-scale eddies responsible for viscous dissipation are not directly resolved, and damping at an intermediate wavelength is imposed through a viscosity correlation that acts as a filter providing dissipation and stability. In this case, the intermediate wavelength should be chosen short enough that the large-scale motion is not affected. The basic reason that this approach may be successful is the existence of the turbulent "energy cascade" in which it is not the short wavelength phenomena that govern the rate of energy dissipation, but rather it is governed by the rate at which energy is injected into the cascade by the long wavelength motion. Thus, it is only necessary to model the dissipation process at a sufficiently small scale so that the mean motion remains unaffected. While the incomplete TFM stability is not so simple, we may proceed with this idea for now and improve upon it later.

2.5.2 First-Order Semi-Implicit Scheme (Inviscid)

We choose a semi-implicit scheme for the FFM that is closest to is the one used in the TFM-TFIT code of Chap. 3. It is a combination of the explicit and implicit

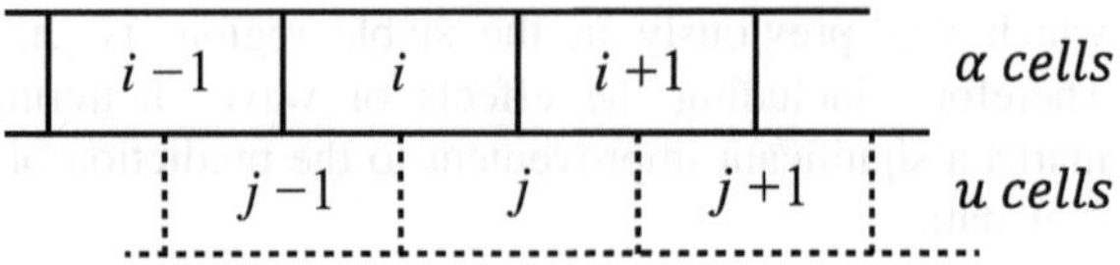

Fig. 2.9 Staggered continuity cells and momentum junctions

schemes presented in Appendices B.4.1 and B.4.2, respectively. The finite difference equations are as follows:

$$\frac{\alpha_i^{n+1} - \alpha_i^n}{\Delta t} + \frac{u_R^{n+1}\hat{\alpha}_R^n - u_L^{n+1}\hat{\alpha}_L^n}{\Delta x} = 0, \tag{2.67}$$

$$\frac{u_j^{n+1} - u_j^n}{\Delta t} + u_j^n \frac{u_j^n - u_{j-1}^n}{\Delta x} - C\frac{\alpha_R^n - \alpha_L^n}{\Delta x} = 0. \tag{2.68}$$

Equation (2.68) is solved first so that the updated velocities can be used in Eq. (2.67). Due to the staggered grid, shown in Fig. 2.9, some "R" and "L" data are available but $\hat{\alpha}_R^n = \alpha_{i+1/2}^n$ does not exist and it is "donored," i.e., for positive SWT wave speed $c = u + \sqrt{-C\alpha}$ and FOU, $\hat{\alpha}_R^n = \alpha_i^n$.

This scheme illustrates the effect that a staggered mesh has on consistency and stability. Equations (2.67) and (2.68) can be rearranged to show that staggering is equivalent to the introduction of higher order difference terms for finite mesh intervals; however, these terms do not always contribute a damping effect. We will use the terminology of "cell" to refer to the center of the control volume for mass, and "junction" to refer to the center of the momentum control volume at which the velocity is evaluated (i.e., the edges of the mass cells).

The stability property of a numerical scheme concerns the manner in which the numerical solution diverges away from the exact solution. In the case of linear systems it is possible to analytically establish the solution divergence and to develop necessary and sufficient conditions to ensure stability. In the more general case of nonlinear systems, it is necessary to linearize the equations in order to obtain analytical solutions for the differential and difference equations. Thus one cannot obtain, in general, neither necessary nor sufficient conditions to ensure a particular type of divergence between the "exact" and numerical solutions. Experience has shown that local stability in the von Neumann sense, Richtmeyer and Morton (1967), is necessary in most cases and in many, sufficient as well. The von Neumann criterion for stability, described in Appendix B.4, requires that the magnitude of the eigenvalues of the amplification matrix, $\boldsymbol{G}$, satisfies the inequality Eq. (B.30):

$$|\boldsymbol{G}|_{\max} \leq 1 + O(\Delta t), \tag{2.69}$$

where $|G|_{max}$ is the maximum eigenvalue of G. The growth matrix G for the difference scheme is defined by the recursion relation

$$\underline{\phi}^{n+1} = G\underline{\phi}^{n}. \tag{2.70}$$

Equation (2.69) is called a stability condition, but in reality it is a requirement that the difference problem be well-posed, i.e., the wave growth rate remains bounded in the limit as the mesh is refined to zero. This is consistent with the well-posedness criterion that the growth rate of the differential model must remain finite at all wavelengths,

$$|G|_{max}^{N} \equiv e^{-I\omega_{I}^{\Delta}t} \cong \prod_{n=0}^{N} \left(1 \pm |\omega_{I}^{\Delta}|\Delta t_{n}\right), \tag{2.71}$$

where the sign depends on whether the *differential problem* is stable or unstable. The eigenvalues of G are functions of the Fourier component wavelength and, for a given node network, the spatial wavelengths of interest range from $2\Delta x$ to infinity. A difference scheme is unconditionally stable if Eq. (2.69) is satisfied or all values of the time step, Δt. For hyperbolic systems without source terms the stability criterion, Eq. (2.69), can be reduced to $|G|_{max} \leq 1 + O(\Delta t)^{m}$, where m is the order of the truncation error and is at least two for any practical numerical scheme. Only when source terms are present, e.g., the kinematic instability discussed in Sect. 2.4.2, it is necessary to reduce the stability requirement to Eq. (2.69). As long as source terms are approximated consistently in the difference scheme, they do not affect stability.

To clarify the relationship between the numerical scheme and the differential equation system, consider Fig. 2.4. Here the growth coefficients of the differential equation, $-I\omega_{I}^{\Delta} > 0$, for the ill-posed ($\varepsilon = \nu = 0$) and well-posed ($\varepsilon, \nu > 0$) cases, are plotted at a fixed time for the range of wavelengths of interest. For all systems the long wavelengths have a growth approaching the one due to the K–H instability alone. However, for the ill-posed case, additional unbounded growth occurs at short wavelengths, whereas for the parabolic case the growth is mitigated due to the $e^{-\frac{1}{2}(\varepsilon+\nu)k^{2}t}$ term from the dispersion relation, Eq. (2.42). Therefore, if the differential equations are hyperbolic or parabolic the growth rate is bounded and the problem is well-posed. Any numerical solution scheme that satisfies Eq. (2.69) will then result in a well-posed numerical problem.

In turn, if the differential equations are ill-posed it is not clear what requirement should be placed on a numerical solution scheme. Such a case may occur when some real physical process has been omitted or modeled inappropriately by filtering relevant mechanisms, e.g., an incomplete TFM. If one is sure that the long wavelength behavior described by the ill-posed system is basically correct and the short wavelength behavior is not crucial, then the system can be modified arbitrarily in the short wavelength range to obtain bounded growth. A numerical scheme for the modified differential system can be required to satisfy Eq. (2.69) and thus result in a well-posed numerical problem. In such case the differential equations could be modified by the inclusion of derivative terms which stabilize the short wavelengths.

The inclusion of physical or artificial stabilization at short wavelength has two effects. The first, mentioned earlier, is to give bounded growth rates at short wavelengths. The second effect occurs in nonlinear equations, and is not revealed by the linear analysis; whereas, the nonlinear terms can cause dispersion in which a solution component, having a growth rate greater than unity, disperses into longer and shorter wavelength components. Eventually, energy is cascaded to a short wavelength where the growth rate is less than unity and is dissipated. Thus, energy is removed from the unstable solution components. By this process the dissipation at short wavelength can limit the growth at longer wavelengths. For example, as shown in Appendix B.3.1, the one-dimensional Burgers' waves dissipate energy at the shock-like fronts.

The von Neumann analysis introduced in Appendix B.4.1 for the one-way wave equation is now extended to the FFM. The growth matrix for a general two-level difference scheme is defined by Eq. (2.70). The explicit form of the amplification matrix for a particular differencing scheme can be obtained by expressing the spatial variation of ϕ about the point ϕ_j^n in terms of a typical Fourier component, i.e.,

$$\phi_l^n = \phi_j^n e^{Ik(l-j)\Delta x}, \tag{2.72}$$

where k is the corresponding wavenumber. Using Eq. (2.72), Eqs. (2.67) and (2.68) may be written as:

$$\boldsymbol{M}\underline{\phi}^{n+1} = \boldsymbol{N}\underline{\phi}^n. \tag{2.73}$$

The corresponding FOU growth matrix then becomes $\boldsymbol{G} = \boldsymbol{M}^{-1}\boldsymbol{N}$ where

$$\boldsymbol{M} = \begin{bmatrix} e^{\frac{Ik\Delta x}{2}} & \dfrac{\alpha\Delta t\left(e^{Ik\Delta x} - 1\right)}{\Delta x} \\ 0 & 1 \end{bmatrix}, \tag{2.74}$$

$$\boldsymbol{N} = \begin{bmatrix} e^{\frac{Ik\Delta x}{2}} - \dfrac{2Iu\Delta t\, sin\left[\dfrac{k\Delta x}{2}\right]}{\Delta x} & 0 \\ \dfrac{2Ic\Delta t\ \sin\left[\dfrac{k\Delta x}{2}\right]}{\Delta x} & 1 - \dfrac{u\Delta t\left(1 - e^{-Ik\Delta x}\right)}{\Delta x} \end{bmatrix} \tag{2.75}$$

and the eigenvalues are as follows:

$$|\boldsymbol{G}| = 1 - \frac{\Delta t\left(1 - e^{-Ik\Delta x}\right)}{2\Delta x^2}\left[C\alpha\Delta t\left(e^{Ik\Delta x} - 1\right) + 2u\Delta x \right. \\ \left. \pm \sqrt{C\alpha}\sqrt{C\alpha\Delta t^2\left(e^{Ik\Delta x} - 1\right)^2 - 4\Delta x[u\Delta t + (\Delta x - u\Delta t)e^{Ik\Delta x}]}\right]. \tag{2.76}$$

The stability of the numerical scheme is compared to the stability of the differential model by evaluating the eigenvalues of the growth matrix for a range of wavelengths. The difference scheme growth rate is defined by Eq. (B.32):

$$\omega_{\mathrm{I}}^{\Delta} = -I\frac{\ln|\boldsymbol{G}|_{\max}}{\Delta t}, \tag{2.77}$$

i.e., the finite difference growth rate is compared to the growth rate of the dispersion relation of the partial differential equations (2.42). A numerical imaginary wave speed, corresponding to the wave growth speed, may also be considered:

$$c_{\mathrm{I}}^{\Delta} = -I\frac{\ln|\boldsymbol{G}|_{\max}}{k\Delta t}. \tag{2.78}$$

Stable model ($C < 0$)

The Courant–Friedrichs–Lewy (CFL) criterion for SWT $Co \equiv \left(u + \sqrt{-C\alpha}\right)\Delta t / \Delta x = 1$ determines the stability limit of the FOU numerical method. The behavior of the semi-implicit scheme for the stable well-posed model, with $Co = 0.5, 1, C = -1, \ \alpha = 0.5,$ and $u = 1.0$, is shown in Fig. 2.9. The growth rate is plotted in the imaginary plane for $0 < k\Delta \mathrm{x} < 2\pi$. There are two eigenvalues and the most important feature of this plot is that $|\boldsymbol{G}| \leq 1$ when $Co = 0.5$ for both of them, whereas $|\boldsymbol{G}| = 1.$ for $Co = 1.0$, i.e., the wave amplitude is constant. This is the well-known stability condition for stable numerical schemes. Figures 2.10 and 2.11 show the Von Neumann analysis results in the growth vs. wavelength plane. When $Co = 0.5$ the growth rates are less than zero for short wavelengths as expected for a scheme with numerical viscosity, but the growth rates become greater than zero when $Co = 1.5$, so the numerical scheme is now unstable, i.e., the CFL condition is exceeded. The next question is what happens when the equations are unstable so that growth occurs in the differential model to begin with?

Unstable model ($C > 0$)

The stability characteristics of the numerical scheme are compared by plotting the growth rates of the finite-difference system and the ill-posed differential system in Fig. 2.12. The character of the difference scheme is contrasted to the differential system by presenting the relative amplification factors to illustrate applicability of the upwind numerical scheme to ill-posed problems. Figure 2.11 shows the results for $\Delta x = 0.005, \ 0.01$ and 0.02 m and the dispersion relation result for the Euler FFM with $Co = 0.5$.

Fig. 2.10 Growth rate map for stable model

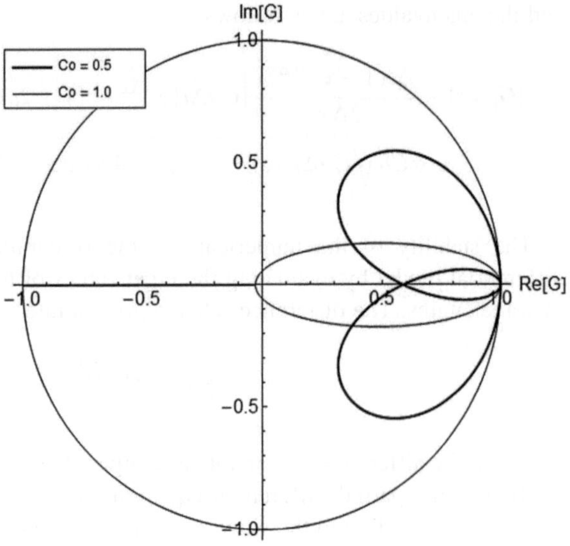

Fig. 2.11 FOU stability, stable model

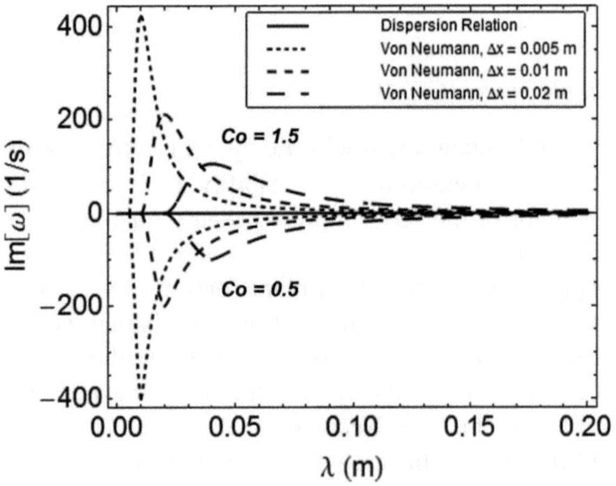

The differential model growth rate is an exponential function and becomes unbounded as the mesh is shrunk to zero ($\Delta x \to 0$), i.e., ill-posed. In order to obtain a well-posed finite-difference problem for finite Δx, the difference scheme should attenuate the exponential growth of the differential system and result in an growth rate less than 0 at the smallest wavelength of interest (i.e., 2 Δx). The semi-implicit scheme with a staggered mesh shows a continual decrease in growth rate for wavelengths $\lambda < 5\Delta x$. This is a convenient engineering solution to an ill-posed problem; however, the cutoff wavelength decreases with mesh size and then the numerical solution does not converge.

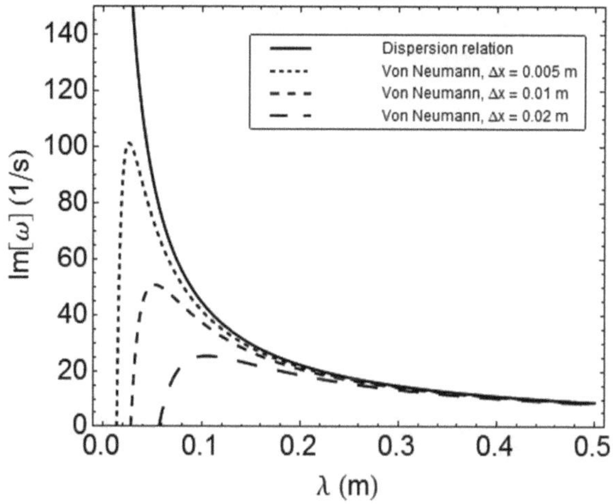

Fig. 2.12 FOU stability, unstable model, $Co = 0.5$

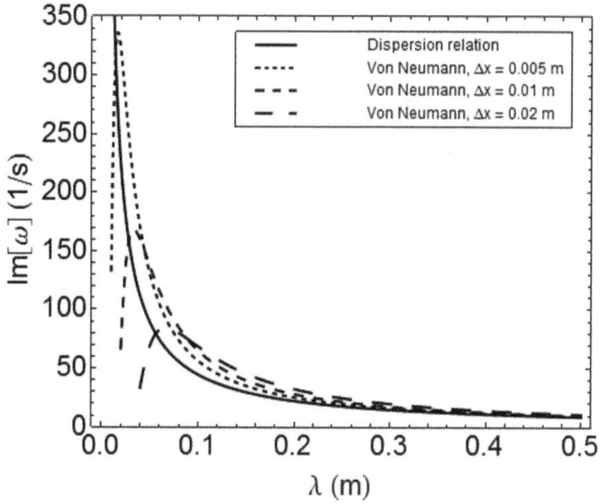

Fig. 2.13 FOU stability, unstable model, $Co = 1.5$

As was shown in the previous section, the numerical method becomes unstable for a Courant number greater than one. For example Fig. 2.13 shows the growth rates with $Co = 1.5$ where the numerical results are *more unstable* than the differential model.

Numerical viscosity is an effective expedient to solve the TFM simply because it damps the short wavelengths of the solution when the TFM becomes KH unstable and ill-posed, as described in section 2.4. However, the numerical model does not converge.

A low pass filter consisting of artificial viscosity terms in the continuity and momentum equations, makes the TFM well-posed and is a better alternative to numerical viscosity, because the numerical model then converges and the cutoff wavelength may be prescribed precisely. The next section illustrates the artificial viscosity solution. The derivation of a low pass filter is presented in Chap. 8.

2.5.3 First-Order Semi-Implicit Scheme (with Viscous Terms)

It was argued in the previous section that the FFM becomes well-posed if a low pass filter, consisting of an artificial viscous term, is added to the two equations. We now write a semi-implicit finite difference model with artificial implicit center-difference viscous terms, i.e., $\varepsilon = \nu$, for a FFM with an arbitrary filter, i.e., $\varepsilon = \nu$:

$$\frac{\alpha_i^{n+1} - \alpha_i^n}{\Delta t} + \frac{u_{\mathrm{R}}^{n+1}\hat{\alpha}_{\mathrm{R}}^{n} - u_{\mathrm{L}}^{n+1}\hat{\alpha}_{\mathrm{L}}^{n}}{\Delta x} = \nu\frac{\alpha_{i+1}^{n+1} - 2\alpha_i^{n+1} + \alpha_{i-1}^{n+1}}{\Delta x^2}, \tag{2.79}$$

$$\frac{u_j^{n+1} - u_j^n}{\Delta t} + u_j^n\frac{u_j^n - u_{j-1}^n}{\Delta x} - C\frac{\alpha_{\mathrm{R}}^n - \alpha_{\mathrm{L}}^n}{\Delta x} = \nu\frac{u_{j+1}^{n+1} - 2u_j^{n+1} + u_{j-1}^{n+1}}{\Delta x^2}. \tag{2.80}$$

This numerical model is still Courant number limited. The von Neumann analysis is performed following the procedure outlined in the previous section. The matrices now are

$$\boldsymbol{N} = \begin{bmatrix} e^{\frac{Ik\Delta x}{2}} - \dfrac{2I\nu\,\Delta t\left(e^{Ik\Delta x} - 1\right)\sin\left[\dfrac{k\Delta x}{2}\right]}{\Delta x^2} & \dfrac{\alpha\Delta t\left(e^{Ik\Delta x} - 1\right)}{\Delta x} \\ 0 & 1 - \dfrac{2\nu\Delta t(\cos[k\Delta x] - 1)}{\Delta x^2} \end{bmatrix}, \tag{2.81}$$

$$\boldsymbol{M} = \begin{bmatrix} e^{\frac{Ik\Delta x}{2}} - \dfrac{2I\Delta tu\sin\left[\dfrac{k\Delta x}{2}\right]}{\Delta x} & 0 \\ \dfrac{2Ic\Delta t\sin\left[\dfrac{k\Delta x}{2}\right]}{\Delta x} & 1 - \dfrac{u\Delta t\left(1 - e^{-Ik\Delta x}\right)}{\Delta x} \end{bmatrix}. \tag{2.82}$$

The equation for the eigenvalues of the growth matrix is more complicated than Eq. (2.76) and is not shown. Figure 2.14 shows the result of the growth rate of the von Neumann analysis compared with the dispersion relation using the same value for the viscosity (i.e., $\nu = 0.01\,\mathrm{m}^2/\mathrm{s}$ and $Co = 0.5$). The result is that the numerical model is also well-posed and the exponential growth of the numerical model is always smaller than that of the differential model, and both approach each other as the mesh resolution increases. It is important to note that the cutoff wavelength decreases with the mesh size, but converges, unlike the inviscid case. For the case with $Co = 1.5$, also shown in Fig. 2.14, the numerical model is consistent but the numerical solutions are more unstable than the exact one as the mesh is refined.

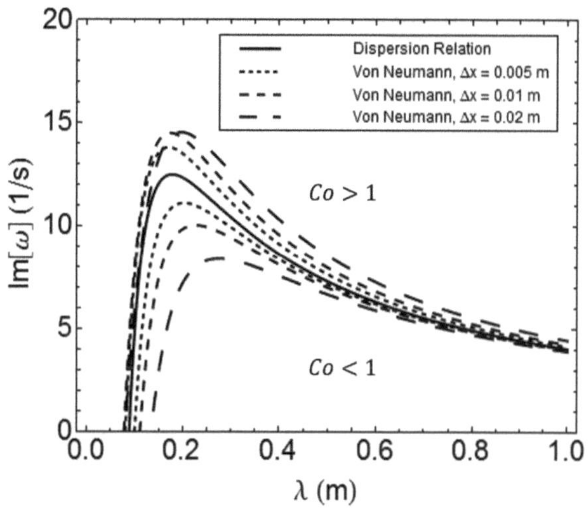

Fig. 2.14 FOU stability, effect of implicit diffusion ($Co = 0.5,\ 1.5$)

The case with explicit viscous terms is of interest because the stability of the model depends on the diffusion number, $R = \frac{\Delta t}{\Delta x^2}$, in addition to the Courant number. The von Neumann analysis matrices of the scheme with explicit viscous terms are as follows:

$$N = \begin{bmatrix} e^{\frac{Ik\Delta x}{2}} & \dfrac{\alpha \Delta t\left(e^{Ik\Delta x} - 1\right)}{\Delta x} \\ 0 & 1 \end{bmatrix}, \tag{2.83}$$

$$M = \begin{bmatrix} e^{\frac{Ik\Delta x}{2}} - \dfrac{2I\Delta t u \sin\left[\frac{k\Delta x}{2}\right]}{\Delta x} + \dfrac{2I\nu\ \Delta t\left(e^{Ik\Delta x} - 1\right) sin\left[\frac{k\Delta x}{2}\right]}{\Delta x^2} & 0 \\ \dfrac{2Ic\Delta t\ sin\left[\frac{k\Delta x}{2}\right]}{\Delta x} & 1 - \dfrac{u\Delta t\left(1 - e^{-Ik\Delta x}\right)}{\Delta x} + \dfrac{2\nu\Delta t(\cos[k\Delta x] - 1)}{\Delta x^2} \end{bmatrix}. \tag{2.84}$$

Figure 2.15 shows the results of the von Neumann analysis for $Co = 0.5$ and $\Delta x = 0.01 - 0.05\,\text{m}$. The corresponding values of the diffusion number are $R = 0.1 - 0.5$. The graphic shows that the numerical model is ill-posed at $\Delta x = 0.01$ m, a result that is well known from the numerical analysis of the one-way wave convection–diffusion equation (Strang 2007). This then poses a more severe constraint on the time step than the Courant criterion as the mesh is refined. Furthermore, there is another constraint with the dispersion of the solution that is determined by the Peclet number which is also addressed by Strang (2007). Because of these difficulties the diffusion terms are usually treated implicitly.

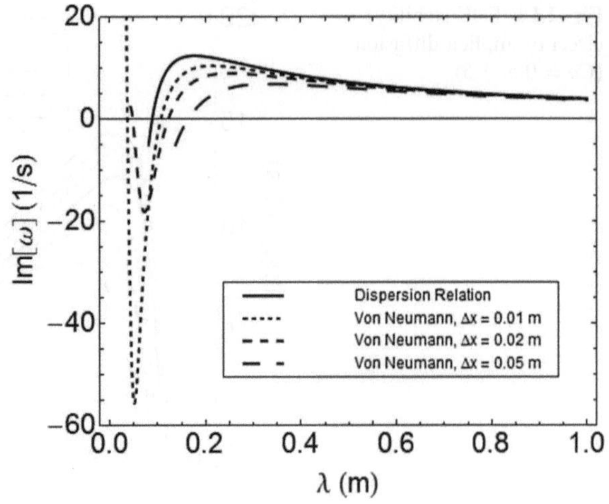

Fig. 2.15 FOU stability, effect of explicit diffusion ($Co = 0.5, R = 0.1 - 0.5$)

2.5.4 *First-Order Fully Implicit Scheme (with Viscous Terms)*

A fully implicit scheme has the advantage that it is not Courant number limited. The first-order implicit scheme for the one-way wave equation presented in Appendix B.4.2 is now applied to the FFM:

$$\frac{\alpha_i^{n+1} - \alpha_i^n}{\Delta t} + \frac{u_R^{n+1}\hat{\alpha}_R^{n+1} - u_L^{n+1}\hat{\alpha}_L^{n+1}}{\Delta x} = \nu \frac{\alpha_{i+1}^{n+1} - 2\alpha_i^{n+1} + \alpha_{i-1}^{n+1}}{\Delta x^2}, \tag{2.85}$$

$$\frac{u_j^{n+1} - u_j^n}{\Delta t} + u_j^{n+1}\frac{u_j^{n+1} - u_{j-1}^{n+1}}{\Delta x} - C\frac{\alpha_R^{n+1} - \alpha_L^{n=1}}{\Delta x} = \nu \frac{u_{j+1}^{n+1} - 2u_j^{n+1} + u_{j-1}^{n+1}}{\Delta x^2}. \tag{2.86}$$

The von Neumann matrices for this scheme are as follows:

$$N = \begin{bmatrix} e^{\frac{Ik\Delta x}{2}} + \frac{2Iu\Delta t \sin\left[\frac{k\Delta x}{2}\right]}{\Delta x} - \frac{2I\nu\Delta t(-1+e^{ik\Delta x})\sin\left[\frac{k\Delta x}{2}\right]}{\Delta x^2} & \frac{\alpha\Delta t(-1+e^{Ik\Delta x})}{\Delta x} \\ -\frac{2IC\Delta t \sin\left[\frac{k\Delta x}{2}\right]}{\Delta x} & 1 + \frac{u\Delta t(1-e^{-Ik\Delta x})}{\Delta x} - \frac{2\nu\Delta t(-1+\cos[k\Delta x])}{\Delta x^2} \end{bmatrix}, \tag{2.87}$$

$$M = \begin{bmatrix} e^{\frac{Ik\Delta x}{2}} & 0 \\ 0 & 1 \end{bmatrix}. \tag{2.88}$$

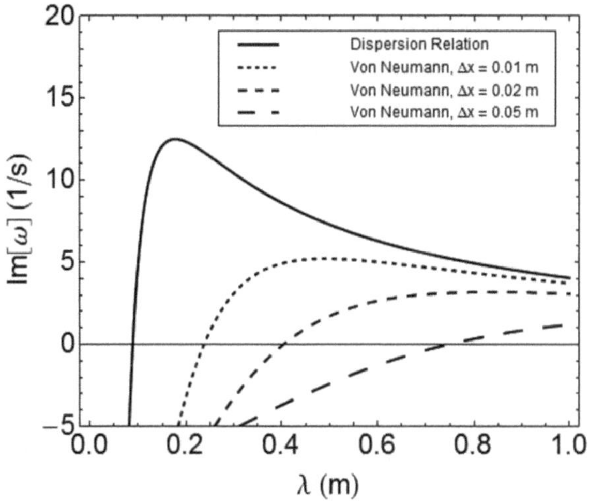

Fig. 2.16 FOU implicit scheme stability, unstable model, $Co = 5$

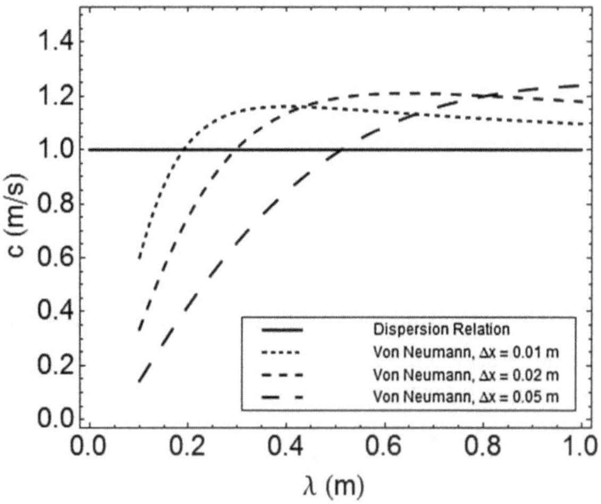

Fig. 2.17 FOU implicit scheme stability, unstable model, $Co = 5$

The main advantage of the implicit scheme is that long time steps may be used. To illustrate this point we consider a time step that is one order of magnitude longer, i.e., $Co = 5$ for the unstable case, i.e., $C = 1$. Figure 2.16 shows that the growth rate is significantly damped compared to the semi-implict scheme, i.e., there is excessive numerical viscosity. Furthermore, Fig. 2.17 shows that there is excessive dispersion too, similar to the results in Appendix. B.4.2. Clearly long time steps deteriorate the accuracy of the wave propagation predictions and should be used with care. Such long time steps are only appropriate when material wave propagation phenomena are not relevant.

2.5.5 Second-Order Semi-Implicit Scheme

Appendix B.4.3 presented the well-known Lax–Wendroff second-order numerical scheme applied to the one-way wave equation. For practical reasons a different second-order numerical scheme is now proposed for the FFM, where the basic underlying structure, utilizing a staggered grid shown in Fig. 2.9, is maintained. This scheme shares similar wave propagation characteristics with the Lax–Wendroff scheme.

Equations (2.37) and (2.38) are arranged into the form

$$\frac{\partial}{\partial t}\underline{\phi} = \underline{f_{\phi}}\left(\alpha^{n\pm k}, u^{n\pm k}, t^{n\pm k}\right), \tag{2.89}$$

where the right-hand side functions, f_{ϕ}, are assumed to be known at each time level $n \pm k$ through a finite-difference scheme, and the time is advanced by treating the PDE as $2N$ ordinary differential equations (ODEs).

The finite-difference functions are given by

$$f_{\alpha,i} = -\overline{u}\frac{\hat{\alpha}_{\mathrm{R}} - \hat{\alpha}_{\mathrm{L}}}{\Delta x} - \alpha_i\frac{u_{\mathrm{R}} - u_{\mathrm{L}}}{\Delta x} + \varepsilon\frac{\alpha_{i+1} - 2\alpha_i + \alpha_{i-1}}{\Delta x^2} + S_{\alpha,i} \tag{2.90}$$

for the α equation at the i location and

$$f_{u,j} = -u_j\frac{\hat{u}_{\mathrm{R}} - \hat{u}_{\mathrm{L}}}{\Delta x} + C\frac{\alpha_{\mathrm{R}} - \alpha_{\mathrm{L}}}{\Delta x} + \nu\frac{u_{j+1} - 2u_j + u_{j-1}}{\Delta x^2} + S_{u,j} \tag{2.91}$$

for the u equation at the j location. The overbar indicates a cell averaged (mean) value. Variables with a hat do not exist at the specified location. The additional subscripts refer to the values to the "right" and "left" of the reference location. For example, in Eq. (2.90) the faces j and $j+1$ surround the reference i position, likewise for Eq. (2.91) the cell centers $i-1$ and i surround the reference j position. When the variables required at the right and left are available, they are used directly, e.g., u exists at j and $j+1$ in Eq. (2.90). When the variables do not exist at the right and left locations as indicated by the hat, they are donored or extrapolated using the values at the neighboring locations. A flux limiter is used for the extrapolation and discussed later. The diffusion terms are discretized with a standard second-order center difference scheme. For those simulations where periodic boundary conditions are applied three "ghost" cells are placed at the beginning and end of the domain that mirror the conditions at the opposite end.

For the extrapolated variables there are many options. Two obvious choices are first-order upwinding (FOU) and second-order center differencing. For the 1D FFM, the previous sections illustrated upwinding with edge values simply donored from adjacent cells in the direction of the SWT wave. The problem is that applying the FOU method to the convection variables adds a significant amount of numerical diffusion and lowers the accuracy of the entire scheme. On the other hand, viewing the FFM equations as simply two PDEs, leaving physics considerations aside, one might opt for a center difference scheme. However, this would leave the overall finite difference scheme with all linear second-order difference stencils, leading to spurious numerical oscillations, which are undesirable for the present application where the underlying governing equations are already known to be unstable.

In order to combine increased accuracy and enhanced stability, a nonlinear flux limiter is used for the extrapolated variables (Drikakis and Rider 2005; Tannehill et al. 1997). This implies some physical interpretation in determining the direction of the "wind" or flow, here given by the sign of $c = u + \sqrt{-C\alpha}$. The general-piecewise limiter (GPL) of Waterson and Deconinck (2007) is used here because it is relatively simple and can easily be tuned to produce a variety of popular schemes.

The structure of the flux limiter for the right face value of the α variable in Eq. (2.90) is,

$$\hat{\alpha}_{\mathrm{R}} = \alpha_i + \frac{\Delta x_i}{2}\Psi(r)\left(\frac{\partial \alpha}{\partial x}\right)_{\mathrm{UD}}, \tag{2.92}$$

where r is the gradient ratio defined by,

$$r = \frac{\left(\frac{\partial \alpha}{\partial x}\right)_{\mathrm{CD}}}{\left(\frac{\partial \alpha}{\partial x}\right)_{\mathrm{UD}}}. \tag{2.93}$$

The subscripts CD and UD indicate the center and upwind differences for the location i. In the present work, a uniform mesh will always be applied so that Eq. (2.93) simplifies to

$$r = \frac{\alpha_{i+1} - \alpha_i}{\alpha_i - \alpha_{i-1}} \tag{2.94}$$

for positive flow, i.e., $c > 0$. The function $\Psi(r)$ is the flux limiter and the GPL scheme is given by

$$\Psi(r) = \max\left[0, min\left\{(2+a)r, \frac{1}{2}(1+k)r + \frac{1}{2}(1-k)r, M\right\}\right], \tag{2.95}$$

where a, k, and M are control parameters. The left values as well as the extrapolated variables of Eq. (2.91) are found in the same manner with the neighboring cell locations shifted appropriately.

The GPL limiter of Eq. (2.95) can be tuned to give several classical flux limiters, e.g., the Minmod scheme of Roe (1986) by setting $a = k = -1$ and $M = 1$ and the MUSCL scheme of van Leer (1979) by setting $a = k = 0$ and $M = 2$. In this work, the GPL scheme will be set to $a = 0$, $k = 1/2$, and $M = 4$, which results in the converted normalized variable SMART scheme of Gaskell and Lau (1988). The SMART scheme performed better for discontinuous solution data and was also slightly higher than second-order accurate for smooth data. The SMART scheme has been selected since it is known in advance that the TFM equations produce solutions with shock-like structures, i.e., small regions of space where the solution changes rapidly. It should be pointed out that unlike Minmod and MUSCL, the limiter form of SMART is not exactly total variation diminishing (TVD). However, it is nonoscillatory in its original normalized variable form which has a rough equivalence to TVD (Drikakis and Rider 2005). The GPL can also be tuned to a few other schemes of less practical interest outlined by Waterson and Deconinck (2007). Additionally, by setting the limiter to the constants $\Psi(r) = 0$ or $\Psi(r) = 1$, the FOU and center difference schemes are recovered, respectively.

For the time advancement a Runge–Kutta method is selected. Rather than the standard fourth-order method that is often used, a strong stability preserving (SSP) third-order method is used in keeping with the theme of increased numerical stability. Essentially SSP schemes are the temporal equivalent to TVD spatial discretizations. The optimal third order, three stage (3-3) SSP Runge–Kutta method of Gottlieb and Shu (1998) is used to approximate Eq. (2.89), which is defined by

$$\begin{aligned} \phi^{(1)} &= \phi^n + \Delta t \cdot f_\phi(\alpha^n, u^n, t^n) \\ \phi^{(2)} &= \frac{1}{4}\phi^{(1)} + \frac{3}{4}\phi^n + \frac{1}{4}\Delta t \cdot f_\phi\left(\alpha^{(1)}, u^{(1)}, t^{n+1}\right) \\ \phi^{(n+1)} &= \frac{2}{3}\phi^{(2)} + \frac{1}{3}\phi^n + \frac{2}{3}\Delta t \cdot f_\phi\left(\alpha^{(2)}, u^{(2)}, t^{n+\frac{1}{2}}\right) \end{aligned}, \tag{2.96}$$

where Δt is the time step size between levels n and $n+1$ and the difference functions are given in Eqs. (2.90) and (2.91).

This method is the most widely used SSP Runge–Kutta scheme since it is relatively inexpensive computationally and has a CFL criterion of unity. Combining the CFL condition with the restriction due to the explicit treatment of the diffusion terms, the resulting heuristic numerical stability constraint is

$$\Delta t \le \min\left[\frac{\Delta x^2}{2\nu}, \frac{2\nu}{u_\infty^2}\right] \tag{2.97}$$

assuming $\varepsilon = \nu$ and $u_\infty = \max|u_j|$ in the domain.

In the end, the global method is a mix of second-order centered differences, fractional-order flux limited differences, and a third-order time marching scheme. The total accuracy of the combined scheme is as yet unknown, although one might expect it to fall somewhere between second- and third-order accuracy, the bounding upper and lower limits of the individual components. To answer this question and to determine if the resulting code has implemented these algorithms correctly, Sect. 2.6.4 will look at the problem of code verification by the method of manufactured solutions.

Section B.4.4 presented a second-order implicit scheme for the one-way wave equation. The major shortcoming is the significant increase in the dispersion of material waves, so the approach is not amenable to the simulation of material waves and is not pursued further since the purpose of higher-order accuracy in this book is precisely that.

2.6 Verification

2.6.1 Kreiss–Yström Equations

The numerical method described in the previous section was implemented using the Fortran programming language. As with any newly developed code, verification exercises should be performed. In general, verification is divided into two components: code verification and solution (or calculation) verification (Roache 1998, 2002). Code verification is the process of determining if the numerical algorithms were implemented correctly (Oberkampf and Roy 2010; Oberkampf et al. 2004) while solution verification deals with quantifying the numerical error of a given solution.

The present work focuses on the Kreiss and Yström (2002) system of partial differential equations (PDEs), which is remarkably similar to the FFM, given by

$$\frac{\partial \alpha}{\partial t} + u\frac{\partial \alpha}{\partial x} + \left[1 + \frac{\alpha}{2}\right]\frac{\partial u}{\partial x} = \varepsilon\frac{\partial^2 \alpha}{\partial x^2} - 2\alpha, \tag{2.98}$$

$$\frac{\partial u}{\partial t} + u\frac{\partial u}{\partial x} = C\frac{\partial \alpha}{\partial x} + v\frac{\partial^2 u}{\partial x^2} \tag{2.99}$$

and hereafter referred to as the KY equations. There are a few slight differences between the way we employ Eqs. (2.98) and (2.99) compared to the authors. First, the parameter C was unity in the original model. The adjustable coefficient C is used here as a means to vary the "degree of instability", because it represents the KH criterion in Eqs. (2.37) and (2.38). Second, the viscosities have been given different symbols although the same value will be used for both.

Among the obvious differences in Eq. (2.98) and the FFM Eq. (2.37), the viscous term is significant because it acts like artificial viscosity in the TFM. Furthermore, there is another significant difference, the constraint $0 \leq \alpha \leq 1$ is not present. Nevertheless, the dynamic similarity of the KY equations and the FFM will be demonstrated in the next two sections and the nonlinear analysis of the KY equations in Ch. 4 is the preliminary step to approach the nonlinear behavior of the more complicated FFM, which is addressed afterward.

2.6.2 Characteristic Analysis

The FFM and the present Kreiss and Yström (2002) system are related in several significant ways. Both models are conditionally ill-posed as an initial and boundary value problem (IBVP) without higher order stabilization, i.e., the first-order systems (may) possess complex characteristics. Both models are predicted to be linearly unstable. The magnitude of the growth rates are linearly proportional to the wavenumber without stabilization and quadratically damped at high wavenumbers with higher order stabilization. Both models have similar dynamics, i.e., they produce similar waveforms. However, the mathematical model of Kreiss and Yström (2002) is simpler in several desirable ways. It does not have a complicated flow-regime map of closure laws for different flow conditions—some of which affect the differential form of the governing equations. The dynamics of the problem are not bounded by physical constraints, e.g., channel height. It does not require special numerical techniques. Finally, and possibly most importantly, the "degree of instability" can be directly controlled by specifying the constant value of C in Eq. (2.86). In turn, in the 1D TFM of Eqs. (2.37) and (2.38) the coefficient is not constant but instead it is determined by the local flow conditions so that the linear stability of the model changes throughout the solution, and may even change from nonhyperbolic to hyperbolic depending on the flow conditions. Therefore, the simplified form and added controllability of the two KY equations make them an ideal first step for the nonlinear studies of Chap. 4.

The KY Eqs. (2.98) and (2.99) can be written in vector form as

$$\boldsymbol{A}\frac{\partial}{\partial t}\underline{\phi} + \boldsymbol{B}\frac{\partial}{\partial x}\underline{\phi} + \boldsymbol{D}\frac{\partial^2}{\partial x^2}\underline{\phi} = \underline{F}, \tag{2.100}$$

where the dependent variable vector is $\underline{\phi} = [\alpha, u]^T$, the source vector is $\underline{F} = [-2\alpha, 0]^T$, and the coefficient matrices are defined by

$$\boldsymbol{A} = \boldsymbol{I}, \quad \boldsymbol{B} = \begin{bmatrix} u & 1 + \frac{\alpha}{2} \\ -C & u \end{bmatrix}, \quad \boldsymbol{D} = \begin{bmatrix} -\varepsilon & 0 \\ 0 & -\nu \end{bmatrix}. \tag{2.101}$$

The eigenvalues of Eq. (2.99) are given by,

$$c = u \pm i\sqrt{C\left(1 + \frac{\alpha}{2}\right)}. \tag{2.102}$$

In the absence of diffusion, i.e., $\varepsilon = \nu = 0$, these eigenvalues define the characteristics of the system. Equation (2.102) shows that for $C > 0$ (and $\alpha > -2$) the characteristics are complex conjugates. Therefore, the system is ill-posed as an initial-boundary value problem. Even when the diffusion matrix is included, the complex characteristics are the central cause of the linear growth discussed later. On the other hand, for $C < 0$ (and $\alpha > -2$) the characteristics are both real and the system is hyperbolic. In this case, the problem simplifies dramatically and the KY model is similar to a modified shallow water model. There is a large body of work concerning the nature of the shallow water equations and so the hyperbolic case is not of primary interest. Equivalent but opposite classifications occur for the case $a < -2$ which is not studied here.

2.6.3 Dispersion Relation

To better highlight the effect of the diffusion terms, the equations are linearized about an initial reference state with an infinitesimally small perturbation superimposed, i.e., $\underline{\phi} = \underline{\phi}_0 + \underline{\phi}'$. This solution is inserted into Eq. (2.100) and three assumptions are applied to greatly simplify the resulting equation: the initial reference state satisfies Eq. (2.100) automatically, products of perturbations are negligible and the reference state is either steady or the length scale of the reference state is considerably larger than that of the imposed perturbation, i.e., $\frac{\partial \phi_0}{\partial x} \ll \frac{\partial \phi'}{\partial x}$. The remaining terms define the linear perturbation equation,

$$\frac{\partial \underline{\phi}'}{\partial t} + \boldsymbol{A}_0 \frac{\partial \underline{\phi}'}{\partial x} + \boldsymbol{D} \frac{\partial^2 \underline{\phi}'}{\partial x^2} - \frac{\partial \underline{F}_0}{\partial \underline{\phi}'^T} \underline{\phi}' = 0, \tag{2.103}$$

where $\boldsymbol{A}_0$ and F_0 are evaluated at ϕ_0. In typical fashion, the perturbation is assumed to be a traveling wave,

$$\underline{\phi}' = \underline{\hat{\phi}}' \mathrm{e}^{i(kx - \omega t)}, \tag{2.104}$$

where $\hat{\phi}'$ is the amplitude, k is the wavenumber, and ω is the angular frequency. The more intuitive variables of wavelength and frequency are related by $\lambda = 2\pi/k$ and $f = \omega/2\pi$, respectively. In general, the angular frequency may be complex and, from Eq. (2.104), when the imaginary component is positive the perturbation will grow exponentially in time. The real component of the angular frequency and the wavenumber defines the wave speed according to $c = \omega_{\mathrm{R}}/k$.

Inserting Eq. (2.104) into Eq. (2.103) gives

$$\left(-i\omega + ik\boldsymbol{A}_0 + (ik)^2 \boldsymbol{D} - \frac{\partial \underline{F}_0}{\partial \underline{\phi}'^T} \right) \underline{\phi}' = 0. \tag{2.105}$$

For a nontrivial solution to exist, the coefficient matrix of Eq. (2.105) must be singular, i.e.,

$$\det \left[\omega - k\boldsymbol{A}_0 + ik^2 \boldsymbol{D} + i \frac{\partial \underline{F}_0}{\partial \underline{\phi}'^T} \right] = 0. \tag{2.106}$$

Solving Eq. (2.106) for the angular frequency gives

$$\omega = ku_0 - \frac{i}{2} \left[(\varepsilon + \nu)k^2 + 2 \pm \sqrt{\left((\varepsilon - \nu)k^2 + 2\right)^2 + (2c_i k)^2} \right], \tag{2.107}$$

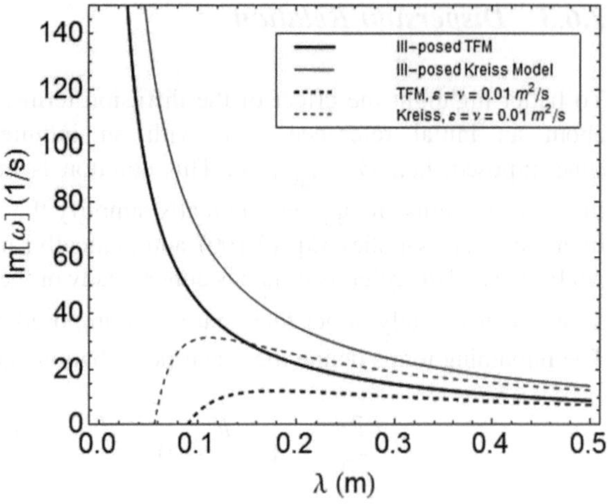

Fig. 2.18 Comparison of FFM and KY dispersion relations

where c_i is the imaginary component of the eigenvalues previously derived in Eq. (2.102). The larger imaginary root of the angular frequency is shown in Fig. 2.18 for a reference state of $\alpha_0 = u_0 = 1$ and $C = 1$. Consider the case $\varepsilon = \nu$:

$$\omega = uk - i\nu k^2 \pm k\sqrt{-C\left(1 + \frac{\alpha}{2}\right)}. \tag{2.108}$$

This dispersion relation is very similar to that of FFM equation (2.42) when $\varepsilon = \nu$ and $\sigma = 0$:

$$\omega = uk - i\nu k^2 \pm k\sqrt{-C\alpha}. \tag{2.109}$$

The model is ill-posed in the inviscid limit, i.e., $\varepsilon = \nu = 0$. Figure 2.18 shows the dispersion relations with $\alpha = 0.5$, $u = 1\,\mathrm{m/s}$, $C = 1$, and $\nu = 0$, $0.01\,\mathrm{m/s^2}$ and the dynamic resemblance with Eqs. (2.37) and (2.38) becomes apparent.

It is readily seen from Eq. (2.108) that when diffusion is neglected, $\varepsilon = \nu = 0$, the growth rate, ω_I, will have a positive component indicating exponential growth. The ill-posed behavior is obvious: an infinitely short wavelength perturbation ($k \to \infty$) has an infinitely large growth rate ($\omega_I \to \infty$). For large k the growth rate is approximately linear with respect to k with a small, constant damping due to the source term. The linear growth rate is also seen in the 1D TFM under certain conditions and is a characteristic feature of the Kelvin–Helmholtz instability that is imbedded within the model equations.

If both viscosities are positive constants the linear stability changes even more dramatically, rather than approaching an asymptote the stability curve bends down and approaches negative infinity as $k \to \infty$. For the special case $\varepsilon = \nu$, the imaginary part of Eq. (2.107) reduces to three components: a negative constant component from the source term, a positive linear component from the underlying ill-posedness, and a quadratic negative term from the diffusion terms. The combined result is that the growth rate is positive for a wavenumber range from $k \in (0, k_0)$ and negative for $k \in (\kappa_0, \infty)$, as shown in Fig. 2.18. The value of the cut off wavenumber is given by $k_0 = \nu^{-1}\sqrt{c_i^2 - 2\nu}$. In the growth region there is a single maximum growth rate, the critical growth rate, given by $\omega_c = \frac{c_i^2}{4\nu} + \frac{\nu}{c_i^2} - 1$ which occurs at the critical wavenumber $k_c = \sqrt{\frac{c_i^2}{4\nu^2} - \frac{1}{c_i^2}}$. While there is still exponential growth, this type of linear instability is physically acceptable. Perturbations will grow if the wavelength is sufficiently long and decay at the short length scales. Lastly, it should be noted that the KY equations are not dispersive in any case since the real part of the angular frequency is always linear in k so that the wave speed is constant, i.e., $c = u$ (Whitham 1974).

The linear stability properties of the KY equations and how it is affected by the viscosities parallel that of the 1D TFM. In particular, to achieve a cutoff wavenumber with decay at short wavelengths, second-order diffusion terms must be added to all equations even when there is no physical justification. It is worth noting that both the 1D TFM and the KY equations can also be regularized with a combination of kinematic or turbulent viscosity and a third-order term related to surface tension. Such an approach introduces dispersion which increases the complexity of the problem and will not be explored here.

2.6.4 Method of Manufactured Solutions

The simplest way to test the code's correctness (errors or bugs) and accuracy (observed convergence rate) is to compare a numerical solution to a known exact analytical solution. Unfortunately the KY equations have no known solution that is sufficiently complex to exercise all of the terms. Certainly $\alpha(x,t) = 0$ and $u(x,t) = A$, where A is any constant, is a solution to Eqs. (2.98) and (2.99) but it provides no useful information for assessing the accuracy of the code (although such trivial solutions may be useful for checking the correctness of the code, i.e., debugging). Fortunately this problem has a clever remedy: simply manufacture a solution without being concerned about whether or not it exactly satisfies the governing PDEs. Then, if the manufactured solution does not exactly satisfy the governing PDEs, it is relatively straightforward to determine the residual source term that would modify the equations to make the manufactured solution an exact solution. This approach is often referred to as the method of manufactured solutions (MMS).

To begin, it helps to write of the KY equations as a pair of operators:

$$L_\alpha(\alpha, u) = \frac{\partial \alpha}{\partial t} + u\frac{\partial \alpha}{\partial x} + \left(1 + \frac{\alpha}{2}\right)\frac{\partial u}{\partial x} - \varepsilon\frac{\partial^2 \alpha}{\partial x^2} + 2\alpha \tag{2.110}$$

and

$$L_u(\alpha, u) = \frac{\partial u}{\partial t} + u\frac{\partial u}{\partial x} - C\frac{\partial \alpha}{\partial x} - \mu\frac{\partial^2 u}{\partial x^2}. \tag{2.111}$$

Then in operator form, the KY equations become $L_\alpha(\alpha, u) = L_u(\alpha, u) = 0$ for some $\alpha(x, t)$ and $u(x, t)$ that are exact solutions to Eqs. (2.98) and (2.99). Following Roache (1998), traveling waves are selected for the manufactured solutions:

$$\widetilde{\alpha} = 2 + 0.5 \sin\left[\widetilde{k}(x - \widetilde{c}t)\right], \quad \widetilde{u} = 1 + 0.5 \sin\left[\widetilde{k}(x - \widetilde{c}t)\right]. \tag{2.112}$$

The manufactured solutions are then inserted into the operators of Eqs. (2.100) and (2.101) giving

$$\begin{aligned} L_\alpha(\widetilde{\alpha}, \widetilde{u}) = \widetilde{k}\left(\frac{3}{2} - \frac{\widetilde{c}}{2}\right) \cos\left[\widetilde{k}(x - \widetilde{c}t)\right] + \left(1 + \frac{\varepsilon k^2}{2}\right) \sin\left[\widetilde{k}(x - \widetilde{c}t)\right] \\ + \widetilde{k}\left(\frac{3}{16}\right) \sin\left[2\widetilde{k}(x - \widetilde{c}t)\right] + 4 = S_\alpha(x, t) \end{aligned} \tag{2.113}$$

and

$$\begin{aligned} L_u(\widetilde{\alpha}, \widetilde{u}) = \left(\frac{\widetilde{k}}{2}\right)\left(1 - \widetilde{c} - C + \frac{1}{2} \sin\left[\widetilde{k}(x - \widetilde{c}t)\right]\right) \cos\left[\widetilde{k}(x - \widetilde{c}t)\right] \\ + \left(\frac{\nu\widetilde{k}^2}{2}\right) \sin\left[\widetilde{k}(x - \widetilde{c}t)\right] = S_u(x, t) \end{aligned} \tag{2.114}$$

which define the source terms, $S_\alpha(x, t)$ and $S_u(x, t)$, necessary to modify the KY equations so that the manufactured solutions of Eq. (2.99) and (2.100) are exact solutions. The procedure could equivalently be called the method of manufactured equations since it is really the governing equations that are changed from their original form, $L_\phi(\alpha, u) = 0$, to a modified form, $L_\phi(\alpha, u) = S_\phi(x, t)$, to allow the handpicked solution to be an exact solution.

The source terms defined in Eqs. (2.113) and (2.114) are used as the sources in the finite difference functions in Eqs. (2.90) and (2.91) at discrete space locations x_i or x_j and time levels t^n. The initial conditions are given by the manufactured solutions at $t = 0$. Periodic boundary conditions are applied so that no special boundary values need to be determined from the manufactured solutions, but the domain or solution should be set so that there are an integer number of periods. The error between the code calculated solution and the manufactured exact solution will be assessed with a global L_2 norm defined by

$$e_2(\phi) = \frac{\sqrt{\sum_i^N \left(\phi_i - \widetilde{\phi}(x_i)\right)^2}}{\sqrt{\sum_i^N \left(\widetilde{\phi}(x_i)\right)^2}}. \tag{2.115}$$

The coarsest grid used for the MMS is $N = 25$ and the nodalization is successively doubled, i.e., Δx is halved, up to $N = 800$. The convergence rate between a grid N and the doubled grid $2N$ is given by

$$O = \frac{\ln\left[\dfrac{e_2(\phi_{2N})}{e_2(\phi_N)}\right]}{\ln\left[\dfrac{\Delta x_{2N}}{\Delta x_N}\right]} \tag{2.116}$$

which gives the order of accuracy of the code for grid sizes in the vicinity of $3N/2$. In the present work the grid refinement is always achieved by doubling the number of nodes so the denominator in Eq. (2.116) is simply ln(1/2). Additionally the time step should be refined consistently with the grid, which can be problematic since the time step is restricted by Δx^2 for small enough Δx. For each case the ratio $r_\Delta = \Delta t/\Delta x$ is set to a constant value so that the numerical stability condition of Eq. (2.97) is satisfied for all grids.

To begin, the simplest waveform is selected, i.e., $\widetilde{k} = \widetilde{c} = 1$ in Eq. (2.112), so that the solution is 2π-periodic in space and time. The domain is given by $x \in [-\pi, \pi]$ and the error will be assessed at $t = 2\pi$ to give one period in both space and time. The time step for each grid is set from the ratio $r_\Delta = \Delta t/\Delta x = 0.0125$. As may be somewhat expected, the MMS does not work and the calculated solution diverges from the manufactured solution. The traditional refinement plot or table are not shown because they offer little information. What is important to demonstrate is not simply that the solution diverges, but how it diverges.

The error for each variable is shown in Fig. 2.19 as a function of time for four different grids. (The two coarsest grids have been left out because the truncation error begins to affect the divergence rate.). The straight line in Fig. 2.19 beyond the finest grid is $e^{\omega_c(t-t^*)}$ with ω_c calculated using the average α of the exact solution

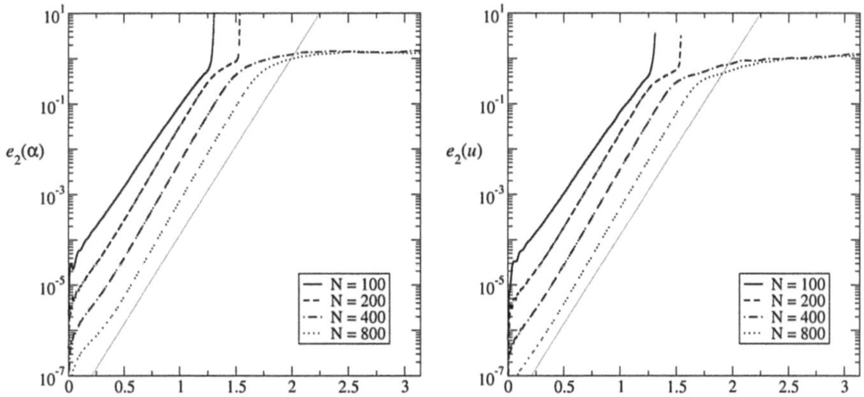

Fig. 2.19 Divergence of numerical method from the exact solution for the case $x \in [-\pi, \pi]$, straight line shows the average critical growth rate predicted from linear theory

and t^* is an arbitrary constant. This shows that for each case, there is an initial transient period after which the error grows exponentially and approximately uniformly at the critical growth rate. Beyond this period the solution either diverges to infinity (an excursion) or reaches an asymptotic value. A discussion on this difference is postponed to the end of Sect. 4.3.4.

Figure 2.18 shows that while the MMS did not work, it failed in a way that was predicted by the linear stability analysis of Sect. 2.6.3. Error is introduced to the calculation through two primary ways: round-off error in the finite precision representation of the source terms and truncation error by the finite difference representation of the differential equations. While these errors are initially very small, the linear stability analysis predicts that even infinitesimally small perturbations will grow exponentially in time. Therefore, some adjustments are needed in order to proceed with the code verification.

The most obvious change is to make the model hyperbolic, either by adjusting the model or adjusting the manufactured solution. This was tested by setting $C = -1$, which makes the model hyperbolic for the manufactured solution of the previous section, and convergence was indeed verified. However, this method of code verification raises several concerns, namely, that such an approach may not be applicable to other similar models, e.g., the 1D TFM.

Without altering the form of the equations or the solution, there is one other option available: shrink the geometry of the problem so that domain simply cannot contain wavelengths with positive growth rate, i.e., $L \leq \lambda_0$. Using the maximum value of α in the manufactured solution, the maximum cutoff wavenumber is $k_0 = 20\sqrt{2.15} \cong 29.3$. Therefore, the domain will be reduced to $x \in [-\pi/30, \pi/30]$ so that the maximum possible wavelength is less than λ_0 and linear growth should not be possible. The wavenumber of the manufactured solution is modified to fit the new domain, $\widetilde{k} = 30$, while the previous wave speed is retained, $\widetilde{c} = 1$, so that the error is still assessed at one temporal period, $t = 2\pi$. The time step for each grid is set by $r_\Delta = \Delta t/\Delta x = 0.004$.

Now the error is reduced as the grid is refined. The error for each grid is shown in Table 2.1 and the convergent behavior is shown in Fig. 2.20. The convergence rate between two grids is also given in Table 2.1 which shows that the order of accuracy of the method is slightly better than second order. There is a noticeable convergence of the rate of convergence itself, which was also reported by others using the MMS (Burg and Murali 2006). For the coarsest grids the refinement brings almost third-

Table 2.1 Error and convergence rate between two successive grids for the case $x \in [-\pi/30, \pi/30]$

	α		u	
N	e_2	O	e_2	O
25	5.860E-02	–	5.187E-02	–
50	8.345E-03	2.81	6.568E-03	2.98
100	1.363E-03	2.61	9.103E-04	2.85
200	2.652E-04	2.36	1.501E-04	2.60
400	5.817E-05	2.19	2.925E-05	2.36
800	1.363E-05	2.09	6.410E-06	2.19

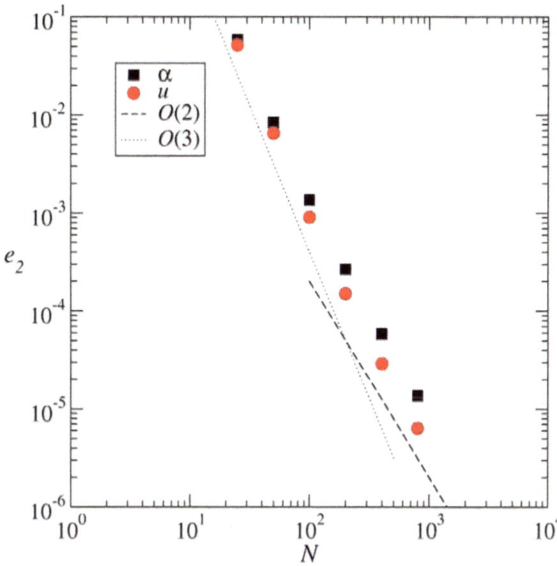

Fig. 2.20 Convergence rates for the second MMS case, $x \in [-\pi/30, \pi/30]$.

order accuracy, which is shown in Fig. 2.20 as the steeper initial line. However, as the grid is refined further the order of accuracy is reduced to near second order which is also shown in Fig. 2.20. It should be noted that this convergence rate is for smooth data (solutions) only. Discontinuous or piece-wise continuous solutions are discussed next.

Using such a small domain may not be physically relevant for some models, but this is not a concern of code verification. This provides a method to further distinguish ill-posed models from regularized models, since for an ill-posed model there will be no domain where the MMS is applicable. Lastly it should be pointed out that we are still left wondering whether or not the numerical method (and really any time marching scheme) is appropriate for this type of problem, but this is not really a question for the MMS or of code verification in general. With code verification one only wants to be sure that the code is working correctly, which has been shown.

2.6.5 *Water Faucet Problem*

The water faucet problem of Ransom (1984), derived in App. B.5.5, is a stable SWT case, i.e., $C = 0$, with a void jump. The purpose of this section is to verify the numerical FFM with the water faucet problem to asses the effect of a discontinuous solution. The uniform initial conditions are $\alpha_{10} = 0.8$ and $u_{10} = 10\,\mathrm{m/s}$. The boundary conditions at the inlet are held constant at α_{10} and u_{10}. The SWT Eqs. (2.47) and (2.48) are solved analytically for $L = 12\,\mathrm{m}$, $\rho_1 = 1000\,\mathrm{kg/m^3}$, $\rho_2 = 1\,\mathrm{kg/m^3}$, $C = 0$ and the source term $F = g$. The solution, derived in Appendix B.5.5, is

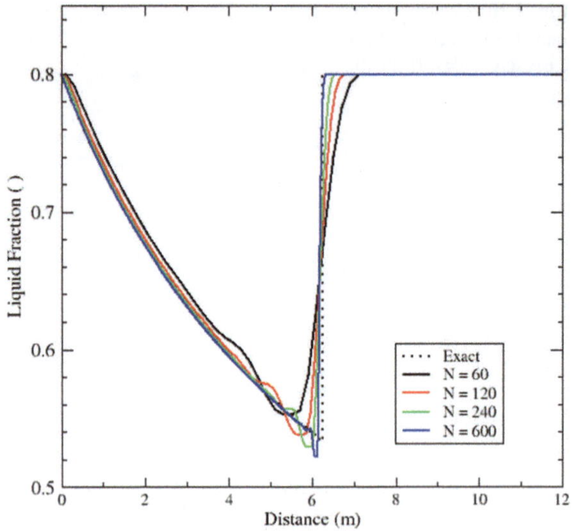

Fig. 2.21 Numerical solutions of the water faucet problem at $t = 0.5$ s with SSP2-SMART scheme

$$x < u_{10}t + \frac{1}{2}gt^2 \quad \rightarrow \quad \alpha_1(x,t) = \frac{\alpha_{10}u_{10}}{\sqrt{u_{10}^2 + 2gx}}, \qquad u_1(x,t) = \sqrt{u_{10}^2 + 2gx} \quad (2.117)$$

or otherwise

$$\alpha_1(x,t) = \alpha_{10}, \quad u_1(x,t) = u_{10} + gt. \quad (2.118)$$

The second-order semi-implicit scheme of Sect. 2.5.5, including the SMART flux limiter, is verified in the presence of a contact discontinuity. The solutions are shown in Fig. 2.21 at $t = 0.5$ s for $Co = 0.5$. There is a small undershoot right before the shock, but otherwise the solution converges satisfactorily.

To quantify the convergence of two second-order schemes the L_1 error for the liquid fraction is calculated as

$$L_1(\alpha) = \frac{1}{N}\sum_{i=1}^{N} |\alpha_i - \alpha_{\text{exact}}(x_i)|. \quad (2.119)$$

The errors as a function of grid number are then fitted with a linear slope on a log–log plot to determine the global order of accuracy in Fig. 2.22. The convergence rates for both higher order schemes are significantly reduced, compared with the continuous case shown in Fig. 2.20. This reduction has been explained by Banks et al. (2008) for linearly degenerate waves. The convergence rates of the FOU, SSP2-MM, and SSP2-SMART schemes shown in Fig. 2.22 are extremely close to the theoretical L_1 errors of Banks et al. (2008), i.e., $O\left(\frac{1}{2}\right)$ for FOU and $O\left(\frac{2}{3}\right)$ for the SSP2-MM scheme. The SSP2-SMART scheme preforms slightly better, i.e., approximately $O(0.77)$.

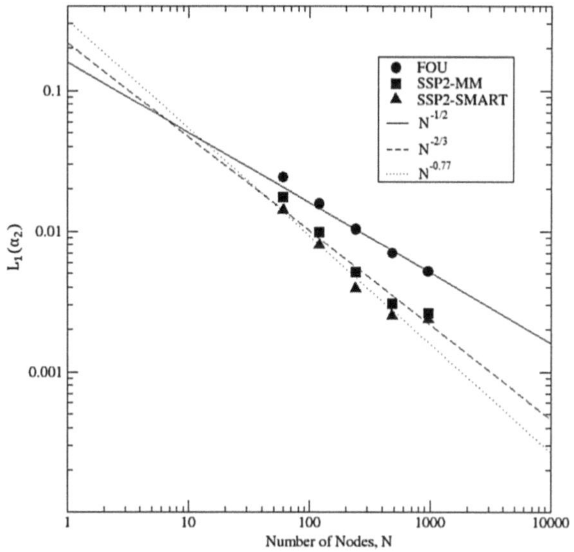

Fig. 2.22 Convergence rates for three different schemes

Finally, it may be appropriate at this point to clarify why the linear theory of Banks et al. (2008) is applicable to the water faucet problem, which is a nonlinear problem. In the vicinity of the void jump, the velocity is continuous according to Eqs. (2.117) and (2.118), although it is discontinuous in the first derivative. Therefore, the velocities on either side of the jump are equal. Since the characteristics are given by the velocity, the jump speed is $u(x_d(t))$ and the water faucet problem describes a contact discontinuity (Lax 1972) which is linearly degenerate, unlike material shocks that have discontinuous velocities for $C > 0$, cf. Sect. B.5.3.

2.7 Kelvin–Helmholtz Instability

The purpose of this section is to show how the Kelvin–Helmholtz (KH) instability behaves in horizontal stratified flow for a well-posed FFM compared to an ill-posed one. The KH instability has been the subject of considerable research within the framework of the 1D TFM. Taitel and Dukler (1976) used it with linear stability theory to predict the flow regime transition from stratified to slug flow. In their simple interpretation, the 1D TFM becomes ill-posed at the transition as was shown in Sect. 2.4.1. However, Ramshaw and Trapp (1978) demonstrated that adding surface tension makes the model well-posed even when unstable. Very little work has been done on the nonlinear stability of the KH unstable TFM and since SWT does not contemplate the case $C > 0$ there are no equivalent analyses either. This topic is addressed in the next two chapters. The dynamically unstable TFM is therefore unique and exhibits, when the model is too incomplete, ill-posed behavior.

A "double wave" simulation, first introduced by Holmås et al. (2008), is considered here. The initial condition is a long wave of relatively large amplitude with a small amplitude high frequency, i.e., short wavelength, disturbance superimposed on top. The length scale of the short wave is arbitrary, i.e., 1 cm, but resembles that of capillary waves. A Gaussian is used for the large wave, instead of the single sine wave used by Holmås et al. (2008) which has a discontinuous derivative at both ends, namely:

$$\alpha_l(x, t=0) = 0.5 + 0.1\mathrm{e}^{-256\left(x-\frac{L}{2}\right)^2} + 0.002 \sin\left(\frac{2\pi x}{0.01024}\right). \tag{2.120}$$

The simulation of the KH unstable FFM, Eqs. (2.37)–(2.38), is obtained for the initial near linear stage of wave growth with $\rho_1 = 1000\,\mathrm{kg/m^3}$, $\rho_2 = 1\,\mathrm{kg/m^3}$, $u_1 = 0.2\,\mathrm{m/s}, C = 1$ (i.e., $u_2 - u_1 = 12.2\,\mathrm{m/s}$), $\varepsilon = \nu = 0, F = 0, H = 0.15$ m, with and without surface tension, i.e., $\sigma = 0.07\,\mathrm{N/m}$ or 0 N/m. The boundary conditions are periodic. According to the dispersion analysis, the short-wavelength ripples will die out and the large wave will grow with time if the model is well-posed.

Figure 2.23 (bottom) shows the simulation of the FFM with surface tension (i.e., the well-posed problem). The high frequency wave is below the cutoff frequency predicted by the linear stability analysis and it is damped out while the large wave grows. On the other hand, Fig. 2.23 (top) shows that when surface tension is not included, i.e., the model is ill-posed, the high frequency wave has a larger growth rate and dominates the solution after a short time.

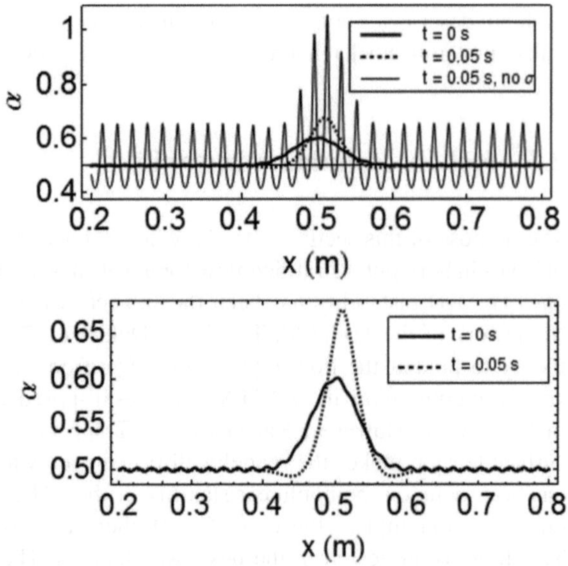

Fig. 2.23 Evolution of FFM wave with very short wavelength perturbation beyond KH threshold. *Thick solid line*: initial condition. *Dashed line*: well-posed, *thin solid line*: ill-posed

2.8 Summary and Discussion

The 1D TFM for horizontal or near-horizontal stratified flow has been simplified to the two-equation FFM to facilitate the local stability analyses of material waves. First, the incompressible assumption was applied to remove the acoustic waves, leaving only the material waves. Second, the fixed flux assumption removed the very long material wave instabilities. Finally the model was reduced to the SWT equations, in the limit $r_\rho \to 0$, which allows the well-known surface wave analysis of Whitham (1974) to be applied to the TFM. Two separate instabilities were identified, the dynamic KH instability and the kinematic SWT instability. Several results were obtained for these, using characteristic and dispersion analyses, under a unified treatment made possible by the simplified FFM equations.

The characteristic analysis shows that the FFM becomes ill-posed when the KH criterion is exceeded. This distinct KH-TFM condition arises because of the jump between the two-phase velocities and is not shared by single-phase flow models except for the exceptional multidimensional Euler vortex sheet problem. The dispersion analysis demonstrates that viscosity or surface tension, both of which stabilize the single-phase interface shear layer problem, make the TFM well-posed and in the process they stabilize it even though the shear layer physics is still absent. Furthermore, they change the nature of the material waves, in the case of viscosity, from hyperbolic to parabolic. Therefore, a well-posed TFM is hyperbolic–parabolic, like the Navier–Stokes equations. This linear result is both welcome and necessary but it is not sufficient, since the TFM must also be Lyapunov stable, i.e., wave growth must be bounded. In Chap. 4 we demonstrate Lyapunov stability with nonlinear simulations of a viscous 1D FFM that results in limit cycles and chaos.

The dispersion analysis also gives rise to the SWT instability of Whitham (1974) which occurs at a lower gas velocity and is used to predict the transition to wavy flow. An analytic expression for the stability boundary of stratified to wavy flow was derived from this condition. Finally the wave sheltering effect was incorporated into the model and good agreement was obtained with experimental data for the flow regime transition of Vallee et al. (2010).

The numerical stability of first-order finite difference schemes was analyzed with the von Neumann method. Even though these schemes have been well understood for a long time, their application presents a unique challenge when the TFM is ill-posed. It was found that while the FOU method may be used to regularize an ill-posed TFM, it also leads to nonconvergence in some cases. Furthermore, while the requirements that the scheme must be stable and nondispersive are easily met, too much stabilization may result not only in the desired dissipation of short waves but of long waves too, which may be detrimental.

A second-order scheme with a flux limiter was developed for a well-posed FFM. Its convergence was verified by the method of manufactured solutions (Roache 1998, 2002) and its capability to handle a contact discontinuity was verified with the water faucet problem. This numerical model is used throughout the book for the simulation of nonlinear behavior of the FFM. Finally a simulation was performed with this model for the initial near linear stage of material wave growth caused by the KH instability. A comparison between a well-posed and an ill-posed FFM illustrates the initial near linear behavior under the ill-posed condition, while nonlinear simulations will be pursued in the next two chapters.

The stability analysis methods developed so far for the FFM will be applied to a full TFM in the next chapter.

References

Anshus, B. E., & Goren, S. L. (1966). A method of getting approximate solutions to Orr–Sommerfeld equation for flow on a vertical wall. *AIChE Journal, 12*(5), 1004.

Arai, M. (1980) Characteristics and Stability Analyses for Two-Phase Flow Equation Systems with Viscous Terms. *Nuclear Science and Engineering, 74*, 77–83.

Banks, J. W., Aslam, T, and Rider, W. J. (2008) On sub-linear convergence for linearly degenerate waves in capturing schemes. *Journal of Computational Physics, 227*, 6985–7002.

Barmak, I., Gelfgat, A., Ullmann, A., Brauner, N., & Vitoshkin, H. (2016). Stability of stratified two-phase flows in horizontal channels. *Physics of Fluids, 28*, 044101.

Barnea, D., & Taitel, Y. (1993). Kelvin–Helmholtz stability criteria for stratified flow: Viscous versus non-viscous (inviscid) approaches. *International Journal of Multiphase Flow, 19*, 639–649.

Barnea, D. & Taitel, Y. (1994) Interfacial and Structural Stability of Separated Flow. *International Journal of Multiphase Flow, 20*, 387–414

Benjamin, T. B. (1959). Shearing flow over a wavy surface. *Journal of Fluid Mechanics, 6*, 161–205.

Brauner, N., & Maron, D. M. (1993). The role of interfacial shear modelling in predicting the stability of stratified two-phase flow. *Chemical Engineering Science, 48*(16), 2867–2879.

Burg, C. O. E., & Murali, V. K. (2006). The residual formulation of the method of manufactured solutions for computationally efficient solution verification. *International Journal of Fluid Dynamics, 20*(7), 521–532.

Cohen, L. S., & Hanratty, T. J. (1965). Generation of waves in the concurrent flow of air and a liquid. *AIChE Journal, 11*(1), 138–144.

Drazin, P. G., & Reid, W. H. (1981). *Hydrodynamic stability*. Cambridge: Cambridge University Press.

Drew, D. A., & Passman, S. L. (1999). Theory of multicomponent fluids. In *Applied mathematical sciences*. Berlin: Springer.

Drikakis, D., & Rider, W. (2005). *High resolution methods for incompressible and low-speed flows*. Berlin: Springer.

Gaskell, P. H., & Lau, A. K. C. (1988). Curvature-compensated convective transport: SMART, a new boundedness-preserving transport algorithm. *International Journal for Numerical Methods in Fluids, 8*, 617–641.

Gidaspow, D. (1974, September 3–7). Round table discussion (RT-1-2): Modeling of two-phase flow. In *Proceedings of 5th International Heat Transfer Conference*, Tokyo, Japan.

Gottlieb, S., & Shu, C.-W. (1998). Total variation diminishing Runge-Kutta schemes. *Mathematics of Computation, 67*, 73–85.

Henry, R. E., Grolmes,M. A., & Fauske, H. K. (1971). *Pressure-pulse propagation in two-phase one- and two-component mixtures. Technical Report ANL-7792*. Argonne National Laboratory.

Holmås, H., Sira, T., Nordsveen, M., Langtangen, H. P., & Schulkes, R. (2008). Analysis of a 1D incompressible Two-Fluid model including artificial diffusion. *IMA Journal of Applied Mathematics, 73*, 651–667.

Ishii, M., & Hibiki, T. (2006). *Thermo-fluid dynamics of two-phase flow* (1st ed.). New York: Springer.

Keyfitz, B. L., Sever, M., & Zhang, F. (2004). Viscous singular shock structure for a non-hyperbolic Two-Fluid model. *Nonlinearity, 17*, 1731–1747.

Kocamustafaogullari, G. (1985). Two-Fluid modeling in analyzing the interfacial stability of liquid film flows. *International Journal of Multiphase Flows, 11*, 63–89.

Kreiss, H.-O., & Yström, J. (2002). Parabolic problems which are ill-posed in the zero dissipation limit. *Mathematical and Computer Modelling, 35*, 1271–1295.

Kushnir, R., Segal, V., Ullmann, A., & Brauner, N. (2014). Inclined two-layered stratified channel flows: Long wave stability analysis of multiple solution regions. *International Journal of Multiphase Flow, 62*, 17–29.

Lax, P. D. (1972) Hyperbolic Systems of Conservation Laws and the Mathematical Theory of Shock Waves, SIAM, Philadelphia, USA

Lighthill, M. J., & Whitham, G. B. (1955). On kinematic waves I. Flood movement in long rivers. *Proceedings of Royal Society of London, 229*, 281.

Lopez de Bertodano, M. A., Fullmer W., Vaidheeswaran, A. (2013) One-Dimensional Two-Equation Two-Fluid Model Stability. *Multiphase Science and Technology, 25*(2), 133–167.

Oberkampf, W. L., & Roy, C. J. (2010). *Verification and validation in scientific computing*. Cambridge: Cambridge University Press.

Oberkampf, W. L., Trucano, T. G., & Hirsch, C. (2004). Verification, validation and predictive capability in computational engineering and physics. *Applied Mechanics Reviews, 57*(5), 345–384.

Ramshaw, J. D., & Trapp, J. A. (1978). Characteristics, stability and short wavelength phenomena in two-phase flow equation systems. *Nuclear Science and Engineering, 66*, 93–102.

Ransom, V. H. (1984). Benchmark numerical tests. In G. F. Hewitt, J. M. Delhay, & N. Zuber (Eds.), *Multiphase science and technology*. Washington, DC: Hemisphere.

Richtmeyer, R. D., & Morton, K. W. (1967). *Difference methods for initial-value problems* (2nd ed.). New York: Interscience.

Roache, P. J. (1998). *Verification and validation in computational science and engineering*. Albuquerque: Hermosa.

Roache, P. J. (2002). Code verification by the method of manufactured solutions. *Journal of Fluids Engineering, 124*, 4–10.

Roe, P. L. (1986). Characteristic-based schemes for the Euler equations. *Annual Review of Fluid Mechanics, 18*, 337–365.

Stadtke, H. (2006). *Gasdynamic aspects of two-phase flow: Hyperbolicity, wave propagation phenomena, and related numerical methods*. Weinheim, Germany: Wiley-VCH.

Strang, G. (2007). *Computational science and engineering*. Wellesley, MA: Wellesley-Cambridge.

Taitel, Y., & Dukler, A. E. (1976). A model for prediction of flow regime transitions in horizontal and near horizontal gas-liquid flow. *AIChE Journal, 22*, 47–55.

Tannehill, J. C., Anderson, D. A., & Pletcher, R. H. (1997). *Computational fluid mechanics and heat transfer*. Washington, DC: Taylor & Francis.

Tiselj, I., & Cerne, G. (2000). Some comments on the behavior of the RELAP5 numerical scheme at very small time steps. *Nuclear Science and Engineering, 134*, 306–311.

Vallee, C., Lucas, D., Beyer, M., Pietruske, H., Schutz, P., & Carl, H. (2010). Experimental CFD grade data for stratified two-phase flows. *Nuclear Engineering and Design, 240*, 2347–2356.

van Leer, B. (1979). Towards the ultimate conservative difference scheme. V. A second-order sequel to Godunov's method. *Journal of Computational Physics, 32*, 101–136.

Wallis, G. B. (1969). *One-dimensional two-phase flow*. New York: McGraw-Hill.

Waterson, N. P., & Deconinck, H. (2007). Design principles for bounded higher-order convection schemes—A unified approach. *Journal of Computational Physics, 224*, 182–207.

Whitham, G. B. (1974). *Linear and nonlinear waves*. New York: Wiley.

Chapter 3
Two-Fluid Model

Abstract The stability analysis of the four-equation incompressible TFM for horizontal stratified flow is presented. Both the characteristic equation and the dispersion relation are obtained. A von Neumann analysis of the semi-implicit First-Order Upwind (FOU) scheme, currently used in industrial reactor safety codes, is performed and compared to the dispersion analysis. A second-order finite difference scheme is then implemented and verified with the modified water faucet problem of Ransom (Course A—Numerical modeling of two-phase flows. Technical Report EGG-EAST-8546, EG&G Idaho, 1989). The difference with the previous chapter is that the flux is not fixed as a spatiotemporal constant. Removing the fixed-flux condition of the previous chapter brings in the implicit pressure Poisson technique which makes the algorithm significantly more elaborate.

This chapter also presents the nonlinear simulations of the four-equation TFM for the experiment of Thorpe (Journal of Fluid Mechanics, 39, 25–48, 1969) that is used to validate the model. The major outcome of this chapter is the implementation of surface tension and of the dissipative Reynolds stresses together into a validated well-posed TFM that has the capability to simulate local instabilities. These mechanisms are neglected in industrial 1D TFM codes but they have a strong impact on the numerical simulations of KH unstable flows for two reasons: they provide physical mechanisms to stabilize the TFM both linearly and nonlinearly and they allow the numerical model to converge in a statistical sense.

3.1 Introduction

The methods presented in the previous chapter for the Fixed-Flux Model (FFM) will now be applied to the four-equation Two-Fluid Model (TFM), focusing on the local instability of material waves and exploring their nonlinear evolution following Fullmer (2014). The four-equation TFM is mathematically more complex than the FFM. While the material wave results are similar, the full model brings the pressure into the equations and allows the simulation of acoustic waves (Stadtke 2006) and system transients. Also, the absence of the Fixed Flux condition allows the solution of global material waves and instabilities such as density wave oscillations which will be covered in Chap. 6.

M.L. de Bertodano et al., *Two-Fluid Model Stability, Simulation and Chaos*,
DOI 10.1007/978-3-319-44968-5_3

First, the local material wave linear stability is analyzed. The only difference with the previous chapter is that the low density ratio assumption is removed. As may be expected the results are similar.

Second, an in-depth assessment of a few finite difference numerical schemes, applied to the full TFM equations, is performed. The main difference with the previous chapter is the pressure Poisson equation technique of Harlow and Welch (1965), which is the method of choice in engineering applications. The first-order upwind (FOU) scheme is emphasized since it is the method currently used in the nuclear reactor safety community. Then, two second-order schemes using different flux limiters are compared. The verification of the numerical schemes is performed with a simple sine-wave test and the nonlinear water faucet problem. The material discontinuity of the water faucet problem is of particular interest since the formation of continuity shocks provide a nonlinear viscous dissipation mechanism identified by Kreiss and Ystrom (2006) that enhances the TFM Lyapunov stability.

Finally, a validation is performed using Thorpe's experiment (1969) for wavy near horizontal flow beyond the KH instability. This validation highlights the nonlinear behavior of the TFM waves and it motivates the more comprehensive nonlinear stability analyses of Chap. 4.

3.2 Incompressible Two-Fluid Model

The incompressible TFM equations that preclude the acoustic waves, cf. Sect. 2.2, are the starting point of the stability of material or void waves for near horizontal flows. A brief review of linear stability theory is needed to motivate the choice of the models that are finally selected. Ramshaw and Trapp (1978) included surface tension into the TFM to make it well-posed. This is a proper physical stabilization of short waves. There are also several artificial ways to regularize the TFM. Holmås et al. (2008) investigated the effect of artificial viscous terms on the stability of the TFM and found that a well-posed model was obtained. However, the latter requires the introduction of a viscous term in the continuity equations, which is not consistent with the rigorous averaging of the TFM but is representative of numerical regularization with the FOU scheme, widely used in industrial codes. There are other types of artificial regularization. Two examples in which artificial differential models are used are the CATHARE code (Bestion 1990) and the RELAP5/MOD3.3 code (Information Systems Laboratories 2003). In CATHARE a sometimes artificial interfacial pressure model is adopted and in RELAP5/MOD3.3 an incomplete virtual mass model which artificially extends the hyperbolic domain of the TFM is applied to all flow regimes.

In this chapter, an incompressible, isothermal well-posed 1D-TFM is applied to horizontal stratified flow in a rectangular channel. In this work, we do not derive the TFM from a local instant model—that is, a model composed of Navier–Stokes equations for each continuous phase separated by an interface—because (a) this has been performed previously by a variety of other authors and, (b) it is not trivial.

Entire texts are dedicated to the complicated averaging (time, volume, and/or ensemble) that must be performed, particularly with respect to the interfacial boundary and jump conditions. At this point, we direct readers that are unfamiliar with two-fluid modeling approaches to other texts, for example, Ishii and Hibiki (2006), Drew and Passman (1999), Morel (2015) or Gidaspow (1994) for authoritative examples of time, ensemble, volume, and kinetic theory paths to the TFM. Interested readers may also find the works of Anderson and Jackson (1967), Drew (1983), Ransom (1989), Zhang and Prosperetti (1994), Jackson (1997), Morel et al. (1999), Lakehal et al. (2002), Pannala et al. (2010, Ch. 1 and 2) worthwhile for additional reading on the two-fluid model approach. Some details of the averaging process from the general, 3D TFM to the 1D model given as follows are presented in Appendix A. Essentially, we make a relatively stringent assumption that the heavier phase is solely below the lighter phase so that the volume fraction at a given location is an indication of the height of the (single) interface between the two phases. The resulting model is therefore quite restrictive and not generally applicable to other geometries, other flow regimes, or even additional physics often associated with wavy stratified flow, e.g., the appearance of bubbles or droplets. Hence, readers are advised that the following results are likely not directly applicable to other two-phase scenarios of interest. However, the approach and the consequences of the general findings are, we feel, widely relevant to 1D TFM modeling, if indirectly.

Some may question why such an intricate study is carried out with such a specific 1D TFM or with a 1D model at all. One reason is that 1D modeling of two-phase flows remains relevant in industrial applications—particularly nuclear reactor safety. Even with the ever increasing power of leadership class supercomputing, 3D simulation of systems on the order of tens of meters or larger remains a challenging problem. Another reason is that we wish to address the ill-posed condition of the TFM. The discovery of the ill-posedness of the 1D TFM (Gidaspow 1974) set off quite a firestorm that has yet to fully die down. [We strongly recommend reading an article by Lyczkowski (2010) for a historical and somewhat personal perspective of the discovery and its aftermath.] In its wake, a fair amount of confusion remained; some believing that additional modeling could fix the problem, others that the 1D TFM is irreparably flawed. The following analysis, and this book as whole, attempts to address this concern or at least to shed some light on this divide. Therefore, without further ado, the incompressible, isothermal 1D-TFM for horizontal stratified flow in a rectangular channel is:

$$\boxed{\frac{D_1\alpha_1}{Dt} + \alpha_1\frac{\partial u_1}{\partial x} = 0,} \tag{3.1}$$

$$\boxed{\frac{D_2\alpha_2}{Dt} + \alpha_2\frac{\partial u_2}{\partial x} = 0,} \tag{3.2}$$

$$\boxed{\begin{aligned}\rho_1\frac{D_1u_1}{Dt} &= -\frac{\partial p_{2i}}{\partial x} - \rho_1 g_y H\frac{\partial \alpha_1}{\partial x} + \sigma H\frac{\partial^3 \alpha_1}{\partial x^3} + \rho_1 g_x - \frac{W+2\alpha_1 H f_1}{\alpha_1 A}\frac{}{2}\rho_1|u_1|u_1 \\ &+ \frac{1}{\alpha_1 H}\frac{f_i}{2}\rho_2|u_2-u_1|(u_2-u_1) + \frac{\rho_1}{\alpha_1}\frac{\partial}{\partial x}\left(\alpha_1\nu_1\frac{\partial u_1}{\partial x}\right),\end{aligned}} \tag{3.3}$$

$$\boxed{\begin{aligned}\rho_2\frac{D_2u_2}{Dt} &= -\frac{\partial p_{2i}}{\partial x} + \rho_2 g_y H\frac{\partial \alpha_2}{\partial x} + \rho_2 g_x - \frac{W+2\alpha_2 H f_2}{\alpha_2 A}\frac{}{2}\rho_2|u_2|u_2 \\ &- \frac{1}{\alpha_2 H}\frac{f_i}{2}\rho_2|u_2-u_1|(u_2-u_1) + \frac{\rho_2}{\alpha_2}\frac{\partial}{\partial x}\left(\alpha_2\nu_2\frac{\partial u_2}{\partial x}\right).\end{aligned}} \tag{3.4}$$

where, α_k, ρ_k, u_k, and p_{2i} are the averaged void fractions, densities, velocities, and the lighter phase interfacial pressure, respectively. The subscripts $k=1, 2$ indicate the heavier phase and lighter phase, as shown in Fig. 2.3. The void fractions are related by the total void fraction condition, $\alpha_1+\alpha_2=1$, which can be rearranged to give derivatives of one phase in terms of the other. The remaining parameters, g, H, σ, and ν_k are the gravity vector, channel height, surface tension, and the effective (material + turbulent) kinematic viscosities. The coefficients f_k and f_i are the phasic wall and interfacial Fanning friction factors, respectively.

3.3 Linear Stability

3.3.1 Characteristics

The characteristic equation corresponding to the first-order system of Eqs. (3.1)–(3.4) is given by Det$[\boldsymbol{B} - c\boldsymbol{A}] = 0$, where c are the characteristic wave speeds. The coefficient matrices are given by,

$$\boldsymbol{A} = \begin{bmatrix} 1 & 0 & 0 & 0 \\ -1 & 0 & 0 & 0 \\ 0 & \rho_2 & 0 & 0 \\ 0 & 0 & \rho_1 & 0 \end{bmatrix} \tag{3.5}$$

and

$$\boldsymbol{B} = \begin{bmatrix} u_2 & \alpha_2 & 0 & 0 \\ -u_1 & 0 & \alpha_1 & 0 \\ -\rho_2 g_y H & \rho_2 u_2 & 0 & 1 \\ -\rho_1 g_y H & 0 & \rho_1 u_1 & 1 \end{bmatrix} \tag{3.6}$$

for a variable vector $[\alpha_2, u_2, u_1, p_{2i}]^T$. The two acoustic characteristics are infinite: $c_{3,4} = \infty, \infty$, due to the assumption of incompressibility, and the two (finite) material characteristics are:

$$c_{1,2} = \frac{\widetilde{\rho u} \pm \sqrt{\alpha_1 \alpha_2 \left[-(u_2 - u_1)^2 \rho_1 \rho_2 + (\rho_1 - \rho_2) \widetilde{\rho} g_y H \right]}}{\widetilde{\rho}}, \tag{3.7}$$

where $\widetilde{\rho} = \alpha_1 \rho_2 + \alpha_2 \rho_1$ and $\widetilde{\rho u} = \alpha_2 \rho_1 u_1 + \alpha_1 \rho_2 u_2$. The term in the square root is the balance of inertia and gravity forces and the characteristics become imaginary, i.e., ill-posed or elliptic, when the Kelvin–Helmholtz stability condition is exceeded, i.e.,

$$(u_2 - u_1)^2 > \frac{(\rho_1 - \rho_2) \widetilde{\rho} g_y H}{\rho_1 \rho_2}. \tag{3.8}$$

This equation reduces to Eq. (2.36) in the limit $\frac{\rho_2}{\rho_1} \to 0$. Figure 3.1 shows the characteristic speeds as a function of void fraction for Thorpe's water–kerosene experiment (1969), cf. Sect. 3.6.1, where the dimensionless characteristic speed is $c^* = \frac{c - u_1}{u_2 - u_1}$. The experimental conditions are as follows: $\alpha_1 = \alpha_2 = 0.5$, $\rho_1 = 1000\ \text{kg/m}^3, \rho_2 = 780\ \text{kg/m}^3$, and $H = 0.03\ \text{m}$. The curves of $c^*_{1,2}$ combine to form closed elongated curves, each root spanning from one major vertex to the other. It is seen that for $u_2 - u_1 = 0.27$ and $0.28\ \text{m/s}$ the model is ill-posed for low void fractions. Also, the model becomes unconditionally ill-posed for greater relative velocities.

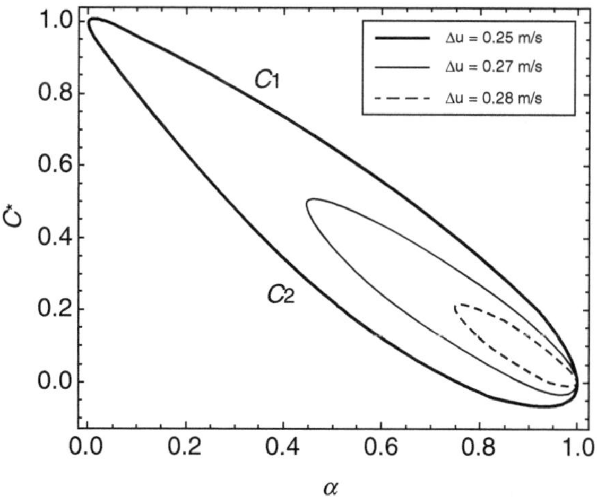

Fig. 3.1 TFM characteristics for the conditions of the Thorpe experiment (1969)

3.3.2 Dispersion Analysis

The equations (3.1)–(3.4) are now written in full matrix form,

$$\boldsymbol{A}\frac{d}{dt}\underline{\phi}+\boldsymbol{B}\frac{d}{dx}\underline{\phi}+\boldsymbol{D}\frac{d^2}{dx^2}\underline{\phi}+\boldsymbol{E}\frac{d^3}{dx^3}\underline{\phi}+\underline{F}=0, \tag{3.9}$$

so that the dispersion relation may be obtained

$$\text{Det}\left[-i\omega\boldsymbol{A}+ik\boldsymbol{B}+(ik)^2\boldsymbol{D}+(ik)^3\boldsymbol{E}+\frac{\partial\underline{F}}{\partial\underline{\phi}}\right]=0. \tag{3.10}$$

To simplify the linear analysis we set $\nu_1=\nu_2=\nu=$ constant. The product derivative viscous stress terms obtained after derivation by parts are linearized with respect to a flat interface and uniform velocities so they vanish. Thus, the diffusivity matrix is

$$\boldsymbol{D}=\begin{bmatrix}0 & 0 & 0 & 0\\ 0 & 0 & 0 & 0\\ 0 & -\rho_2\nu & 0 & 0\\ 0 & 0 & -\rho_1\nu & 0\end{bmatrix} \tag{3.11}$$

and the stiffness matrix is

$$\boldsymbol{E}=\begin{bmatrix}0 & 0 & 0 & 0\\ 0 & 0 & 0 & 0\\ 0 & 0 & 0 & 0\\ H\sigma & 0 & 0 & 0\end{bmatrix} \tag{3.12}$$

and $\boldsymbol{A}$ and $\boldsymbol{B}$ are given by Eqs. (3.5) and (3.6). The dispersion relation for the case without algebraic terms, i.e., $\frac{\partial\underline{F}}{\partial\underline{\phi}}=0$, that precludes the kinematic instability, can now be obtained resulting in:

$$\frac{\omega}{k}=\frac{\widetilde{\rho u}}{\widetilde{\rho}}-\frac{i}{2}\nu k$$
$$\pm\frac{\sqrt{\alpha_1\alpha_2\left[-(u_2-u_1)^2\rho_1\rho_2+(\rho_1-\rho_2)\widetilde{\rho}g_yH-\left(\frac{\widetilde{\rho}^2\nu^2}{4\alpha_1\alpha_2}-H\widetilde{\rho}\sigma\right)k^2\right]}}{\widetilde{\rho}}. \tag{3.13}$$

Surprisingly this equation is not much more complicated than the FFM Eq. (2.43) and reduces to it as the density ratio approaches zero. Beyond the KH criterion the model remains stable for short wavelengths, i.e., $k\rightarrow\infty$, so that waves do not grow below a critical wavelength and hence the TFM is well-posed. Below the KH criterion there is no wave growth at any wavenumber so the equations are well-posed and stable when friction forces are neglected, i.e., the kinematic instability that arises due to frictional forces discussed in Sect. 2.4.2 has been precluded. This is consistent with the statement by Ramshaw and Trapp (1978) that "the fact that the basic equation system constitutes an improperly posed problem is caused by a physical deficiency in the model—the failure to account for surface tension (or other short-wavelength effects)." Similar conclusions may be reached for other flow regimes, for example, bubbly flows described in Chap. 5. Therefore, it may be stated with some generality that the TFM is ill-posed only when it is too incomplete.

3.3.3 KH Instability

The present analysis follows closely the more complete multidimensional analysis of Thorpe (1969). In it he demonstrated that two major effects neglected so far in the 1D analysis only play a minor role for this particular case: acceleration and the velocity distributions around the interface, i.e., velocity distribution parameters (shape factors).

The Kelvin–Helmholtz instability condition is defined by Eq. (3.8). For the present analysis the variables are selected to match the conditions of Thorpe's water–kerosene experiment (1969) that is described in Sect. 3.6.1: $\alpha_1 = \alpha_2 = 0.5$, $\rho_1 = 1000\ \mathrm{kg/m^3}$, $\rho_2 = 780\ \mathrm{kg/m^3}$, effective viscosity for both phases $\nu = 0.0001\ \mathrm{m^2/s}$, $\sigma = 0.04\ \mathrm{N/m}$, and $H = 0.03$ m. The major difference with Chap. 2 is the density ratio, which now gives a relative velocity of $(u_1 - u_2)_i \cong 0.27$ m/s for the critical KH stability criterion, from Eq. (3.8). The phasic velocities, $u_2 = 0.2$ m/s and $u_1 = -0.2$ m/s, are chosen so that the problem is KH unstable, i.e., growth is expected at certain wavelengths. The wave growth rates under various assumptions are shown in Fig. 3.2 and are qualitatively similar to the results for the two-equation model of Chap. 2.

The basic 1D TFM corresponds to the 1D Euler equations of single-phase flow and is obtained from Eqs. (3.1) to (3.4) by setting $\nu = \sigma = 0$. The Euler TFM is ill-posed, i.e., growth rates increase unboundedly as the wavelength approaches zero. The zero wavelength growth rate is infinite for any relative velocity except for the trivial cases of homogeneous flow (i.e., zero relative velocity) or single-phase flow. In spite of this deficiency this TFM is often used in industrial codes, but it must be regularized in order to prevent high frequency waves from destroying the solution.

When kinematic viscosity is added to the TFM, i.e., $\nu = 0.0001\ \mathrm{m^2/s}$, the resulting growth rate increases as the wavelength decreases but it is finite in the zero limit so the maximum growth rates still occurs at zero wavelength, i.e., $\omega_i = \frac{\alpha_1 C}{\nu}$, and is very high which is unphysical. Therefore, while such a model

Fig. 3.2 Dispersion analysis of the 1D TFM for a fluid–fluid system

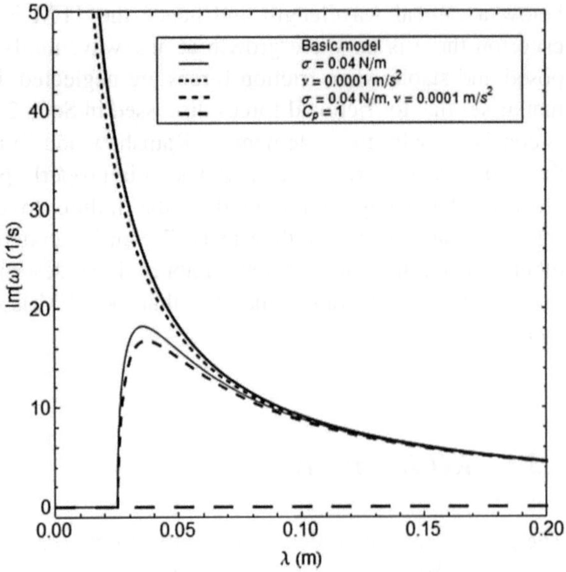

may not be ill-posed strictly speaking, it appears to offer little practical benefits over the ill-posed model. This result was first obtained by Arai (1980).

Considering now only the effect of surface tension, $\sigma = 0.04\ \mathrm{m/s}$, the model becomes well-posed because of the stabilizing effect of the capillary force at short wavelengths. However, even though physical stabilization is desirable, it is not used in industrial codes because its scale is usually well below the scales of engineering significance and larger scale artificial regularization is finally employed, as discussed in Section 2.4.1. Figure 3.2 also shows the combined result of using $\nu = 0.0001\ \mathrm{m^2/s}$ and $\sigma = 0.04\ \mathrm{m/s}$. The combined effect is practically a linear superposition of the individual effects, viscosity playing an insignificant role compared to surface tension. Surface tension stabilization (Ramshaw and Trapp 1978) is the earliest published demonstration that the unstable 1D TFM may be rendered well-posed by including short wavelength physics. Nevertheless, even when the model is well-posed, there is still strong exponential wave growth which presents a serious challenge to practical applications of the 1D TFM. In Section 3.6 and in the next chapter, the analysis of nonlinear simulations that overcome this difficulty is presented. It will be shown that viscosity is the key mechanism of nonlinear stability and that Lyapunov stability is indeed possible because the Lyapunov coefficient close to the nonlinear invariants is one order of magnitude smaller than the linear growth rate.

One means of artificial regularization is to add the interfacial pressure as an interfacial force,

$$M_1^{\mathrm{p}} = C_{\mathrm{p}}\rho_1(u_2 - u_1)^2\frac{\partial \alpha_1}{\partial x} \tag{3.14}$$

to the liquid momentum equation. Such a model has a physical interpretation for bubbly flow but it may be applied to all flow regimes simply because it hyperbolizes the TFM unconditionally when $C_p \geq 1$, so that the growth rate becomes zero for all wavelengths, as shown in Fig. 3.2. Since unstable stratified flows produce waves that grow, unconditional stability implies that the material wave dynamics of the model have been removed. This may be a desirable engineering compromise in TFM codes. However, removing the KH instability does remove significant physics that should be eventually recovered by ad hoc means, which are flow regime specific. For example, regularized 1D TFM industrial codes replace the material wave dynamics with static flow regime maps out of this necessity.

3.4 Numerical Stability

Before choosing a numerical scheme we need to keep in mind that the transient two-phase flow problem includes phenomena which have large variations in the associated time constants. The smallest time constant is associated with the inter-phase energy and mass transfer. These processes are closely coupled to the pressure and energy fields and to a large extent behave as source terms in the equations. Since the instabilities considered in this book are determined by the momentum equations, the TFM is simplified by applying the isothermal condition which removes the energy equations and these very short time scales.

The next fastest time constant is associated with relaxation of acoustic waves which are propagated relative to the fluid at the sound speed, but our focus is on material waves which may be captured with the incompressible TFM. Therefore, we treat the pressure implicitly and solve it with the well-known pressure Poisson equation using time steps much larger than those prescribed by the acoustic speed Courant limit (Harlow and Welch 1965).

Finally, the slowest time constants are associated with mass convection propagating with the material velocity. These time constants affect the numerical simulations in slightly different ways. The propagated effects result in material Courant-type time step limits for explicit differencing of the governing terms of the equations. For the purpose of this book which is the stability of the material waves it is desirable to model these waves explicitly. Thus, the semi-implicit finite difference schemes used in this chapter ensure this property and the time step is thus limited only by the material Courant limit.

If desired, this process can be carried one step further, and even the material Courant limit can be eliminated using more implicit schemes. In this regard several methods exist which are not material Courant time step limited, for example, the nearly implicit schemes, Mahaffy (1982); and the fully implicit schemes, Patankar (1980). However, using the long time steps associated with a fully implicit scheme has a severe impact on the prediction of the material waves, as presented in Appendix. B.4.2. Therefore, such schemes are not considered in this chapter.

3.4.1 TFIT Two-Fluid Model

The 1D TFM presented in Eqs. (3.1)–(3.4) is a rather complicated system of partial differential equations and analytical solutions to all but the most simple problems are impossible. In general, solutions of the system require numerical methods. Typically for the 1D TFM this has been achieved with either finite difference or finite volume schemes. The starting point of this work was a pilot code developed by Ransom (2000) from the earlier DISCON code (Trapp and Mortensen 1993) developed at the Idaho National Laboratory. The pilot 1D TFM code named TFIT uses a first-order upwind finite difference scheme very similar to the nuclear thermal-hydraulics code RELAP5 (ISL 2003), which was also developed at the INL with major contributions from the authors of TFIT.

A major feature of TFIT is the use of FOU damping to stabilize an ill-posed TFM, generally regarded as an acceptable approach to solve two-phase flow problems. TFM numerical stabilization has its roots in the single-phase flow scheme developed by Harlow and Welch (1965). The main source of damping in this scheme is the use of a donor formulation for the momentum flux term of the momentum equation, also known as upwind differencing. This method is the basis for the so-called semi-implicit scheme which was applied to two-phase problems by Lile and Reed (1978) as the basis for the original TRAC code numerical scheme.

The overly dissipative nature of the FOU scheme is actually a desirable feature for an ill-posed numerical 1D TFM. This is applicable to thermal-hydraulics codes for large-scale systems where short wavelength physical mechanisms, and in particular the KH instability, often occur one or more orders of magnitude below the scales of engineering interest. Thus, the numerical regularization of the FOU scheme has been widely accepted in industry and therefore warrants its own study. The FOU scheme is analyzed in depth in Sects. 3.4.2–3.4.6.

However, in this chapter short wavelength physics, i.e., surface tension and Reynolds stresses, have been included in the governing Eqs. (3.1)–(3.4) with the explicit objective of analyzing the KH instability. Therefore, the numerical diffusion needs to be reduced as much as possible. In Sect. 3.4.7 a higher order shock capturing scheme is proposed with an order of accuracy near two, i.e., $O(1.7)$, that effectively doubles the accuracy of the original FOU scheme.

3.4.2 Staggered Cell Structure

The governing equations are discretized with a finite difference scheme on a staggered grid shown in Fig. 3.3. With a staggered grid, the velocity and pressure are stored at separate locations (as opposed to a colocated grid, where all variables are located at the same positions). TFIT's staggered grid structure is similar to that of RELAP5 (Information Systems Laboratory 2003). The channel is divided into a

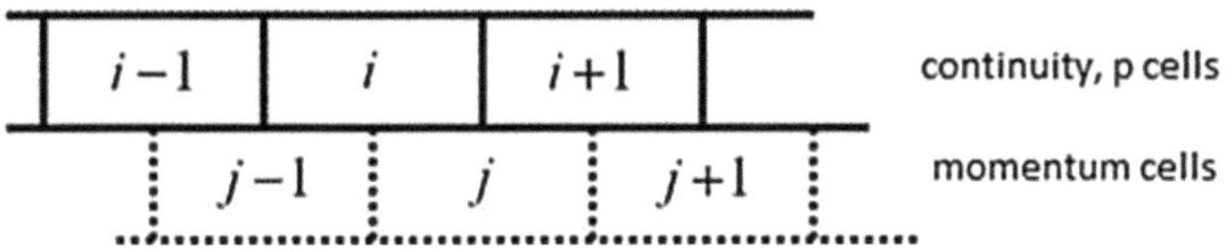

Fig. 3.3 Sketch of the staggered grid used in the TFIT code

number of finite volumes or cells with centers at the midpoints where the pressures, void fractions, densities, and temperatures are stored and junctions where the phasic velocities are stored. The cell centers can also be viewed as junction faces since they are located at the boundaries of momentum volumes. The shorthand R and L will be used to denote junctions or cell centers to the right and left of a reference location, e.g., the R, L junctions of cell center i are j, $j-1$, the R, L cell centers of junction j are $i+1$, i, and so on, as sketched in Fig. 3.3. Currently, only uniform grids will be used so that $x_{j+1} - x_j = x_{i+1} - x_i = \Delta x$ is constant. Additionally, the time is also discrete with $t = n\Delta t$ corresponding to the known time level n, i.e., the old time, and $n+1$ the unknown, new time level. In the equations that follow, the finite spatial location of each variable will be added as a subscript and the time level as a superscript. Note that the relative velocity is, for now, indicated with a lower case r to avoid confusion with the right cell indicator.

The conservative discrete forms of Eqs. (3.1)–(3.2) are given by,

$$\left(\frac{\partial \alpha_1}{\partial t}\right)_i + \frac{\hat{\alpha}^n_{1,R}\overset{*}{u}_{1,R} - \hat{\alpha}^n_{1,L}\overset{*}{u}_{1,L}}{\Delta x} = 0, \tag{3.15}$$

$$\left(\frac{\partial \alpha_2}{\partial t}\right)_i + \frac{\hat{\alpha}^n_{2,R}\overset{*}{u}_{2,R} - \hat{\alpha}^n_{2,L}\overset{*}{u}_{2,L}}{\Delta x} = 0. \tag{3.16}$$

The momentum equations (3.3) and (3.4) are multiplied by their respective void fractions and then discretized,

$$\begin{aligned}\overline{\alpha}^n_{1,j}\rho_1\left[\left(\frac{\partial u_1}{\partial t}\right)^n_j + u^n_{1,j}\frac{\overset{*}{u}_{1,R} - \overset{*}{u}_{1,L}}{\Delta x}\right] = &-\overline{\alpha}^n_{1,j}\frac{\overset{*}{p}_{2\text{int},R} - \overset{*}{p}_{2\text{int},L}}{\Delta x} + \overline{\alpha}^n_{1,j}\rho_1 g_y H\frac{\alpha^n_{1,R} - \alpha^n_{1,L}}{\Delta x}\\ &+ \overline{\alpha}^n_{1,j}\sigma g_y H\frac{\alpha^n_{1,i+1} - 3\alpha^n_{1,i} + 3\alpha^n_{1,i-1} + \alpha^n_{1,i-2}}{\Delta x^3}\\ &+ \frac{(\alpha_1\mu_1)^n_R\left(u^n_{1,j+1} - u^n_{1,j}\right) - (\alpha_1\mu_1)^n_L\left(u^n_{1,j} - u^n_{1,j-1}\right)}{\Delta x^2}\\ &- F^n_{w1,j}u^n_{1,j}\left(2u^{n+1}_{1,j} - u^n_{1,j}\right) - F^n_{\text{int}1,j}u^n_{r,j}\left(2u^{n+1}_{r,j} - u^n_{r,j}\right)\\ &+ \overline{\alpha}^n_{1,j}\rho_1 g_x,\end{aligned} \tag{3.17}$$

$$\overline{\alpha}_{2,j}^{n}\rho_2\left[\left(\frac{\partial u_2}{\partial t}\right)_j^n + u_{2,j}^n\frac{\overset{*}{u}_{2,\mathrm{R}} - \overset{*}{u}_{2,\mathrm{L}}}{\Delta x}\right] = -\,\overline{\alpha}_{2,j}^{n}\frac{\overset{*}{p}_{2\mathrm{int,R}} - \overset{*}{p}_{2\mathrm{int,L}}}{\Delta x} + \overline{\alpha}_{2,j}^{n}\rho_2 g_y H\frac{\alpha_{2,\mathrm{R}}^n - \alpha_{2,\mathrm{L}}^n}{\Delta x}$$
$$+\frac{(\alpha_2\mu_2)_\mathrm{R}^n\left(u_{2,j+1}^n - u_{2,j}^n\right) - (\alpha_2\mu_2)_\mathrm{L}^n\left(u_{2,j}^n - u_{2,j-1}^n\right)}{\Delta x^2}\overline{\alpha}_{2,j}^{n}$$
$$-F_{\mathrm{w}2,j}^n u_{2,j}^n\left(2u_{2,j}^{n+1} - u_{2,j}^n\right) - F_{\mathrm{int}2,j}^n u_{\mathrm{r},j}^n\left(2u_{\mathrm{r},j}^{n+1} - u_{\mathrm{r},j}^n\right)$$
$$+\,\overline{\alpha}_{2,j}^{n}\rho_2 g_x. \tag{3.18}$$

Furthermore, the volume fraction constraint is required:

$$\alpha_{1,i}^{n+1} + \alpha_{2,i}^{n+1} = 1. \tag{3.19}$$

The continuity equations are centered about a cell center, i, and the momentum equations centered about a junction, j. There are several new symbols which require additional explanation. The superscript * does not denote specific time level by itself, but is related to the pressure–velocity coupling which is discussed separately in Sect. 3.4.4.

If every superscript * were expressed as n, the scheme would be explicit, however, through the pressure–velocity coupling, the scheme is referred to as semi-implicit or pressure implicit. Additionally, the 'overbar' and 'hat' operators have been introduced for variables that are not defined at the denoted locations. In general, variables that are not related to derivatives are simply taken as the average of the two adjacent locations and denoted with the overbar, e.g., the void fractions multiplying the material derivatives on the LHS of Eqs. (3.17) and (3.18) are given by $\overline{\alpha}_{k,j}^{n} = \frac{\left(\alpha_{k,\mathrm{R}}^n + \alpha_{1,\mathrm{L}}^n\right)}{2}$. Variables that are related to convection that do not exist at a specified location are treated with a donoring or extrapolation scheme. The donoring scheme plays an important role in the accuracy and stability which is discussed separately in the following section. It is important to note that some variables at the R and L locations do exist due to the staggered grid. For example, in the spatial derivatives of Eqs. (3.15) and (3.16), the void fractions need to be donored while the velocities are known because the right and left faces of cell centers are junctions. Similar situations arise in the discrete momentum equations.

Rather than using only one void fraction, both are used and the sum of the void fractions, Eq. (3.19), is solved at every time step and coupled to the two continuity equations to ensure that void fraction continuity is preserved in the new time solution. This helps ensure that the void fraction of each phase is bounded between 0 and 1 and is important for the water packing algorithm.

The wall and interface drag terms, that are nonlinear functions of the velocity, have been linearized about the old time level. A general source term, S, that is a nonlinear function of u_k can be linearized as

$$S^{n+1} = S^n + \left(\frac{\partial S}{\partial u_k}\right)^n \left(u_k^{n+1} - u_k^n\right). \tag{3.20}$$

The semi-implicit quadratic wall and interfacial shear models given in Eqs. (3.17) and (3.18) reduce from Eq. (3.20), where the bulk coefficients are given by

$$F_{\mathrm{w}k,j}^n = \frac{W + 2\overline{\alpha}_{k,j}^n H}{2A} f_{\mathrm{w}k,j}^n \, \rho_k \mathrm{sgn}\left(u_{k,j}^n\right) \tag{3.21}$$

and

$$F_{\mathrm{int}\,k,j}^n = \frac{1}{2H} f_{i,j}^n \, \rho_2 \mathrm{sgn}\left(u_{k,j}^n - u_{nk,j}^n\right). \tag{3.22}$$

This allows the drag models to have an implicit evaluation without requiring computationally expensive algorithms to solve an otherwise coupled nonlinear problem.

All of the derivatives related to the staggered grid are second-order accurate center differences. The viscous terms are treated with a standard second-order finite difference stencil. The third derivative owing to surface tension is also a second-order accurate center difference even though, due to the staggering, the form of this term looks uncommon. All numerical terms described thus far are second-order accurate. It then falls to the donoring, time stepping, and pressure–velocity coupling to determine whether or not the scheme remains second order or reduces to first-order accuracy.

3.4.3 First-Order Semi-Implicit Scheme

The semi-implicit FOU numerical method for the TFM is similar to that already discussed in Sect. 2.5.2 for the two-equation FFM. In this section, the semi-implicit numerical method will be described as it is implemented in the TFIT code, developed by Ransom (2000). The spatial derivatives of the mass fluxes in the continuity equations are evaluated by using donor old-time void fractions and evaluating the velocity with which they are fluxed implicitly. In addition, the only spatial derivative in the momentum equations that are evaluated semi-implicitly are the pressure gradients leading to the pressure Poisson equation. This permits the momentum equations to be solved for the velocities in terms of the adjoining cell pressures at each velocity node.

The original TFIT code utilizes first-order upwinding (FOU) for all donored variables. The FOU method incorporates a physical interpretation into the finite difference scheme: the donored value is that of the adjacent cell or junction in the direction from which the information (flow) is propagating. Therefore,

$$\hat{\alpha}_{k,\mathrm{R}} = \begin{cases} \alpha_{k,i}, & \text{if } u_{k,\mathrm{R}} > 0 \\ \alpha_{k,i+1}, & \text{if } u_{k,\mathrm{R}} < 0 \end{cases} \tag{3.23}$$

for void fraction and

$$\hat{u}_{k,\mathrm{R}} = \begin{cases} u_{k,j}, & \text{if } \overline{u}_{k,\mathrm{R}} > 0 \\ u_{k,j+1}, & \text{if } \overline{u}_{k,\mathrm{R}} < 0 \end{cases} \tag{3.24}$$

for velocity donoring. The same scenarios are applied at the L sides shifted by one spatial location. Combining both donoring schemes into the difference equations, it is easy to see that the resultant finite difference schemes are first-order backward or forward differences depending on whether the flow is positive or negative, respectively. The leading order truncation error is proportional to Δx or $O(1)$. Additionally, the $O(1)$ truncation term is a coefficient of a second derivative which results in numerical diffusion. That is, even without a viscous term in the governing equations, the discrete equations will tend to diffuse or smear sharp gradients similarly to physical viscous diffusion. The simplicity, physical interpretation, and (most importantly) robust stability of the FOU scheme are a mainstay of all industrial nuclear reactor safety codes solving the 1D TFM.

As with most IBVP codes, the time marching was originally first-order backward Euler. For void fraction this is given by

$$\left(\frac{\partial \alpha_k}{\partial t}\right)_i = \frac{\alpha_{k,i}^{n+1} - \alpha_{k,i}^{n}}{\Delta t}, \tag{3.25}$$

and for velocity by

$$\left(\frac{\partial u_k}{\partial t}\right)_j = \frac{u_{k,j}^{n+1} - u_{k,j}^{n}}{\Delta t}, \tag{3.26}$$

where Δt is the time step size.

3.4.4 *Implicit Pressure Poisson Equation*

In the solution of the discretized equation system, a pressure correction algorithm based on the pressure Poisson equation, which is similar to the method first employed by Harlow and Welch (1965), is used to decouple the solution of the momentum and continuity equations. This and the staggered grid are the foundations of the semi-implicit schemes employed in this chapter. To quote Harlow (2004): "…the principal challenge for developing a numerical technique to solve incompressible fluid flows using the primitive variables is to find a way to ensure that the divergence of the velocity field vanishes. For the first version of the MAC method we attempted to describe the fields of velocity and pressure with co-located (cell-centered) values in a mesh of rectangular cells. The procedure was to be completely Eulerian, with the fluid flowing through the stationary mesh. We knew that a crucial requirement was to achieve the incompressibility constraint through the determination of pressures in each computational cycle consistent with the desired objective. In effect the achievement of that goal also will remove the sound-speed restrictions on the time step per cycle; sound signals could propagate to all parts of the mesh instantaneously. The only problem with the idea soon turned out to be the difficulty in expressing the divergence constraint with cell-centered placement of velocities. To overcome this impediment we changed to placement of velocity components at the edges of each computational cell... To derive the equation for pressure we used the... trick of taking the divergence of the moment (um) equation, and thus obtained a Poisson equation for the pressure..."

The solution procedure for a new time step follows three stages. First the momentum equations are solved with the old time pressure so that $p^{*}_{2\text{int}}$ becomes $p^{n}_{2\text{int}}$ in Eqs. (3.17) and (3.18). The momentum solution with the old time pressure gives an approximate, or pressure explicit, velocity prediction. Then, the new time velocities are linearized about the pressure explicit velocity in terms of the pressure gradient,

$$u^{n+1}_{k,j} = u^{*}_{k,j} + \left(\frac{\partial u_k}{\partial \Delta p_{2\text{int}}}\right)^{n}_{j} \left(\frac{\partial \Delta p_{2\text{int}}}{\partial t}\right)^{n+1}_{j}, \tag{3.27}$$

where $\Delta p_{2\text{int},j} = p_{2\text{int},R} - p_{2\text{int},L}$. Therefore, in the continuity Eqs. (3.25) and (3.26) the velocities are expanded in terms of the approximate velocity, the old time pressure gradient and, most importantly, the new time pressure gradient. This results in a pressure Poisson equation that couples the spatial domain at the new time level requiring matrix inversion, i.e., an implicit solution. This strategy is referred to as a semi-implicit or pressure-implicit method which is very important in slightly compressible flows to eliminate the need to satisfy acoustic Courant limits. Finally, after solving the coupled continuity equations for $p^{n+1}_{2\text{int}}$, the new time velocity can be corrected using Eq. (3.27).

3.4.5 von Neumann Analysis

The dispersion analysis has been used to study stability properties of the differential TFM, but ultimately the model is solved numerically. To determine whether the conclusions drawn from the dispersion analysis are still valid for a numerical TFM, the frequency response of the discrete equations must also be determined. Such an analysis is commonly referred to as a von Neumann stability analysis (von Neumann and Richtmyer 1949).

The von Neumann analysis is a form of linear stability analysis, therefore the products of independent variables need to be linearized, which generally consists of dropping the space–time dependence of the coefficients of the discrete TFM defined by Eqs. (3.15)–(3.18). Only the original FOU numerical scheme will be analyzed this way, since the second-order scheme would produce a quadratic matrix equation. Similarly to the dispersion analysis, the phase-1 volume fraction is written as a function of the phase-2 volume fraction and the remaining truly nonlinear terms are dropped. Then the von Neumann analysis is carried out on the linearized equations by considering the evolution of a single Fourier mode, from old time to new time, decomposing the discrete variable vector into

$$\underline{\phi}(x,t) = \hat{\underline{\phi}}(t)\mathrm{e}^{ikx}. \tag{3.28}$$

For example, the linearized momentum convection terms become:

$$\begin{aligned} \alpha_k\rho_k u_k \frac{u_{k,j}^n - u_{k,j-1}^n}{\Delta x} &= \frac{\alpha_k\rho_k u_k}{\Delta x}\left(u_k^n \mathrm{e}^{ikx_j} - u_k^n \mathrm{e}^{ik(x_j-\Delta x)}\right) \\ &= \frac{\alpha_k\rho_k u_k}{\Delta x}\left(1 - \mathrm{e}^{-ik\Delta x}\right)u_k^n \mathrm{e}^{ikx_j} \end{aligned}, \tag{3.29}$$

When the Fourier decomposition is applied to the linearized equations, the common terms e^{ikx_i} and e^{ikx_j} can be canceled from the mass and momentum equations, respectively. Then moving the new time variables to the left-hand side and the old time variables to the right, cf. Sect. 2.5.2, results in the linear system

$$\boldsymbol{M}\underline{\phi}^{n+1} = \boldsymbol{N}\underline{\phi}^{n}. \tag{3.30}$$

The nonzero (row, column) components of the new time matrix $\boldsymbol{M}$ are as follows:

$$\begin{aligned} M_{1,1} &= 1, \\ M_{1,2} &= i2(1-\alpha_2)r_\Delta \sin(k\Delta x/2), \\ M_{2,1} &= -1, \\ M_{2,3} &= i2\alpha_2 r_\Delta \sin(k\Delta x/2), \\ M_{3,2} &= (1-\alpha_2)\rho_1 + 2\Delta t\, F_{\mathrm{w}1}u_1 + 2\Delta t\, F_{i1}u_{\mathrm{R}}, \\ M_{3,3} &= -2\Delta t\, F_{i1}u_{\mathrm{R}}, \\ M_{3,4} &= i2(1-\alpha_2)r_\Delta \sin(k\Delta x/2), \\ M_{4,2} &= 2\Delta t\, F_{i2}u_{\mathrm{R}}, \\ M_{4,3} &= \alpha_2\rho_2 + 2\Delta t\, F_{\mathrm{w}2}u_2 - 2\Delta t\, F_{i2}u_{\mathrm{R}}, \\ M_{4,4} &= i2\alpha_2 r_\Delta \sin(k\Delta x/2) \end{aligned}$$

and the components of the old time matrix $\boldsymbol{N}$ are as follows:

$$
\begin{aligned}
N_{1,1} &= 1 - u_1 r_\Delta\left(1 - \mathrm{e}^{-ik\Delta x}\right),\\
N_{2,1} &= u_2 r_\Delta\left(1 - \mathrm{e}^{-ik\Delta x}\right) - 1,\\
N_{3,1} &= i2(1-\alpha_2)\rho_1 g_y H r_\Delta \sin\left(k\Delta x/2\right),\\
N_{3,2} &= (1-\alpha_2)\rho_1\left[1 - u_1 r_\Delta\left(1 - \mathrm{e}^{-ik\Delta x}\right) + 2\nu_1\left(\Delta t/\Delta x^2\right)\left(\cos\left(k\Delta x\right) - 1\right)\right]\\
&\quad + \Delta t\, F_{\mathrm{w}1} u_1 + \Delta t\, F_{i1} u_{\mathrm{R}},\\
N_{3,3} &= -\Delta t\, F_{i1} u_{\mathrm{R}},\\
N_{4,1} &= i2\alpha_2\rho_2 g_y H r_\Delta \sin\left(k\Delta x/2\right) + i(1-\alpha_1)\sigma H\left(\Delta t/\Delta x^3\right)\\
&\quad \left[6\sin\left(k\Delta x/2\right) - 2(3k\Delta x/2)\right],\\
N_{4,2} &= \Delta t\, F_{i2} u_{\mathrm{R}},\\
N_{4,3} &= \alpha_2\rho_2\left[1 - u_2 r_\Delta\left(1 - \mathrm{e}^{-ik\Delta x}\right) + 2\nu_2\left(\Delta t/\Delta x^2\right)\left(\cos\left(k\Delta x\right) - 1\right)\right]\\
&\quad + \Delta t\, F_{\mathrm{w}2} u_2 - \Delta t\, F_{i2} u_{\mathrm{R}},
\end{aligned}
$$

where $r_\Delta = \Delta t/\Delta x$ is the ratio of the time step to the grid spacing. All other components that are not explicitly given earlier are zero. Multiplying Eq. (3.30) by $\boldsymbol{M}^{-1}$ gives the recursion relation,

$$\underline{\phi}^{n+1} = \boldsymbol{M}^{-1}\boldsymbol{N}\underline{\phi}^{n} = \boldsymbol{G}\underline{\phi}^{n} \tag{3.31}$$

which relates the new time variables to the old time variables through the growth or amplification matrix, $\boldsymbol{G}$. For a matrix system, the amplification factor, ζ, can be determined from the eigenvalues of $\boldsymbol{G}$

$$\zeta = \max\left(\sqrt{\xi_i \xi_i^*}\right), \tag{3.32}$$

where ξ_i^* is the complex conjugate and ξ satisfies

$$\det(\boldsymbol{N} - \xi\boldsymbol{M}) = 0, \tag{3.33}$$

In other words, ζ is the modulus of the maximum eigenvalue of the amplification matrix. The amplification factor determines the growth or decay from one time step to another about a local node. Now, since *some* growth is expected, the amplification factors need to be converted into growth rates for direct comparison to the dispersion analysis of the differential equations. In the dispersion analysis it was assumed that the evolution of the perturbation in time could also be described by a Fourier mode, i.e., $\zeta = \mathrm{e}^{-i\omega^\Delta \Delta t}$. The imaginary component of the discrete growth rate is then,

$$\omega_{\mathrm{I}}^\Delta = \ln(\zeta)/\Delta t \tag{3.34}$$

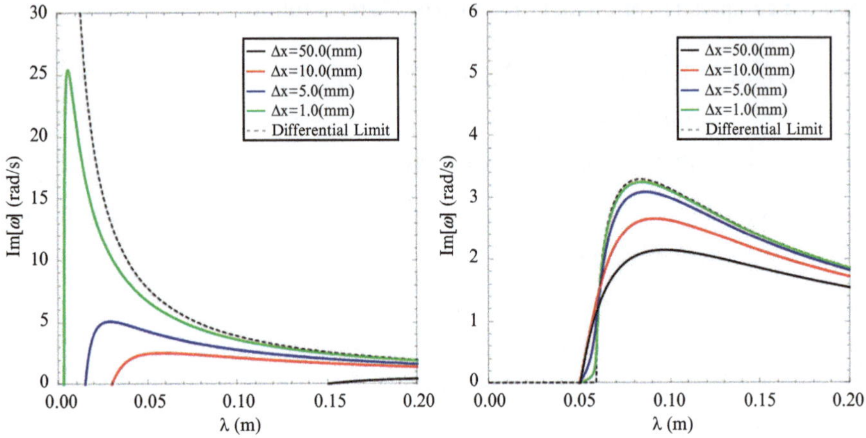

Fig. 3.4 Comparison of the differential and discrete growth rates for the ill-posed Euler TFM (*left*) and the well-posed TFM with surface tension (*right*)

which can be compared directly to the differential model growth rate, ω_I, obtained from the TFM dispersion analysis, Eq. (3.13).

The values of the discrete growth rates from the von Neumann analysis are compared in Fig. 3.4 to the differential growth rates that were calculated previously for the physical, material, and flow properties of Thorpe's experiment, cf. Sect. 3.6.2. Additionally, for the von Neumann analysis, appropriate mesh and time step information is also required. Here several grid sizes have been evaluated while keeping a constant Courant number of unity. An additional stability requirement exists when surface tension and viscosity are included since they are solved explicitly in TFIT as shown in Eq. (3.27). This stability requirement is neglected, without loss of generality, by simply moving the higher order terms to the new time matrix, which is very simple within the confines of the von Neumann analysis. The discrete growth rates are plotted beginning at a wavelength of $2\Delta x$, the smallest representable wavelength for a given grid. In general, the discrete results confirm what has already been discussed in the context of the dispersion analysis. However, the two are not entirely equivalent and two major topics warrant further analysis: numerical regularization of an ill-posed differential model and numerical stability requirements of an unstable, well-posed differential model.

3.4.6 Numerical Regularization

The finite difference equations are equivalent to the original differential equations with an additional infinite series of derivative terms of increasing order and decreasing magnitude (referred to as a modified equation). The leading-order truncation terms for the FOU method are second derivatives which represent numerical diffusion. The coefficient of the diffusion is proportional to Δx and is called the numerical viscosity. Such terms appear in all of the numerical equations—including the continuity equations. Therefore, the ill-posed TFM in discrete form is equivalent to an artificial diffusion model in differential form, cf. Sect. 2.4.1. The only difference between artificial and numerical viscosity is that numerical viscosity is grid and time step dependent.

Figure 3.4 (left) shows that the discrete growth rates of the ill-posed TFM, i.e., neglecting surface tension and turbulent viscosity, exhibit a critical and a cutoff wavelength, i.e., there is no growth at the $2\Delta x$ wavelength. In other words, the ill-posed differential model has been rendered well-posed by numerical regularization. This regularization has been the engineering artifice of choice in the nuclear reactor safety codes that would otherwise be plagued with intractable stability issues.

The rather obvious problem of using numerical regularization to achieve a well-posed model is that in the limit of $\Delta x \to 0$ the cutoff and most dangerous wavelengths, λ_0^Δ and $\lambda_i^\Delta \to 0$. This is a property of numerical methods that are consistent; the differential model is recovered in the limit, which for the present case is ill-posed. The grid dependent cutoff and critical wavelengths result in a differential model that is nonconvergent. Every time the mesh is refined a new critical wavelength appears on the scale of a few Δx. Nonconvergence issues in reactor modeling typically have been avoided by "user guidelines" that place limits on mesh refinement, e.g., $\Delta x \geq D_\mathrm{H}$. While limiting the grid size to a characteristic dimension may have some physical justification due to the area averaging, a nonconvergent numerical model remains troublesome in a mathematical sense, i.e., if the model does not converge then a true verification test of the code cannot be performed.

It is also of interest to determine how the finite difference scheme affects the growth rates when the governing differential model is well-posed yet unstable. In this case, the surface tension force has been included into the von Neumann analysis and compared to the previous result from the dispersion analysis in Fig. 3.4 (right). The general trend is the same as that of the Euler TFM: numerical viscosity reduces the growth rate and it approaches the differential result as $\Delta x \to 0$. The trend of $\omega_\mathrm{I}^\Delta \to \omega_\mathrm{I}$ as $\Delta x \to 0$ for constant r_Δ shows that the scheme is *consistent*, i.e., the discrete equations approach the differential equations in the limit, which should not be confused with convergence. In fact, Lax's equivalence theorem[1] cannot be used to prove convergence for this type of problem because it does not satisfy the traditional numerical stability condition, i.e., $\omega_\mathrm{I}^\Delta \leq 0$. However, in this case a verification test may be performed since the cutoff wavelength is independent of the discretization.

[1]Lax's Equivalence Theorem: Given a properly posed initial value problem and a finite-difference approximation to it that satisfies the consistency condition, stability is the necessary and sufficient condition for convergence.

While $\omega_{\rm I}^{\Delta} \leq 0$ cannot be applied here due to the inherently unstable (yet well-posed) differential model, there needs to be some additional criteria to assess the validity of the numerical method. Here the criteria proposed by Hwang (2003) has been adopted,

$$\max \omega_{\rm I}^{\Delta} \leq \max \omega_{\rm I}. \tag{3.35}$$

Equation (3.35) simply states that the maximum growth rate of the numerical model should not exceed what would be expected if the governing differential equations could be solved exactly. Figure 3.4 (right) also shows that $\lambda_i^{\Delta} \rightarrow \lambda_i$ and $\omega_{\rm I}^{\Delta}(\lambda_i^{\Delta}) \rightarrow \omega_{\rm I}(\lambda_i)$ as $\Delta x \rightarrow 0$. It is rather straightforward to show that the scheme is similarly consistent if $\mathrm{CFL}_k > 1$; however, the numerical stability requirement, Eq. (3.35), is violated as $\omega_{\rm I}^{\Delta}(\lambda_i^{\Delta})$ approaches $\omega_{\rm I}(\lambda_i)$ from above.

From the previous assessment of the regularized ill-posed differential model, an additional requirement is that the cutoff wavelength of a well-posed model should not vanish as the mesh is refined. With the FOU numerical method the discrete cutoff wavelength, λ_0^{Δ}, is slightly less than the differential cutoff, λ_0. However, this slight discrepancy is acceptable so long as the discrete cutoff wavelength remains constant as $\Delta x \rightarrow 0$, which is indeed the case.

3.4.7 Second-Order Semi-implicit Scheme

A higher order method for the 1D TFM cannot be arbitrarily selected. For instance, using a second-order linear upwind method everywhere has difficulties near sharp gradients. In nuclear reactor applications, regions of large gradients can be created by simulated breaks, valve openings, unramped flow regime transitions, or simply by the nonlinear behavior of the governing equations themselves. Therefore, a nonlinear scheme should be used to specify the degree of accuracy depending on the local solution conditions. One of the simplest nonlinear convections schemes is the flux limiter approach (Tannehill et al. 1997).

Another question is what order the method should be. This issue has not been addressed yet for 1D TFM solvers based on the pressure Poisson equation. Studies of the 2D incompressible Navier–Stokes equations, however, suggest that numerical schemes based on the pressure Poisson approach may experience unexpected challenges exceeding second-order accuracy (Guermond and Shen 2003; Guermond et al. 2006). Therefore, there may be no benefit in attempting to use a higher order method for convection if the global accuracy of the scheme is limited by other considerations.

Following the flux limiter approach of Sect. 2.5.5, if the local velocity field is positive, as defined by Eq. (3.23), the void fraction at the R face relative to the reference location i is donored by

$$\hat{\alpha}^{\,n}_{k,\mathrm{R}} = \alpha^n_{k,i} + \frac{\Delta x_i}{2}\Psi(r)\left(\frac{\partial \alpha_k}{\partial x}\right)_{\mathrm{UD}}, \tag{3.36}$$

where the local upwind derivative is given by

$$\left(\frac{\partial \alpha_k}{\partial x}\right)_{\mathrm{UD}} = \frac{\alpha^n_{k,i} - \alpha^n_{k,i-1}}{\Delta x}, \tag{3.37}$$

where the variable r is the derivative ratio of the center difference approximation to the upwind approximation. In the case of a uniform grid this reduces to

$$r = \frac{\alpha^n_{k,i+1} - \alpha^n_{k,i}}{\alpha^n_{k,i} - \alpha^n_{k,i-1}}. \tag{3.38}$$

The function $\Psi(r)$ is the flux limiter function, specified later. The exact same scheme of Eqs. (3.36)–(3.38) is used for velocity donoring, with each spatial location shifted by one half. In the case that the flow is negative, the void fraction is donored by

$$\hat{\alpha}^{\,n}_{k,\mathrm{R}} = \alpha^n_{k,i+1} + \frac{\Delta x_i}{2}\Psi(r)\left(\frac{\partial \alpha_k}{\partial x}\right)_{\mathrm{UD}}, \tag{3.39}$$

where the local upwind derivative is given by

$$\left(\frac{\partial \alpha_k}{\partial x}\right)_{\mathrm{UD}} = \frac{\alpha^n_{k,i+2} - \alpha^n_{k,i+1}}{\Delta x} \tag{3.40}$$

and the derivative ratio by

$$r = \frac{\alpha^n_{k,i+1} - \alpha^n_{k,i}}{\alpha^n_{k,i+2} - \alpha^n_{k,i+1}}. \tag{3.41}$$

Otherwise, for negative flow, the velocity is donored in the same manner as Eqs. (3.39)–(3.41) with a corresponding shift by one half of the grid spacing.

There are many potential candidates for the flux limiter function Ψ. One particularly attractive scheme is the general-piecewise limiter (GPL) of Waterson and Deconinck (2007), which is adopted here. The GPL scheme is appealing because it is relatively simple yet can easily be tuned to produce several popular schemes found in the literature and is given by

$$\Psi(r) = \max\left[0, \min\left\{(2+a)r, \frac{(1+k)r}{2} + \frac{(1-k)r}{2}, M\right\}\right]. \tag{3.42}$$

The classical Minmod flux limiter of Roe (1986) can be recovered by setting $a = k = -1$, $M = 1$. The Minmod limiter is one of the oldest and most widely used and is total-variation diminishing (TVD). However, Minmod is also one of the lowest order-of-accuracy flux limiter schemes but is still considered higher order compared to the linear FOU scheme. The popular MUSCL scheme of van Leer (1979) is obtained by the choice of $a = k = 0, M = 2$. The MUSCL scheme is also TVD and was found by Waterson and Deconinck (2007) to give the highest order of accuracy, $O(2.22)$, of all possible GPL schemes for smooth solutions. Another scheme produced by $a = 0$, $k = 1/2$, $M = 4$ is the converted normalized variable (NV) SMART scheme of Gaskell and Lau (1988). This scheme was found to perform better for discontinuous solutions and also retain $O(>2)$ for smooth solutions. It should be pointed out that, unlike the other schemes, the limiter form of SMART is not exactly TVD. However, it is nonoscillatory in its original NV form which has a rough equivalence to TVD (Drikakis and Rider 2005).

The GPL can be tuned to a few other schemes of less practical interest outlined by Waterson and Deconinck (2007). Also, it can be seen that if $\Psi(r) = 0$ or $\Psi(r) = 1$, then Eqs. (3.36) and (3.39) reduce to the original FOU scheme and the second-order center difference scheme, respectively.

Now that the donored variables can be calculated with higher than first-order accuracy, it is desirable to also upgrade the time marching algorithm. Higher order time advancement schemes typically take one of two forms: multistage methods or multistep methods. In the present TFIT code, the continuity and energy equations are solved together, which is important for problems involving phase change, but essentially eliminates the option for multistage methods which use several stages per time step, e.g., the Runge–Kutta method of Sect. 2.5.5. A multistep method eliminates the need for multiple stages by incorporating older-time data into the solution; this however is done at the cost of storing more information than just the old and new time variables.

One multistep method which is particularly attractive for the 1D TFM is the optimal second-order strong-stability preserving (SSP2) scheme of Gottlieb and Shu (1998). SSP schemes for time advancement are essentially a TDV counterpart for initial value problems. The optimal SSP2 multistep scheme is given by

$$\left(\frac{\partial \alpha_k}{\partial t}\right)_i = \frac{4\alpha_{k,i}^{n+1} - 3\alpha_{k,i}^{n} - \alpha_{k,i}^{n-2}}{6\Delta t} \tag{3.43}$$

for void fraction and

$$\left(\frac{\partial u_k}{\partial t}\right)_j = \frac{4u_{k,j}^{n+1} - 3u_{k,j}^{n} - u_{k,j}^{n-2}}{6\Delta t} \tag{3.44}$$

for velocity. This scheme only requires roughly double the amount of storage and memory issues were not encountered for the range of problem sizes considered herein. The form of Eqs. (3.43) and (3.44) is also desirable because it does not require retaining or rebuilding the RHS, i.e., everything not related to the time derivative, and can be readily implemented into existing Euler time stepping codes.

The Courant limit, i.e., $\mathrm{CFL}_k = u_k \Delta t / \Delta x$, for this scheme is reduced to one half, instead of one for the Euler method, but compared to the benefits of the scheme the reduced Courant limit is only a minor inconvenience. It should also be noted that the current form of Eqs. (3.43) and (3.44) is only valid for a constant time step, Δt.

3.5 Verification

3.5.1 Sine Wave

The first benchmark considered for code verification will only test the void fraction advection scheme, not the full TFM. There is a simple trick to reduce the TFM into two decoupled linear wave equations for void fraction. By considering an inviscid, horizontal flow with periodic boundary conditions, initially uniform pressure and initially uniform and homogeneous velocities, i.e., $u_1 = u_2 = U$, the velocity–pressure coupling of the TFM is removed. Therefore, the code only solves the significantly reduced system of a pair of one-way wave equations, cf. Appendix B.2.1,

$$\frac{\partial \alpha_k}{\partial t} + U \frac{\partial \alpha_k}{\partial x} = 0, \quad k = 1, 2. \tag{3.45}$$

The solution of Eq. (3.45) is simply $\alpha_k(x,t) = f(x - Ut)$ where $f(t=0) = \alpha_k(x, t=0)$ is the initial condition. In what follows $U = 1$ m/s and the computational domain is 1 m long. The simulations are run for a time of 1 s, which corresponds to one 'lap' around the domain, and the final result is compared to the initial condition. The time step for each grid is set by specifying $\mathrm{CFL} = 0.1$. Grid sizes range from $N = 20$ to 1000. To test the schemes against smooth initial data, a single sine wave initial condition is used,

$$\alpha(x, 0) = 0.5 + 0.3 \sin(2\pi x). \tag{3.46}$$

The three different schemes tested are the original first-order method (FOU) and the higher order method using the Minmod (SSP2-MM) and SMART (SSP2-SMART) reductions of the GPL scheme. Numerical solutions calculated using the three methods are compared to the exact solution in Figs. 3.5 and 3.6.

It is clear that the higher order schemes exhibit significantly less numerical diffusion compared to FOU and therefore conform better to the exact solutions. While the FOU scheme looks diffused throughout the domain, the higher order methods exhibit a slight deformation near the maximum and minimum of the sine wave. This is a result of the nonlinear flux limiter which changes the numerical scheme based on the local solution. More specifically, both flux limiter schemes reduce to the FOU donoring method about a local extremum. The superior performance of the SMART scheme, which is also the case for discontinuous initial data as reported by Waterson and Deconinck (2007), will be demonstrated in the next section.

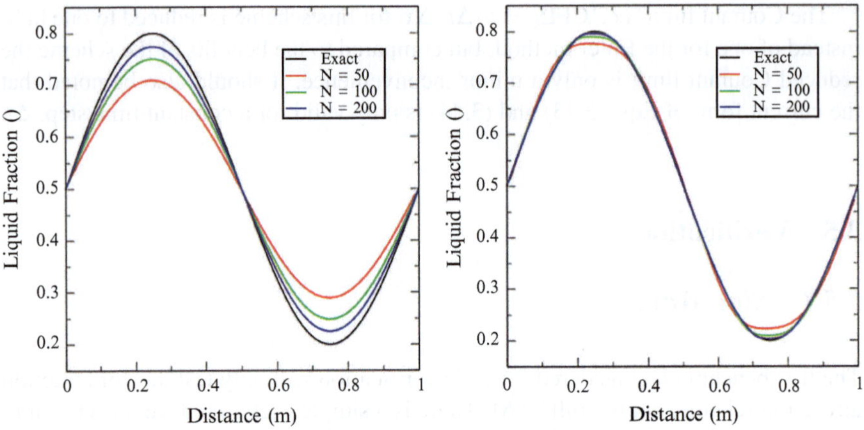

Fig. 3.5 FOU solutions of the linear wave equation compared to the exact solution (*left*), SSP2-MM solutions of the linear wave equation compared to the exact solution (*right*)

Fig. 3.6 SSP2-SMART solutions of the linear wave equation compared to the exact solution

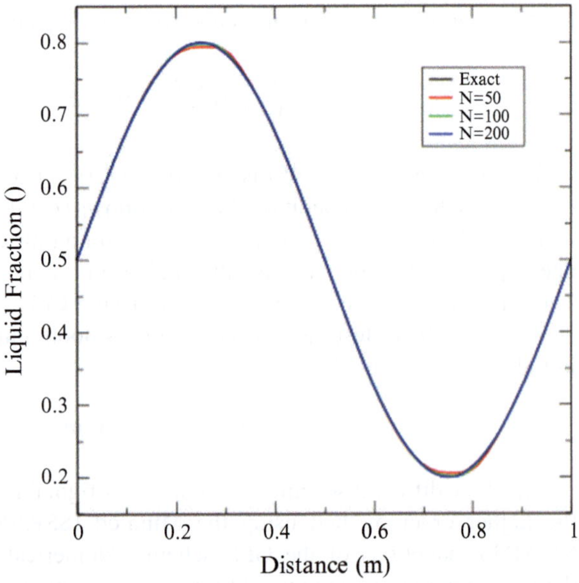

3.5.2 *Water Faucet Problem*

The SWT water faucet problem of Appendix B.5.5 has become a standard benchmark problem for 1D TFM codes. This variation of the Riemann problem is of particular interest because it is one of a few SWT cases for which there exists an exact and analytical discontinuous solution. It must be recognized that there is a fundamental difference between the four-equation TFM and SWT, i.e., the SWT water faucet problem is unconditionally KH stable because the condition $C = 0$ is prescribed whereas the four-equation TFM problem becomes KH unstable and ill-posed at the discontinuity. Therefore, one should expect beforehand that differences may exist between the exact solution of the SWT problem and the numerical TFM simulation. In order to highlight this difference, a few four-equation TFM solutions are benchmarked against the exact solution of the SWT water faucet problem, i.e., Eqs. (2.117) and (2.118). This may be compared with the FFM solutions presented in Sect. 2.6.5. It should be mentioned that an analytical solution of the multiphase water faucet problem at steady state has been recently advanced (Zou et al. 2016). This new solution is exact to the TFM rather than the SWT and should also prove useful to future TFM code developers.

The three numerical methods outlined in the previous section will be assessed. The numerical results provided here will use the liquid fraction profile which clearly presents the challenging problem of a growing discontinuity. Several grid resolutions have been studied ranging from $N = 12$ to 600 nodes, or $\Delta x = 1.0$ to 0.02 m., respectively. For all simulations the time step is given by $r_\Delta \equiv \Delta t / \Delta x = 0.01$ s/m to maintain a constant initial liquid Courant number of $\text{CFL}_l = 0.1$. The numerical solutions will be compared to the quasi-analytical solution at 0.5 s, when the discontinuity is roughly half way through the pipe.

The solution of the water problem with the FOU numerical method, shown in Fig. 3.7 (left), is relatively well behaved for $N < 600$ and looks similar to several published results that also utilized first-order methods, e.g., Ransom and Mousseau (1991). While the liquid fraction jump is diffused by numerical viscosity, the simulations remain stable. In the highest refinement case, a slight instability is seen downstream of the smeared jump. However, when one of the higher order schemes is used, oscillations in the solution begin to appear for resolutions of $N > 60$. Figure 3.7 (right) shows the SSP2-MM solution up to $N = 240$, beyond which point the overshoot becomes so severe that it fills the channel.

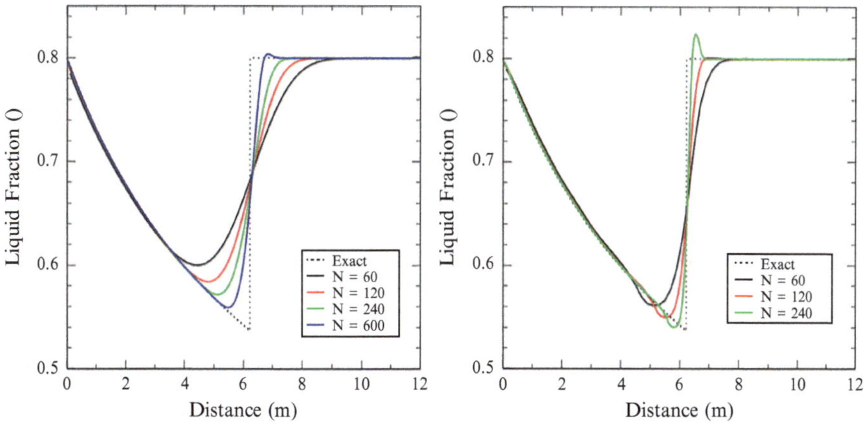

Fig. 3.7 Simulations of the water faucet problem at $t = 0.5$ s with FOU (*left*) and SSP2-MM (*right*) schemes vs. exact SWT solution

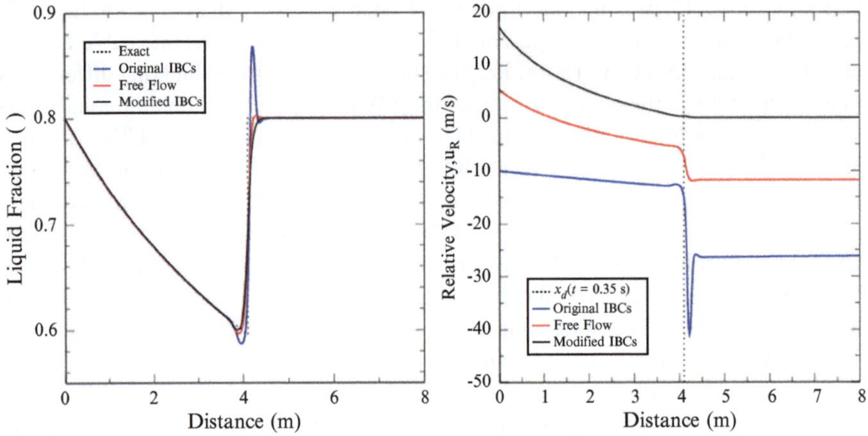

Fig. 3.8 SSP2-MM solutions of the water faucet problem for $N = 600$ at time $t = 0.35$ s showing the effect of different IBCs on the liquid fraction (*left*) and relative velocity (*right*)

The appearance of the oscillations is a result of the numerical viscosity being reduced approximately by an order of magnitude. The ill-posed nature of the differential problem, i.e., the KH instability, makes the frequency of the oscillations mesh dependent. As $\Delta x \rightarrow 0$ there is no limiting or cutoff wavelength because the model has not been made well-posed, i.e., surface tension and viscosity have not been modeled since they are not part of the original SWT formulation of the water faucet problem. In Fig. 3.8 (left) the case $N = 600$ for the SSP2-MM scheme is shown at an earlier time. It is apparent that the ill-posed model is the reason for the increasing instability.

The approach proposed to resolve the overshoot is not to add short wave physics or unphysical closure relations, but to modify the initial and boundary conditions (IBCs) of the water faucet problem to approximate, as closely as possible, the SWT equations for which the exact solution exists. Figure 3.8 (right) shows that the relative velocity upstream of the jump for the original problem is primarily due to the velocity of the falling water column alone but downstream it increases to over 40 m/s. This is due to the necking of the water column and the original boundary condition of zero gas flow at the inlet. As the column of water thins, air is required to fill the growing void and, since it cannot come from the inlet, it is drawn in from the outlet creating a countercurrent flow. The large relative velocity at the discontinuity causes the KH ripple to form, which grows due to the ill-posed nature of the problem further increasing the relative velocity. On the other hand, the analytical SWT solution does not have the gas velocity and therefore the relative velocity is essentially zero. The approach then becomes clear: to reduce the relative velocity—especially at the discontinuity—by modifying the IBCs.

3.5.3 Modified Water Faucet Problem

While the differences between the numerical solutions with higher order methods and the exact SWT solution may be used to justify adding hyperbolizing terms to the TFM, in lieu of making the TFM well-posed hyperbolic we prefer to modify the water faucet problem initial and boundary conditions to minimize the ripple growth due to the KH instability. While the ill-posedness of the differential equations certainly presents a problem for higher order methods, it should not be a justification for adding unphysical constitutive models. Physically, a falling column of water of the specified magnitude will exhibit interfacial instability and a host of other phenomena that are not modeled, such as ripple breaking and atomization, etc. In other words, the water faucet problem is not a validation problem but a verification problem, so it is desirable to distinguish between physics and mathematics and focus on the later.

One method to modify the problem is to simply change the inlet air boundary condition from a zero inflow BC to a free flow BC. With a free flow BC the filling air can now come from either end of the channel. Figure 3.8 (right) shows a significant reduction in the magnitude of the relative velocity and the corresponding reduction in the void oscillation near the jump (left) using a free flow BC. However, since the ill-posed water faucet 1D TFM is only stable for the homogeneous case, $u_R = 0$, the IBCs may be modified even further. First, the initial gas velocity is changed from $u_2(x,0) = 0$ m/s to $u_2(x,0) = 10$ m/s so that the problem is initially homogeneous everywhere. Second, the inlet boundary condition of the gas phase is changed to a function of time given by

$$u_2(0,t) = u_1(x_d(t)) \left[\frac{\alpha_2(x_d(t))}{\alpha_2(t_0)} \right], \tag{3.47}$$

where $x_d(t)$ is the location of the discontinuity. The modified inlet gas velocity given by Eq. (3.47) accounts for the acceleration of the liquid velocity at the jump (first RHS term) and the increasing gas volume that must be filled due to the necking of the water column (second RHS term). The modified IBCs result in a near-homogeneous condition around the jump even though the relative velocity leading up to the jump is now greater. Figure 3.8 (left) shows that the oscillation in the liquid fraction on the downstream side is thus completely eliminated and only a small bump remains on the upstream side. The complete results of the modified problem at $t = 0.5$ s are shown in Fig. 3.9 for the FOU (left) and SSP2-MM (right) schemes. The results for the SSP2-SMART scheme are shown in Fig. 3.10.

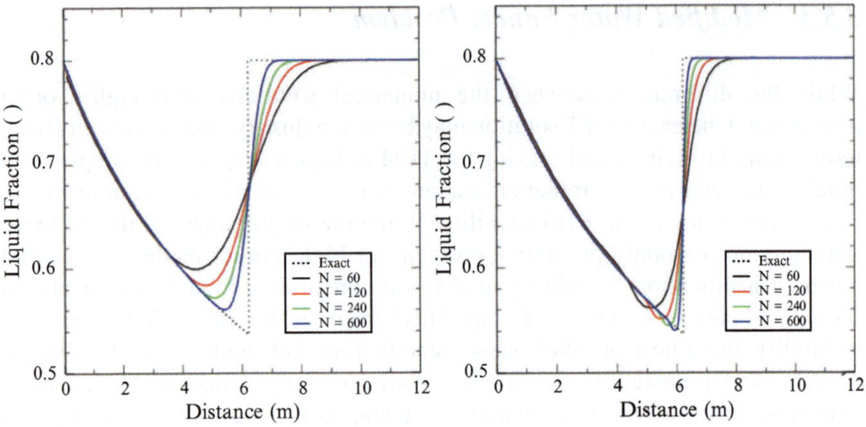

Fig. 3.9 Simulations of the modified water faucet problem at $t = 0.5$ s with FOU (*left*) and SSP2-MM (*right*) vs. exact SWT solution

Fig. 3.10 Simulations at $t = 0.5$ s with the SSP2-SMART scheme vs. exact SWT solution

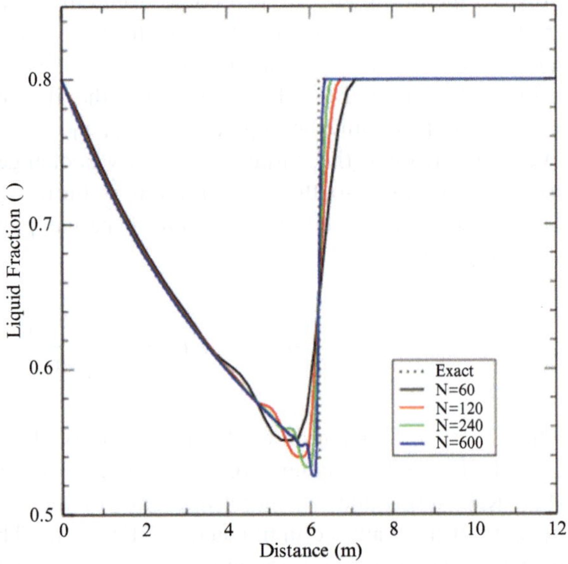

3.5.4 *Convergence*

Next, the TFIT code needs to be verified for convergence. In general, code verification is used to determine two things: whether or not the numerical model matches the differential equations correctly and, if it does, the global order of accuracy of the numerical method. The accuracy of the numerical solutions will be compared to the exact SWT solution, Eqs. (2.117) and (2.118), using the L1-norm defined by

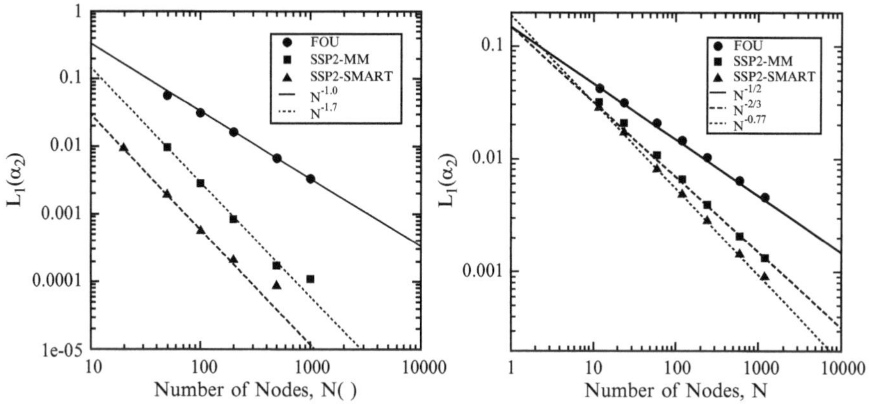

Fig. 3.11 Convergence rates of the three numerical schemes for sine wave (*left*) and water faucet (*right*) problems

$$L_1(\alpha) = \frac{1}{N}\sum_{i-1}^{N} |\alpha_i - \alpha_{\text{exact}}(x_i)|, \tag{3.48}$$

where N is the number of cells, α_{exact} is the exact SWT solution evaluated at x_i discrete locations, and α_i is the corresponding numerical solution.

Convergence tests for both cases and all three schemes are first carried out by calculating the L1 error, Eq. (3.48). The errors as function of grid number are then represented on a log–log plot to determine the global order of accuracy. Figure 3.11 shows a summary of the analyses for the sine wave and water faucet problems. Several key features are as follows: (a) the FOU scheme converges at exactly $O(1)$ for smooth initial data; (b) both higher order schemes converge at approximately the same rate, $O(1.7)$ (c) the convergence rates for the discontinuous initial data are significantly reduced compared to the sine wave, all with $O(< 1)$, for reasons already discussed in Sect. 2.6.5. As before convergence rates of the FOU and SSP2-MM schemes shown in Fig. 3.11 are extremely close to the theoretical values of $O(1/2)$ and $O(2/3)$, while the SSP2-SMART scheme is approximately $O(0.77)$.

3.6 Nonlinear Simulations

3.6.1 Thorpe Experiment

We select the water–kerosene experiment of Thorpe (1969) to validate the nonlinear TFM because it has several advantages over air–water flows. The first advantage of Thorpe's experiment is that, unlike air–water cases, the effect of the kinematic instability may be shown to be negligible so the KH instability may be analyzed in isolation.

The second advantage is that the flow is laminar so the interfacial stability may be analyzed separately from turbulence. Thorpe argues that the Tollmien–Schlichting instability near the walls is precluded because "the maximum Reynolds number, based on the displacement thickness of the boundary layers, which occurs in the experiments before the onset of instability, is 183, which is somewhat below the critical value of 575, estimated by Schlichting (1955) for the onset of instability near a wall. No similar conclusion can be reached about the instability at the interface, which may be characterized by a smaller critical Reynolds number." However, Barmak et al. (2016) recently demonstrated that the simple TFM stability analysis agrees with a more complete Orr–Sommerfeld analysis of the interface stability for laminar stratified flow under similar conditions to Thorpe's experiment. Therefore, a separate treatment of the interfacial instability from turbulence is supported by linear stability theory. Furthermore, Thorpe showed that what amounts to the current TFM stability analysis is in adequate agreement with the experimental results, adding validation to the theory.

The third advantage is that wave breaking is not significant, at least during the initial stages of wave growth. Therefore, the viscous TFM nonlinear simulations, including surface tension, may be performed for a case where one simple constitutive relation for the Reynolds stresses generated by the waves appears to be sufficient.

Thorpe's test section is a rectangular channel of dimensions $H = 0.03$ m, $W = 0.1$ m, and $L = 1.83$ m, schematically shown in Fig. 3.12. The channel is filled with equal volumes of water and a kerosene–carbon tetrachloride mixture with densities $\rho_1 = 1000\ \text{kg/m}^3$ and $\rho_2 = 780\ \text{kg/m}^3$. The remaining material properties are $\mu_1 = 0.001\ \text{Pa} \cdot \text{s}$, $\mu_2 = 0.0015\ \text{Pa} \cdot \text{s}$, and $\sigma = 0.04\ \text{N/m}$. The channel is initially at rest horizontally so the mixture is allowed to completely separate and stratify. Then the channel is suddenly tilted to a specified small angle. A countercurrent flow pattern evolves as the water rushes down, pushing the kerosene up. The time of the onset of instability is measured and the dominant wavelength is captured using a camera. For an angle of tilt of 4.1°, a series of photographs of the flow viewed from the side were published. The onset of the instability occurred at 1.85 s after inclination with an uncertainty of about a quarter of a second—half the estimated time to fully tilt the channel. The dominant wavelength measured after the onset of wave formation is approximately 3.13 ± 0.91 cm which is similar to the prediction of linear stability theory shown in Fig. 3.2.

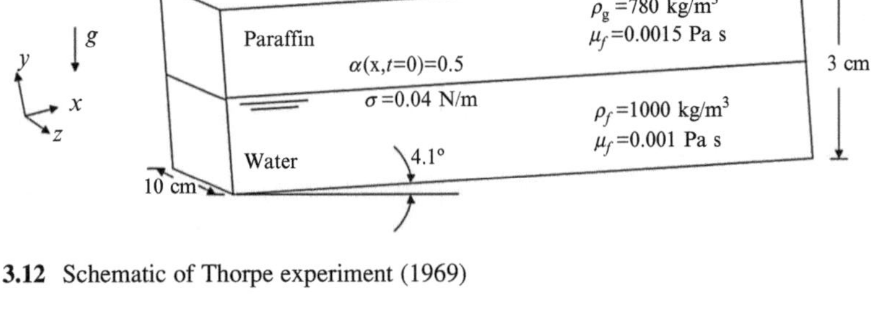

Fig. 3.12 Schematic of Thorpe experiment (1969)

3.6.2 Viscous Stresses

Reynolds stresses represent missing physics, i.e., vorticity, and they play the key role on the nonlinear stabilization of the current TFM Eqs. (3.1)–(3.4). Furthermore, they enable a higher order scheme with reduced numerical dissipation. The implementation of the Reynolds stresses is not trivial because of all the physics that is lost in the averaging. In addition to the pseudo-turbulent viscosity corresponding to the eddies caused by the waves, the covariance of the one-dimensional and three-dimensional strain rates needs to be considered. From a mathematical point of view, the Reynolds stresses are sufficient to demonstrate the role of the viscous nonlinear mechanism on the stability of the model, as shall be demonstrated in the next chapter.

A complete model for the wave induced effective viscosity is not presently available so a simple order-of-magnitude model is proposed, which we reckon sufficient for the numerical simulations herein. Since the phase densities of the Thorpe experiment (1969) are quite close, a first-order approximation is to neglect the damping action of the interface and treat the two-phase flow as a single-phase mixing layer. Even though the conditions of the Thorpe experiment would produce laminar flow in the absence of an interface, the interfacial waves give rise to eddies and velocity fluctuations that may be considered as a pseudo-turbulence. The turbulent viscosity of the mixing layer can be derived analytically (Pope 2009),

$$\nu_{\mathrm{T}} = 0.39^2\, S\, \delta(t) \Delta U, \tag{3.49}$$

where S is the spreading rate, $\delta(t)$ is the mixing layer thickness, and ΔU is the absolute value of the velocity difference between the two streams. The derivation of Eq. (3.49) is made by assuming that the mixing layer is self-similar, i.e., it is only valid if $S \equiv \frac{1}{\Delta U}\frac{d\delta}{dt}$ is constant, which has been experimentally verified for the case for single-phase flow. In the Thorpe problem, the velocity difference is initially increasing linearly so that Eq. (3.49) holds only if the mixing layer thickness expands quadratically in time. A CFD-VOF simulation of the Thorpe experiment performed by Fullmer et al. (2011) was used to verify that this is approximately true with $S \approx 0.0137$–0.0252. Finally, rather than using a time-dependent mixing layer thickness, an estimate of the maximum value is taken, i.e., $\delta(t) \approx H/2$, resulting in the simple turbulent viscosity model,

$$\nu_{\mathrm{T}} = 0.0015\, H|u_2 - u_1|. \tag{3.50}$$

Fullmer et al. (2011) then proposed a 1D Reynolds stress term that produces the same amount of dissipation as the Reynolds stresses obtained with the multidimensional simulation. It is constituted in terms of the turbulent viscosity and the covariance between the CFD and the 1D strain rate tensors. The asymptotic covariance thus obtained was approximately 7, yielding an effective viscosity of the order of $\nu_1 = \nu_2 = \nu_{\mathrm{eff}} = 0.0001\ \mathrm{m}^2/\mathrm{s}$ for the case of Thorpe, which is the value used for our numerical simulations in this chapter.

Finally, additional constitutive equations are required for the closure of the wall and interfacial shear stresses. In a multidimensional model these would be treated as differential terms but in a 1D model they are simple algebraic terms with friction factors.

3.6.3 Wall Shear

The wall shear terms are closed with a simple Darcy-type model,

$$\tau_{k\text{w}} = \frac{1}{2} f_k \rho_k |u_k| u_k, \tag{3.51}$$

where f_k is the phasic wall friction factor. It should be noted that for the wall shear model of Eq. (3.51), f_k is the Fanning friction factor rather than the Darcy friction factor, which is most often used in single-phase flow analysis. The two friction factors are related by $f_{\text{Darcy}} = 4 f_{\text{Fanning}}$ and the Fanning factor is used here simply for consistency with the interfacial drag models, below. The Blasius equation is used in the turbulent friction factor which is made continuous with its laminar counterpart by simply taking the larger of the two values:

$$f_k = \max\left[\frac{16}{Re_k}, 0.0791\, Re_k^{0.25}\right], \tag{3.52}$$

where Re_k is the phasic Reynolds number. The hydraulic diameters are defined by

$$D_{\text{H1}} = 4\frac{\alpha_1 A}{P_{\text{w1}}}, \quad D_{\text{H2}} = 4\frac{\alpha_2 A}{P_{\text{w2}} + P_{\text{i}}}, \tag{3.53}$$

where P_{i} is the length of the interface which is assumed to be equal to the width of the channel. The practice of using P_{i} in the definition of D_{H2} but not in D_{H1} appears to have originated with Agrawal et al. (1973) and persisted for most stratified two-phase flow analyses. The fraction of the wall wetted by each phase also follows directly from the assumed flow geometry,

$$P_{\text{w}k} = W + 2\alpha_k H \tag{3.54}$$

so long as each phase continues to wet the wall. In the limit of $\alpha_k \to 1$, $P_{\text{w}k} = 2W + 2H$ and in the limit $\alpha_k \to 0$, $P_{\text{w}k} = 0$.

3.6.4 Interfacial Shear

In the 1D two-fluid model, the most important of the interfacial momentum transfer forces is the interfacial drag. Drag is not only the most dominant of the interfacial forces but it is responsible for the strong coupling of the phasic momentum equations. The total drag force is decoupled into two components: the skin drag in τ_{ki} and the form drag. For a perfectly smooth interface, only the skin drag exists and is modeled with a Darcy-type correlation similar to the wall shear,

$$\tau_{ki} = \frac{1}{2} f_{\mathrm{i}} \rho_2 |u_{\mathrm{R}}| (u_k - u_{nk}), \tag{3.55}$$

where $u_{\mathrm{R}} = (u_2 - u_1)$ is the relative velocity and u_{nk} is the velocity of the other phase.

If the interface is roughened by small amplitude ripples the form drag may also be important. Like the interfacial shear, the form drag is also modeled with a drag coefficient correlation that would depend on the frontal area of the ripples projected in the streamwise direction. However, the drag coefficient factors determined experimentally are not separated into skin and form drag components. Therefore, the interfacial friction factor, f_{i} in Eq. (3.55), must be interpreted as a 'bulk' friction factor-drag coefficient for both skin and form drag. The ratio of P_{i}/A is the 2D equivalent of the interfacial area concentration, a_{i}. For the assumed quasi-2D flow geometry, the interfacial perimeter is simply the width of the channel and therefore $a_{\mathrm{i}} = H^{-1}$. Following Taitel and Dukler (1976a, b), the interfacial friction factor is proportional to the lighter phase wall friction factor $f_{\mathrm{i}} = C_{\mathrm{i}} f_2$ where the proportionality constant C_{i} accounts for interface roughness. Several values for this term have been proposed varying from 1 up to 15. In this work, $C_{\mathrm{i}} = 1$ is used for liquid–liquid flows, which was the value originally suggested by Taitel and Dukler (1976a, b). For air–water flows, $C_{\mathrm{i}} = 5$ has been suggested by Hurlburt and Hanratty (2001).

3.6.5 Single Nonlinear Wave

To illustrate the nonlinear wave dynamics, a Gaussian solitary wave similar to the one of Sect. 2.8 is simulated under the Thorpe experiment conditions. In Sect. 2.8 the role of physical short wave stabilization was illustrated during the near-linear stage, now we are interested in the nonlinear evolution of the "well-posed" wave. The length of the domain is set to $L = 0.5$ m. The higher order numerical method of Sect. 3.4.7 is used with $\Delta x = 1$ mm, $r_\Delta = 0.5$ s/m, and periodic boundary conditions. The initial conditions are $u_2 = -u_1 = 0.2$ m/s and

$$\alpha_1(x, t = 0) = 0.5 + 0.1\mathrm{e}^{-256(x - L/2)^2}. \tag{3.56}$$

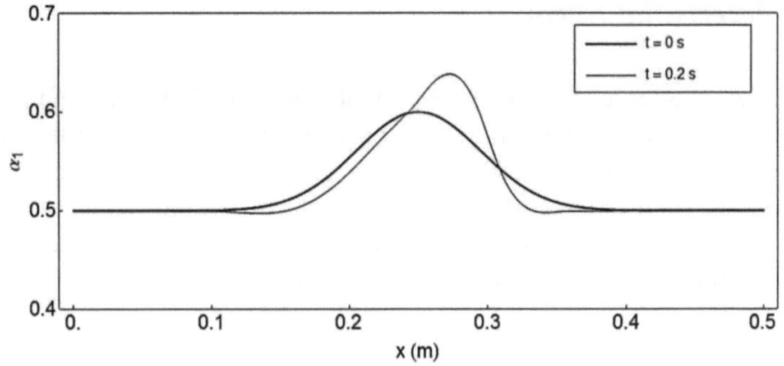

Fig. 3.13 Nonlinear evolution of a Gaussian wave

Figure 3.13 shows the wave profiles of the initial condition and the solution at $t = 0.2$ s considering viscous stresses and surface tension. At the final time the wave has grown due to the KH instability in a similar way to Burgers' equation waves, cf. Section B.3.1. The consequence of this wave evolution is the eventual formation of a continuity shock which provides a key nonlinear viscous dissipation mechanism first identified by Kreiss and Ystrom (2006) that limits the growth of the unstable waves.

3.6.6 Thorpe Experiment Validation

The Thorpe experiment is simulated with the TFIT code using the higher order SSP2-SMART scheme developed in Sect. 3.4.7. Three numerical grids are used in this section, $N = 900$, 1800, and 3600 corresponding to $\Delta x = 2.0$, 1.0 and 0.5 mm. The time step for the first two grids is set by $r_\Delta = 0.5$ s/m and for the higher resolution grid by $r_\Delta = 0.2$ s/m. The complete 1D TFM of Sect. 3.2 is used. Like the Thorpe experiment, the simulation is 'tilted' by multiplying the axial component of the gravity term by a factor of tanh(100 t). The ramping up of the axial gravity minimizes initial transients and allows the hydrostatic pressure profile to develop progressively. This tilting is performed more rapidly[2] than in the experiment—which was estimated to take up to half of a second—reaching 99 % by 0.0265 s.

The development of the waves in the center of the domain with $\Delta x = 1.0$ mm, reported by Fullmer et al. (2011), is compared to the photographs of Thorpe in Fig. 3.14. The numerical results of the height of the interface, $h_1 = \alpha_1 H$, are shown next to the photographs. The scale of the plots is approximately equal to the size of the domain shown in the photographs. It can be seen that the number and the general

[2]While this is a rapid ramping up of the axial force, it is a smooth transition. Previous simulations that did not include this tilting essentially impose a step change in the gravity vector at time zero.

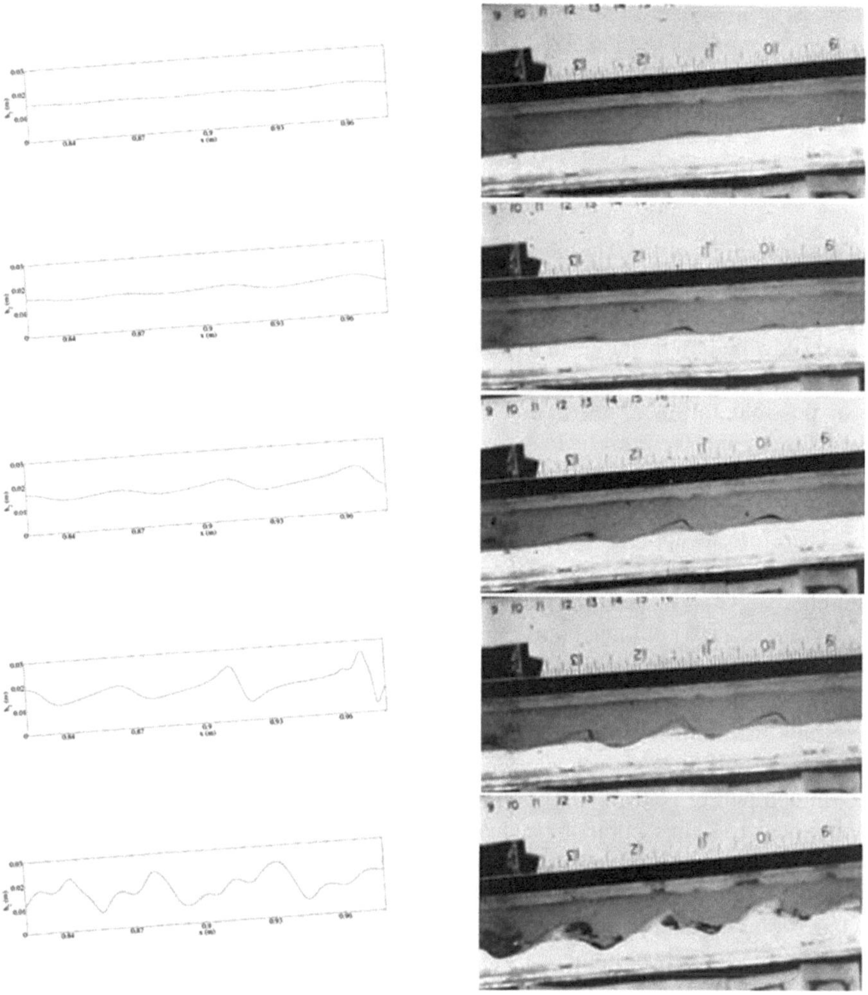

Fig. 3.14 Comparison of the numerical solution (*left*) to the experiment (*right*). Numerical solutions at times 3.23, 3.29, 2.35, 3.41, and 3.59 s. Experimental photographs at 2.0, 2.06, 2.12, 2.18, and 2.35 s. [Photographs reprinted from Thorpe (1969) with permission from Cambridge University Press]

development of the waves appears to be similar between the experiment and the numerical solutions. The wave forms that develop in the numerical solution are nonsymmetric and exhibit the same nonlinear behaviour previously seen for the hypothetical problem in the previous section. There are some discrepancies in the critical wavelength and time of transition between the solution and the data which is quantified below in Sect. 3.6.7. However, the most troubling issue with the results in Fig. 3.14 was that the numerical solution was not unique. When the grid was refined the solutions did not converge—at least once the waves developed.

This issue is investigated further in the following section, leading to the chaotic results set forth by the idealized nonlinear analyses of Chap. 4.

3.6.7 Convergence

The simulations of the Thorpe experiment have shown reasonable agreement with the experimental data but suffered from the problem of convergence. As the phases accelerate countercurrently, eventually the KH criterion is exceeded and a chaotic-wavy flow pattern develops in the center of the domain. In the next chapter, it will be shown that chaos is the root cause of nonconvergence. A full chaos analysis is not possible for the Thorpe case for two reasons: (a) the flow is continually developing and therefore it may be inappropriate to compare the Fourier spectra at different times and (b) the chaotic, wavy, supercritical KH flow only lasts for a few seconds before the fluids separate into the final, steady condition.

Therefore, an approximate Fourier approach is attempted for this spatiotemporal transient problem consisting of the ensemble average of Fourier transforms for a number of slightly perturbed simulations. The different cases are obtained by assuming a $\pm 10\ \%$ uncertainty in the inclination angle θ and each of the three friction factors, f_{i}, f_1, and f_2. Random multipliers of each variable, $\chi_{f_{\mathrm{i}}}$, χ_{f_1}, χ_{f_2}, and χ_θ, are obtained for 20 cases. Each multiplier value was selected from a uniform random distribution and normalized for the desired uncertainty range. Each of the 20 cases results in a distinct solution; all of which were run on three different grids.

In computing void fraction Fourier transforms, only the liquid fraction data in the center 0.512 m section of the channel is considered, i.e., $x \in 0.9 \pm 0.256$ m. The mean amplitude spectrum of the void fraction in wavenumber space is given in Fig. 3.15 (left) at $t = 4.0$ s for the three grids. The ensemble averaged spectra increase in wavenumber resolution with increasing mesh refinement.

Now the converged numerical results can be compared to the experimental data. The $\Delta x = 1.0$ mm grid is chosen because it is shown to resolve up to 99 % of the energy in the spectrum before deviating significantly from the $\Delta x = 0.5$ mm spectrum. On the other hand, the coarser 2.0 mm grid is only able to resolve about 90 % of the energy spectrum before deviating from the finer solutions. Two methods are used to extract information from the 20 simulations and compare with the measured wavelength.

The first method for determining the wavelength is simply to use the Fourier spectra directly. Now it is the long-wavelength (low wavenumber) portion of the spectrum that is important. Figure 3.15 (right) shows the development of the mean void fraction wave amplitude in time as a function of the wavelength. The predominant wavelength first appears in the range of 4–8 cm which is considerably longer than the experimental value of 3.13 $\pm$0.91 cm. As the simulation proceeds the maximum amplitude of the spectrum increases in magnitude and decreases in

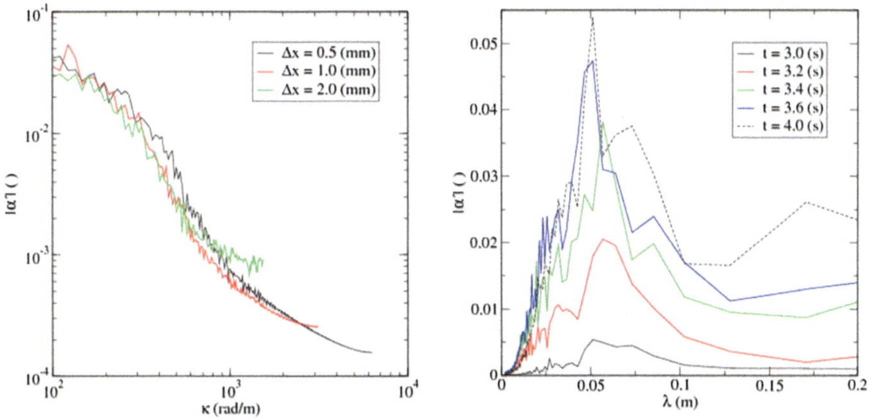

Fig. 3.15 Mean amplitude spectrum of the simulation water fraction in the center 0.512 m of the domain at $t = 4.0$ s with different grid resolutions (*left*) and development of the mean amplitude spectrum of the water fraction in the center 0.512 m of the channel with the $\Delta x = 1.0$ mm grid (*right*)

wavelength, down to approximately 5 cm, as the waves grow and steepen. At approximately 3.6 s the predominant wavelength becomes relatively constant as the waves have filled the center of the channel in all simulations.

One of the drawbacks with using the spectral method to characterize the wavelength of the void fraction variation is that it relies on the Fourier transform which is a linear transform. A Fourier transform does an excellent job at differentiating waveforms that are superimposed on one another, but it does a rather poor job of determining an average wavelength when waves of different wavelengths are adjacent to one another. Therefore, a second method for approximating the wavelength is proposed by simply counting the number of waves in the center 0.512 m of the domain. An efficient, albeit crude, approach is taken to identify a wave as a pair of extrema. Then the average wavelength can be approximated as

$$\lambda_n = \frac{L}{N_{\mathrm{ex}}/2}, \tag{3.57}$$

where $L = 0.512$ m is the test length (i.e., $x = 0.9 \pm 0.256$ m) and N_{ex} is the number of local maxima and minima. The n subscript is used to distinguish the wavelength determined from this integer counting method from the continuous wavelength of the Fourier analysis.

The major drawback of the second method is that equal weight is given to each identified wave, i.e., although the solution has small ripples that are 'riding on the back' of longer waves of larger amplitude, the ripples are counted with equal weight as the large waves. Therefore, the wavelength approximated in this manner is likely to give a significantly smaller wavelength than the spectral method. Figure 3.16 (left) shows that this is indeed the case.

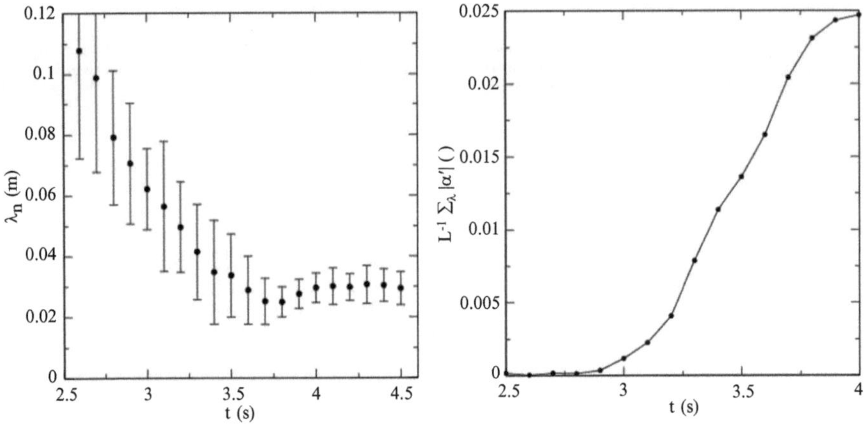

Fig. 3.16 Evolution of the approximate wavelength in the center 0.512 m of the channel with the $\Delta x = 1.0$ mm grid (*left*) and integral of the mean amplitude spectrum, as a function of time (*right*)

The wavelength shown in Fig. 3.16 (left) is the average calculated using Eq. (3.57) for the 20 different cases that were computed, with the error bars giving the standard deviation. At the onset of the instability the wavelength is quite large, approximately 10 cm, but decays to an equilibrium value of $\lambda_n = 3 \pm 0.5$ cm at $t = 4$ s. The smooth trend displayed is somewhat misleading. In actuality, each case decays smoothly to approximately $\lambda_n = 5$ cm and then undergoes an abrupt drop down to 3 cm. The time of this abrupt change varies for each case which is obscured by the averaging. Nonetheless, the overall trend of a longer initial wavelength with a period of reduction followed by a steady period is similar to the spectral results shown in Fig. 3.15.

While the computed predominant wavelength agrees well with the reported data (considering that this is a 1D model) Fig. 3.16 (right) shows that the time of transition is significantly delayed. In this representation, the onset of instability occurs at $t_0 = 2.8$ s. This time is obtained integrating the spectra (by finite summation) over all wavelengths at various times, normalized by the test length, analogous to the Fourier analysis shown in Fig. 3.15 (right).

The calculated time of onset of instability is delayed by nearly a full second from the reported experimental data,[3] $t_{onset} = 1.85 \pm 0.25$ s. The delay of the simulations also disagrees with the linear stability analysis which also predicted that a 10 cm wavelength disturbance would transition faster. One difference between the simulations and the linear stability analysis is the development of the relative velocity. The linear stability analysis assumed an inviscid linear acceleration, whereas the simulations include wall and interfacial drag. Actually, the delay

[3]It should be pointed out that the reported time of onset of instability (Thorpe 1969) had much smaller error bars. However, it should also include "half the time taken to tilt the tube" which was reported to be "usually about 0.25 s." Therefore, this value is used as the uncertainty.

considering the interfacial and wall shear is much closer to the experimentally measured onset of instability. Another discrepancy is that the simulation does not become unstable when linear stability predicts it should, because an initial perturbation is not present in the center of the domain. In the simulations the waves develop from the ends inward. The emptying and filling ends of the domain provide the necessary disturbance for wave formation, but the time for them to reach the center of the domain causes a significant time delay. The time for the onset of instabilities can be improved by introducing a perturbation to the void fraction and/or velocity field just after the center of the domain reaches the critical velocity (Fullmer et al. 2010). However, the introduced perturbations then strongly influence the spectra in the unstable region. Of the two quantities, we feel it is more important to let the spectra develop naturally than predict the transition time more accurately with an imposed perturbation.

3.7 Summary and Discussion

Linear analyses and nonlinear simulations of the stable and KH unstable1D TFM for stratified flows were performed. Furthermore, the stability of first- and second-order accurate finite difference schemes, based on the pressure Poisson equation approach, were analyzed. These methods are employed in current industrial TFM codes and furthermore they are needed for the nonlinear simulations performed in this chapter.

An overview of well-known TFM linear stability analysis was presented. The basic 1D TFM is ill-posed when the relative velocity exceeds the Kelvin–Helmholtz instability criterion. Inclusion of surface tension produces a system that is physically well-posed, i.e., with a cutoff wavelength below which there is no wave growth.

The numerical stability of a first-order finite difference scheme for an ill-posed and a well-posed model, including a pressure Poisson solver, was analyzed with the von Neumann method. The results are similar to those of the previous chapter. Then a second-order scheme with a flux limiter was developed for a well-posed model, again following the procedure of the previous chapter. Its convergence was verified for a sinusoidal void wave and it is $O(1.7)$. However, the convergence rates for the water faucet problem are practically equal to those obtained in Chap. 2, after the water faucet problem was modified to remove the KH instability that arises at the discontinuity when the ill-posed TFM is employed.

A physically well-posed TFM with surface tension alone is dispersive but not dissipative and unbounded wave growth at wavelengths beyond the cutoff occurs, making the model Lyapunov unstable. Therefore, a physical dissipation mechanism is required to saturate wave growth. The nonlinear interaction of viscous stresses with the steep wave fronts, first identified by Kreiss and Ystrom (2006), is the chosen mechanism in this book, although other mechanisms such as wave breaking are also possible. The experiment of Thorpe (1969) for KH unstable waves was used to validate the numerical TFM. The comparison of the nonlinear simulations with the data was satisfactory and the wave growth was bounded, i.e., Lyapunov stability was observed but not demonstrated.

In general, this chapter confirms the results of the previous chapter for the case of the full TFM. Furthermore, the simulations were extended to the nonlinear regime where the results appear to be chaotic. This is the motivation for the more rigorous FFM simulations and nonlinear analyses of the next chapter where Lyapunov stability will be demonstrated.

References

Agrawal, S. S., Gregory, G. A., & Govier, G. W. (1973). An analysis of horizontal stratified two-phase flow in pipes. *Canadian Journal of Chemical Engineering, 51*, 280–286.

Anderson, T. B., & Jackson, R. (1967). Fluid mechanical description of fluidized beds. Equations of motion. *Industrial & Engineering Chemistry Fundamentals, 6*(4), 527–539.

Arai, M. (1980). Characteristics and stability analyses for two-phase flow equation systems with viscous terms. *Nuclear Science and Engineering, 74*, 77–83.

Barmak, I., Gelfgat, A., Ullmann, A., Brauner, N., & Vitoshkin, H. (2016). Stability of stratified two-phase flows in horizontal channels. *Physics of Fluids, 28*, 044101.

Bestion, D. (1990). The physical closure laws in the CATHARE code. *Nuclear Engineering and Design, 124*, 229–245.

Drew, D. A. (1983). Mathematical modelling of two-phase flow. *Annual Review of Fluid Mechanics, 15*, 261–291.

Drew, D. A., & Passman, S. L. (1999). *Theory of multicomponent fluids. Applied mathematical sciences*. Berlin: Springer.

Drikakis, D., & Rider, W. (2005). *High resolution methods for incompressible and low-speed flows*. Berlin: Springer.

Fullmer, W. (2014). *Dynamic simulation of wavy-stratified two-phase flow with the one-dimensional two-fluid model*. Ph.D. Thesis, Purdue University, West Lafayette, IN.

Fullmer, W., Lopez De Bertodano, M., & Ransom V. H. (2011). The Kelvin-Helmholtz instability: Comparisons of one and two-dimensional simulations. In *The 14th International Topical Meeting on Nuclear Reactor Thermal Hydraulics (NURETH-14)*, Toronto, ON, Sept 25–29, 2011.

Fullmer, W., Prabhudharwadkar, D., Vaidheeswaran, A., Ransom V. H., & Lopez-de-Bertodano, M. (2010). Linear and nonlinear analysis of the Kelvin-Helmholtz instability with the 1D Two Fluid model. In *Proceedings of the 7th International Conference on Multiphase Flow*, Tampa, FL, May 30–June 4, 2010.

Gaskell, P. H., & Lau, A. K. C. (1988). Curvature-compensated convective transport: SMART, a new boundedness-preserving transport algorithm. *International Journal for Numerical Methods in Fluids, 8*, 617–641.

Gidaspow, D. (1974). Round table discussion (RT-1-2): Modeling of two-phase flow. In *Proceedings of the 5th International Heat Transfer Conference*, Tokyo, Japan, Sept 3–7, 1974.

Gidaspow, D. (1994). *Multiphase flow and fluidization: Continuum and kinetic theory descriptions*. San Diego, CA: Academic.

Gottlieb, S., & Shu, C.-W. (1998). Total variation diminishing Runge-Kutta schemes. Mathematics of Computation, 67, 73–85.

Guermond, J. L., Minev, P., & Shen, J. (2006). An overview of projection methods for incompressible flows. *Computer Methods in Applied Mechanics and Engineering, 195*, 6011–6045.

Guermond, J. L., & Shen, J. (2003). A new class of truly consistent splitting schemes for incompressible flows. *Journal of Computational Physics, 192*, 262–276.

Harlow, F. H. (2004). Fluid dynamics in group T-3 Los Alamos National Laboratory (LA-UR-03-3852). *Journal of Computational Physics, 195*(2), 414–433.

Harlow, F. H., & Welch, E. J. (1965). Numerical calculation of time-dependent viscous incompressible flow of fluid with free surface. *Physics of Fluids, 8*(12), 2182–2189.

Holmås, H., Sira, T., Nordsveen, M., Langtangen, H. P., & Schulkes, R. (2008). Analysis of a 1D incompressible two fluid model including artificial diffusion. *IMA Journal of Applied Mathematics, 73*, 651–667.

Hurlburt, E. T., & Hanratty, T. J. (2001). Prediction of the transition from stratified to slug and plug flow for long pipes. *International Journal of Multiphase Flow, 28*, 707–729.

Hwang, Y.-H. (2003). Upwind scheme for non-hyperbolic systems. *Journal of Computational Physics, 192*, 643–676.

Information Systems Laboratories. (2003). *RELAP5/MOD3.3 Code manual*, Vol. 1: Code structure, system models, and solution methods. NUREG/CR-5535/Rev P3-Vol I.

Ishii, M., & Hibiki, T. (2006). *Thermo-fluid dynamics of two-phase flow*. New York: Springer.

Jackson, R. (1997). Locally averaged equations of motion for a mixture of identical spherical particles and a Newtonian fluid. *Chemical Engineering Science, 52*(15), 2457–2469.

Kreiss, K. O., & Ystrom, J. (2006). A note on viscous conservation laws with complex characteristics. *BIT Numerical Mathematics, 46*, S55–S59.

Krishnamurthy, R., & Ransom, V. H. (1992). A non-linear stability study of the RELAP5/MOD3 two-phase model. In *Proceedings of the Japan-US Seminar on Two-Phase Flow*, Berkeley, CA.

Lakehal, D., Smith, B. L., & Milelli, M. (2002). Large-eddy simulation of bubbly turbulent shear flows. *Journal of Turbulence, 3*, N25.

Lile, D. R., & Reed, W. H. (1978). A semi-implicit method for two-phase fluid dynamics. *Journal of Computational Physics, 26*, 390.

Lyczkowski, R. W. (2010). The history of multiphase computational fluid dynamics. *Industrial & Engineering Chemistry Research, 49*, 5029–5036.

Mahaffy, John H (1982). A stability-enhancing two-step method for fluid flow calculations. *Journal of Computational Physics 46*(3): 329–341.

Morel, C. (2015). *Mathematical modeling of disperse two-phase flows*. New York: Springer.

Morel, C., Goreaud, N., & Delhaye, J.-M. (1999). The local volumetric interfacial area transport equation: Derivation and physical significance. *International Journal of Multiphase Flow, 25* (6), 1099–1128.

Pannala, S., Syamlal, M., & O'Brien, T. J. (Eds.). (2010). *Computational gas-solids flows and reacting systems: Theory, methods and practice*. Hershey: IGI Global.

Patankar, Suhas (1980). Numerical heat transfer and fluid flow. CRC press.

Pope, S. B. (2009). *Turbulent flows*. Cambridge: Cambridge University Press.

Ramshaw, J. D., & Trapp, J. A. (1978). Characteristics, stability and short wavelength phenomena in two-phase flow equation systems. *Nuclear Science and Engineering, 66*, 93.

Ransom, V. H. (1989). *Course A—Numerical modeling of two-phase flows*. Technical Report EGG-EAST-8546, EG&G Idaho, Idaho Falls, ID.

Ransom, V. H., & Mousseau, V. (1991). Convergence and accuracy of the RELAP5 two-phase flow model. In *ANS International Topical Meeting: Advances in Mathematics, Computations and Reactor Physics*, Pittsburgh, April 28–May 2, 1991.

Ransom, V. H. (2000) Summary of research on numerical methods for two-fluid modeling of two-phase flow, Consulting and Information Systems Laboratories.

Roe, P. L. (1986). Characteristic-based schemes for the Euler equations. *Annual Review of Fluid Mechanics, 18*, 337–365.

Schlichting, H. (1955). *Boundary layer theory*. London: Pergamon.

Stadtke, H. (2006). *Gasdynamic aspects of two-phase flow: Hyperbolicity, wave propagation phenomena, and related numerical methods*. Weinheim, Germany: Wiley-VCH.

Taitel, Y., & Dukler, A. E. (1976a). A model for prediction of flow regime transitions in horizontal and near horizontal gas-liquid flow. *AIChE Journal, 22*, 47–55.

Taitel, Y., & Dukler, A. E. (1976b). A theoretical approach to the Lockhart-Martinelli correlation for stratified flow. *International Journal of Multiphase Flow, 2*, 591–595.

Tannehill, J. C., Anderson, D. A., & Pletcher, R. H. (1997). *Computational fluid mechanics and heat transfer*. Boca Raton: CRC.

Thorpe, J. A. (1969). Experiments on the instability of stratified shear flow: Immiscible fluids. *Journal of Fluid Mechanics, 39*, 25–48.

Trapp, J. A. and Mortensen, G. A. (1993) A discrete particle model for bubble-slug two-phase flows. *Journal of Computational Physics, 107*, 367–377.

van Leer, B. (1979). Towards the ultimate conservative difference scheme. V. A Second-order sequel to Godunov's method. *Journal of Computational Physics, 32*, 101–136.

von Neumann, J., & Richtmyer, R. D. (1949). A method for the numerical calculation of hydrodynamic shocks. *Journal of Applied Physics, 21*, 232–237.

Waterson, N. P., & Deconinck, H. (2007). Design principles for bounded higher-order convection schemes—A unified approach. *Journal of Computational Physics, 224*, 182–207.

Zhang, D. Z., & Prosperetti, A. (1994). Averaged equations for inviscid disperse two-phase flow. *Journal of Fluid Mechanics, 267*, 185–220.

Zou, L., Zhao, H., & Zhang, H. (2016). New analytical solutions to the two-phase water faucet problem. *Progress in Nuclear Energy, 91*, 389–398.

Chapter 4
Fixed-Flux Model Chaos

Before the seas, and this terrestrial ball
And Heav'n's high canopy, that covers all,
One was the face of Nature; if a face:
Rather a rude and indigested mass:
A lifeless lump, unfashion'd, and unfram'd,
Of jarring seeds; and justly Chaos nam'd.

Ovid, Metamorphoses

Abstract The material wave growth of a well-posed Two-Fluid Model (TFM) beyond the Kelvin–Helmholtz (KH) instability has been approached in the previous chapter. The question that remains is what happens to the nonlinear evolution of the waves after their initial growth. Whitham (Linear and Nonlinear Waves, Wiley, New York, 1974) obtained a set of nonlinear solutions for Shallow Water Theory (SWT) consisting of shocks and expansion waves and identified the kinematic SWT instability. In Chap. 2 it was shown that the TFM may be approximated with the Fixed-Flux Model (FFM), which is equivalent to SWT up to the KH instability. Beyond that the FFM is unique and its nonlinear behavior is now addressed.

The chapter begins with the Kreiss–Yström equations (KY). It has already been shown in Sect. 2.6.3 that the linear stability of the KY equations is remarkably similar to the FFM with artificial viscosity. We apply some of the standard tools of system dynamics and chaos theory to the KY equations to obtain the largest Lyapunov Exponent and the fractal dimension. Fixed point, limit cycles and strange attractors are encountered on the route to chaos. We then proceed to the more difficult case of the FFM without artificial viscosity. A well-posed FFM is validated with a new experiment that is similar to Thorpe's (Journal of Fluid Mechanics, 39, 25–48, 1969) but focuses on the chaotic behavior of the waves past their initial growth, which Thorpe didn't report. Then, long-term simulations of the FFM with periodic boundary conditions are performed to obtain the largest positive Lyapunov exponent and the fractal dimension. The Largest Lyapunov exponent turns out one

Electronic supplementary material: The online version of this chapter (doi: 10.1007/978-3-319-44968-5_4) contains supplementary material, which is available to authorized users. Videos can also be accessed at http://link.springer.com/book/10.1007/978-3-319-44968-5_4

M.L. de Bertodano et al., *Two-Fluid Model Stability, Simulation and Chaos*,
DOI 10.1007/978-3-319-44968-5_4

order of magnitude smaller than the linear counterpart and eventually diverging trajectories become bounded by a strange attractor, i.e., Lyapunov stability.

Ultimately, the FFM chaotic behavior newly encountered differs significantly from the well-known linear theory. Therefore, it is important to distinguish between linear stability, which only determines whether the unstable TFM blows up instantaneously (ill-posed) or exponentially (well-posed) and which is valid for a very short interval of time, and nonlinear stability, which determines whether the problem is bounded in the long term.

4.1 Introduction

It has been established that a well-posed 1D TFM may be obtained once appropriate short wavelength physics are incorporated, a necessary condition that is not sufficient for understanding TFM stability without considering nonlinear dynamics. The initial nonlinear stages of the unstable behavior of the TFM beyond the KH instability, addressed in the previous chapter, do not indicate what happens to the nonlinear evolution of the material waves beyond their initial growth and saturation. We now perform long-term nonlinear simulations and analyses to explore what happens next and encounter chaos and Lyapunov stability.

Whitham (1974) performed a set of nonlinear solutions for SWT consisting of shocks and expansion waves and identified the kinematic SWT instability. In the linear analyses of Chap. 2 it was shown that the FFM is equivalent to SWT up to the KH instability. Beyond that the FFM nonlinear behavior is unique and previous research is very sparse. Kreiss and Yström (2002) (KY) and Keyfitz et al. (2004) were the first to analyze two-equation models mathematically similar to the FFM beyond the KH instability. They obtained shocks and expansion waves similar to SWT and further observed that the nonlinear interaction of the steep wave fronts with viscosity bounds the growth of the waves. Other than that Fullmer et al. (2014) extended the work of KY to the domain of chaos theory. It has already been shown in Sect. 2.6.3 that the linear stability of the KY equations is remarkably similar to the FFM with artificial viscosity. In this chapter, we apply some of the standard tools of system dynamics and chaos theory, introduced in Appendix B.6, to the KY equations and then proceed to the more difficult case of the FFM without artificial viscosity.

The nonlinear behavior of a well-posed TFM for the experiment of Thorpe (1969) was presented in Sect. 3.6. Those simulations had the shortcoming that the experiment was too short to achieve the stationary state required for the analysis of chaos. In this chapter, a new longer term experiment is used to validate a FFM for long-term chaotic simulations. The largest Lyapunov coefficient and the fractal dimension are then obtained with the validated FFM. The issue of numerical convergence is also addressed, considering that the model is chaotic.

4.2 Chaos and the Kreiss and Yström Equations

4.2.1 Nonlinear Simulations

The verified numerical method of Sect. 2.6 may now be used to solve the KY Eqs. (2.98) and (2.99) with $C = 1$ and $\varepsilon = \nu = 0.025$. The first and most obvious test is to simulate the original problem of KY and compare our finite difference simulations to their results obtained with a spectral method. The initial conditions are Gaussians of different widths defined by $\alpha_0 = \mathrm{e}^{-2x^2}$ and $u_0 = \mathrm{e}^{-4x^2}$, which are periodic about the origin. The domain is $x \in [-\pi, \pi]$ and the grid is divided into $N = 512$ uniform nodes and the time step is set to $\Delta t = 0.0002$. Comparisons to the published solutions of KY are given in Fig. 4.1.

From the initial condition to $t = 1$ the two numerical solutions are nearly indistinguishable and at $t = 4$ only relatively minor differences are observed. However at $t = 40$ the two solutions are entirely different, even though the features of the two solutions are similar.

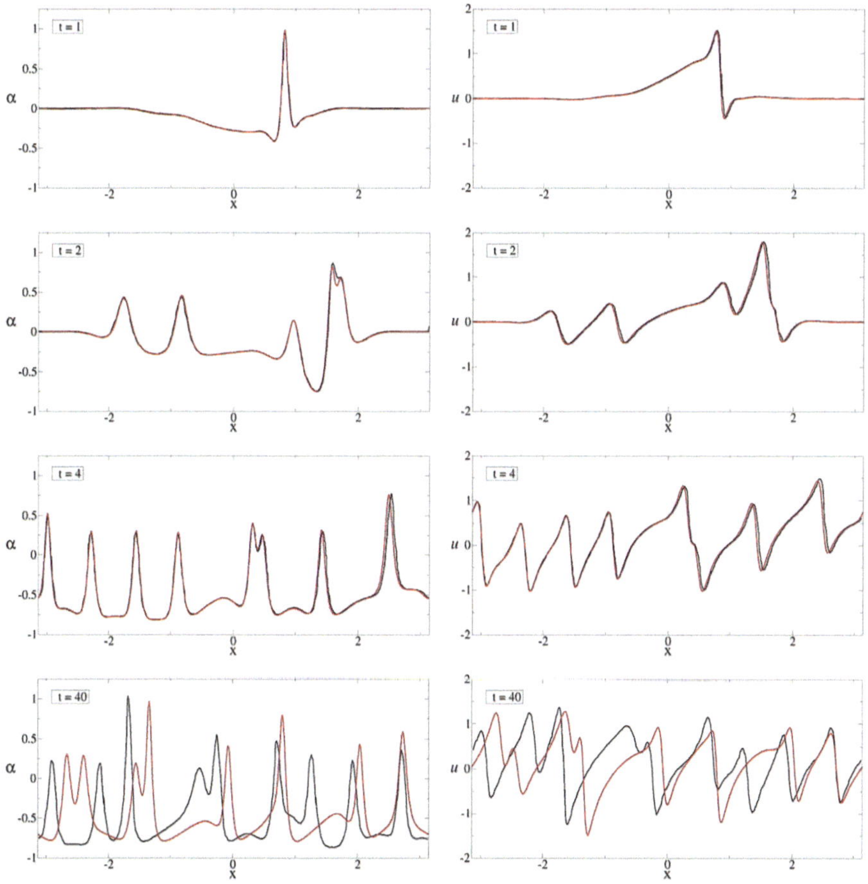

Fig. 4.1 Numerical solution to the KY equations (*red*) compared to the original results (*black*), reprinted from Fullmer et al. (2014), with permission from Elsevier

On the surface the differences don't appear to be so troublesome. After all, the numerical solution of KY was obtained using a higher resolution pseudo-spectral method which they showed to converge up to $t = 8$. Demonstrating that the numerical solution is fully converged, or grid independent, ensures that the numerical solution is an accurate approximation to the system of PDEs. Quantifying the accuracy of a given solution is the process of solution verification or numerical error estimation. Solution verification can take on several forms: ranging from estimating the exact solution from a series of numerical solutions using the method of Richardson (1926) to a more practical approach of refining the grid until two successive solutions become (relatively) indistinguishable. The exact method for solution verification seems to be irrelevant here because any method is predicated on the simple fact that the solutions on successively finer grids are in fact converging. The solutions at $t = 40$ are shown in Fig. 4.2 for successively finer grids from $N = 512$ to $N = 4096$. It can be seen that even with nearly an order of magnitude

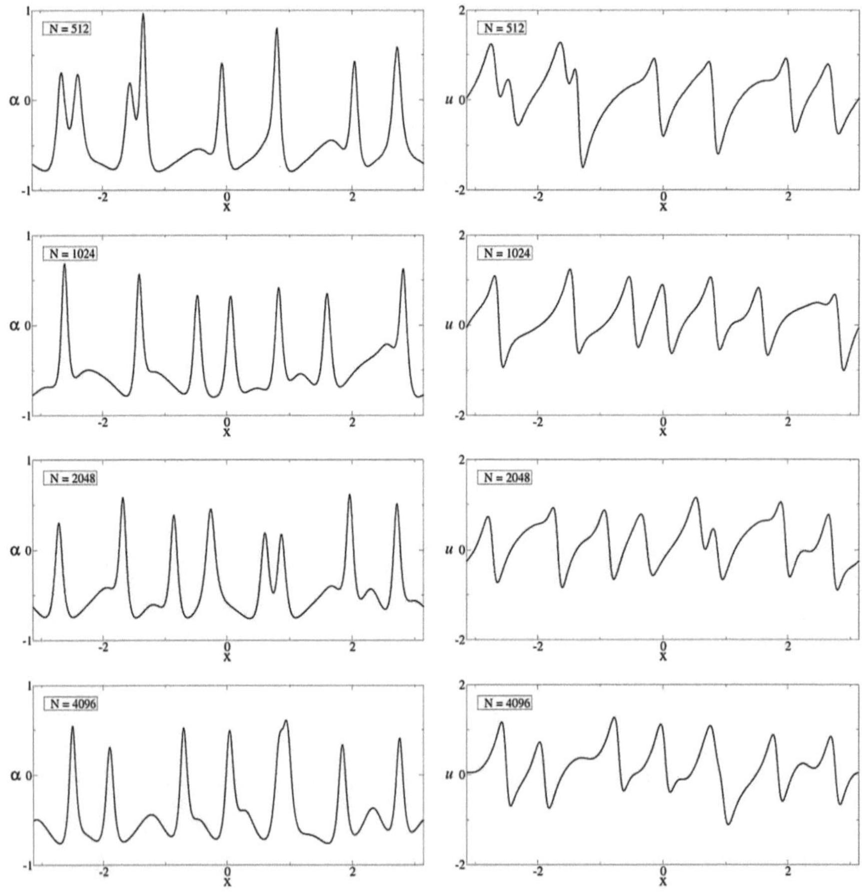

Fig. 4.2 Numerical solutions of the KY equations at $t = 40$ on four different grids, reprinted from Fullmer et al. (2014), with permission from Elsevier

more grid points, the solutions are different from KYs and different from each other. While the features of the solutions are again quite similar, what is so unsettling about Fig. 4.2 is that none of these features seem to be changing systematically with the grid refinement: amplitudes and frequencies are not increasing, slopes are not steepening, and waves are not translating from dispersive errors. Each solution appears to be a different snapshot of a single grid, or at least similar grids, at different times. Such a fundamental lack of convergence renders direct solution verification impossible; at least for this code in this time domain.

4.2.2 Sensitivity to Initial Conditions

While it may be impossible, or at least impractical, to complete the solution verification via error estimation at this point, it is important to understand why this lack of convergence is happening. The answer is similar to what happened with the MMS results shown in Fig. 2.19 on the larger grids: very small errors are growing exponentially. Even though each case starts with the same initial condition, or at least as similar as it can be represented on a finite difference grid, the truncation errors are different for each grid. This can be shown clearly using a single grid ($N = 512$) and introducing a very small perturbation while tracking the evolution of this difference in time.

The initial conditions will now be defined by $\widetilde{\alpha}_0 = \mathrm{e}^{-(2+\delta_\alpha)x^2}$ and $\widetilde{u}_0 = \mathrm{e}^{-(4+\delta_u)x^2}$ so that perturbations are included in the widths of the Gaussians. The original, unperturbed case is recovered by setting $\delta_\alpha = \delta_u = 0$. Following Sprott (2003), the magnitude of the perturbations will be taken to be on the order of the square root of the precision of the floating point numbers being used, double precision in this case, so that $|\delta_\alpha| = |\delta_u| = \delta_0 = 10^{-8}$. Perturbations to the initial condition will be solved in time simultaneously with the unperturbed case. The difference will be treated as an error and quantified with the L_2 norm, Eq. (2.115). Four different perturbations will be considered, each with the magnitude given by δ_0 and the signs of $(\delta_\alpha, \delta_u)$ given alternately by (+ +), (+ −), (− +), and (− −).

Since the behavior of each case is qualitatively similar, the norms of the variables are summed together in Fig. 4.3 to give a total error. Except for a very small time domain initially, the solutions diverge much slower than the linear result, i.e., $\mathrm{e}^{0.38(t-t^*)}$ shown by the thin gray lines in Fig. 4.3, whereas Eq. (2.109) results in a maximum growth of $\mathrm{e}^{10(t-t^*)}$. The value of the coefficient in the exponent will be derived more rigorously in the next section. The slow divergence rate, compared to linear theory, is a result of the nonlinearity and was also identified by KY. While the logarithm of this growth rate is an order of magnitude smaller than predicted of linear theory, the divergence eventually leads to entirely different solutions and the total error reaches an asymptote around $t = 50$. This shows that the solutions are very sensitive to the initial conditions, i.e., small changes in the initial state can eventually lead to large changes in later states, but the difference is bounded. This is an indication of chaos.

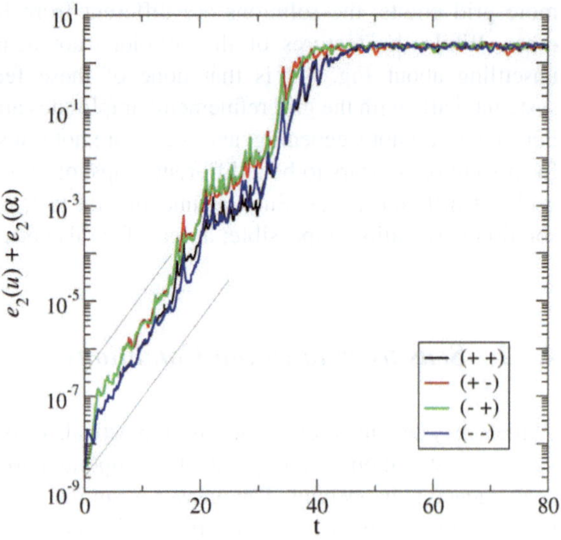

Fig. 4.3 The divergence of four solutions with slightly perturbed initial conditions, reprinted from Fullmer et al. (2014), with permission from Elsevier

4.2.3 Lyapunov Exponent

While there may not be a formal mathematical proof of chaos, a standard test to quantify chaos is with the Lyapunov exponents, cf. Appendix B.6.5, more specifically the Largest Lyapunov Exponent, ω_L. The Lyapunov exponent is the nonlinear equivalent of the linear growth rate obtained from the dispersion relation. For a finite dimensional system of ODEs there are as many exponents as there are dimensions. For a system of PDEs, the number of exponents is infinite. Luckily, the calculation of the largest Lyapunov exponent is tractable, its sign determines whether a system is chaotic and its magnitude quantifies the degree of chaos. More specifically, ω_L represents the average rate at which two nearby states converge to or diverge from one another. ω_L is defined by Eq. (B.100),

$$\omega_L = \lim_{\Delta t \to \infty}\left(\lim_{\delta \to 0}\frac{1}{\Delta t}\ln\frac{\delta(t)}{\delta_0}\right), \tag{4.1}$$

where δ_0 is some initial separation between two trajectories, i.e., two solutions, and $\delta(t)$ is the separation at some later time. Note that δ_0 is not necessarily introduced at $t = 0$. A positive ω_L indicates that the solutions are diverging and that the system is chaotic. The magnitude of ω_L approximates the rate at which predictability of the system is lost (Sprott 2003). If the solutions are diverging, the limits in Eq. (4.1) seem to contradict one another. Therefore, in practice, the two trajectories are never allowed to diverge too far from one another before being reset. ω_L is calculated and averaged over many periods to obtain satisfactory statistics.

The FFM system of PDEs is treated as a $2N$-dimensional system of ODEs for which the general procedure is relatively straightforward. First the problem is solved for a very long period of time to allow any transients to settle out. Then at some specified time a minor perturbation is introduced and this second solution is also followed. The two solutions are tracked for many short intervals of time and the perturbed solution is compared to the original unperturbed solution to determine the evolution of the separation. Then the perturbed solution is renormalized against the unperturbed solution so that the difference between the two solutions does not grow too large and the procedure is repeated as illustrated in Fig. B.40.

The perturbation should be introduced in the direction of the maximum divergence, which may not be easy to evaluate. Therefore, a simple general perturbation is introduced and the system is given a period of time, here 1000 time units, for the perturbation to orient itself in the direction of maximum expansion (Sprott 2003). The first imposed perturbation is specified as $\widetilde{\alpha} = \alpha - \text{sgn}(x)\Delta_0$ and $\widetilde{u} = u - \text{sgn}(x)\Delta_0$ with tilde overbars again signifying the perturbed solution. It should be noted that the perturbation slightly breaks periodicity equally at the edge and center of the domain. This is so that there is no integral change to the perturbed solution which would slightly alter its equilibrium position. The magnitude of the separation in both variables at each spatial location, i.e., in each of the 2N-dimensions, is the same as that used previously, i.e., $\Delta_0 = 10^{-8}$. The total separation is given by the 2N-dimensional Euclidean norm,

$$\delta_k = \sqrt{\sum_i^N (\widetilde{\alpha}_i - \alpha_i)^2 + (\widetilde{u}_i - u_i)^2}. \tag{4.2}$$

The subscript k is used to indicate that ω_L is in general not the same as the iteration number n unless the perturbation is introduced at time zero. The initial separation is then $\delta_0 = \sqrt{2N}\Delta_0$. ω_L is calculated after each iteration so that the separation does not get too large and then the perturbation is adjusted (see Sect. B.6.6 for a more detailed description of the procedure) so that the perturbed orbit (i.e., orientation from the reference solution) is preserved while renormalizing the net separation to be δ_0 again, i.e.,

$$\widetilde{\phi}_i \leftarrow \phi_i + \frac{\delta_0\left(\widetilde{\phi}_i - \phi_i\right)}{\delta_k}. \tag{4.3}$$

Finally ω_L is given by the logarithm of the ratio of the net separation at time k to $k - 1$, and since the starting separation is always renormalized to the initial separation, it is simply

$$\omega_\text{L} = \frac{1}{k} \sum_{k=1}^{k \to \infty} \ln\left(\frac{\delta_k}{\delta_0}\right). \tag{4.4}$$

Generally this calculation needs to be carried out for very long periods of time to determine ω_L with a high degree of precision. However, the primary concern is simply to determine whether or not ω_L is positive and to obtain an approximate value.

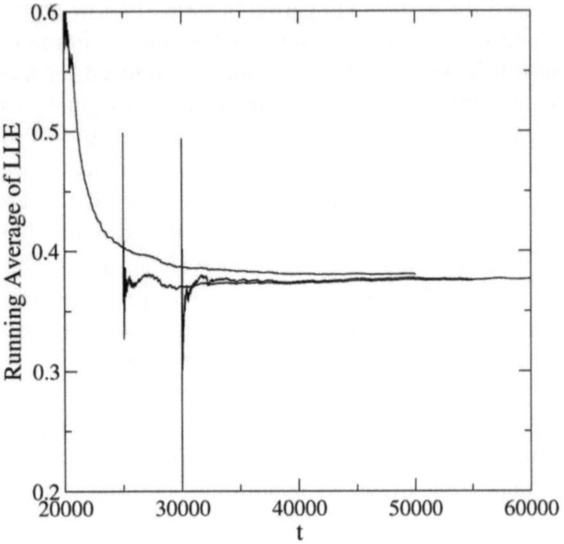

Fig. 4.4 Running average of three ω_L calculations started at different times, reprinted from Fullmer et al. (2014), with permission from Elsevier

The ω_L calculation was started at three different times, $t = 2 \cdot 10^4, 2.5 \cdot 10^4$, and $3 \cdot 10^4$. Each calculation started with the initial perturbation described previously, which was introduced 10^3 time units before the calculation began to allow the perturbed solution to orient itself. Each calculation was carried out for $3 \cdot 10^4$ time units. The running average of each calculation is shown in Fig. 4.4. Even at $t = 2 \cdot 10^4$ there is still some of the initial transient in the solution, which becomes apparent in the running average of the first ω_L calculation. However, the average eventually settles down when the number of samples becomes large and agrees with the other two calculations. Taking an average of the three final values and using $1/\sqrt{f\,t}$ to estimate the order of accuracy (where f is the characteristic frequency and t is the calculation time), ω_L is determined to be 0.38 ± 0.05. This measure quantifies the nonlinear rate of the divergence of the solutions with the slightly perturbed initial conditions and agrees with the estimate shown in Fig. 4.3.

4.2.4 *Fractal Dimension*

Beyond determining whether ω_L is positive, the second most important feature of chaos is the fractal dimension of the dynamics, cf. Appendix B.6.7. The fractal dimension represents the "strangeness" of the attractor (Grassberger and Procaccia 1983). There are several techniques to measure the dimensionality, each of which produces slightly different results. Here, the correlation dimension is used because it is easier to calculate than other measures and converges rapidly (Abarbanel 1996).

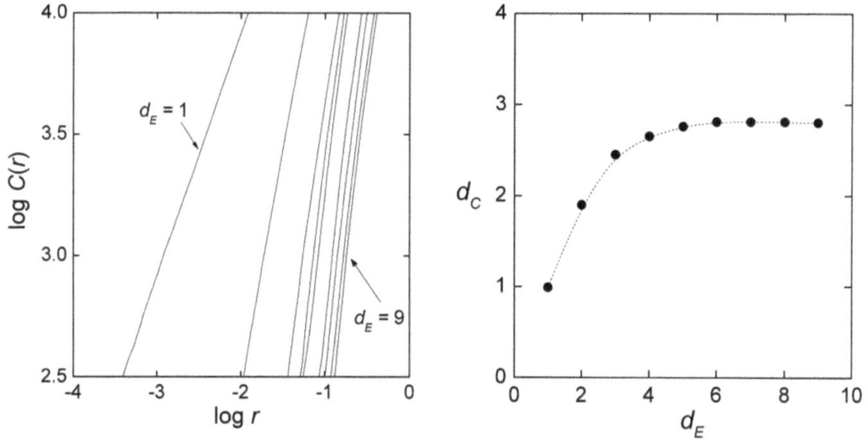

Fig. 4.5 Calculation of the correlation dimension as a function of the embedded dimension. The *left panel* shows the correlation sum as function of the radius: the slope gives the correlation dimension and the separate curves show different embedded dimensions (min and max labeled). The correlation vs. imbedded dimension information is then summarized in the *right panel*, reprinted from Fullmer et al. (2014), with permission from Elsevier

In general, the correlation dimension is calculated by shrinking a hypersphere of radius r in d_E-dimensional space and counting the number of points inside to obtain the correlation sum, $C(r)$. As the total number of points tends to infinity, the correlation sum converges to the correlation integral and behaves as Eq. (B.101),

$$C(r) \propto r^{d_c}, \tag{4.5}$$

as the radius tends to zero. Therefore, the slope of $C(r)$ vs. r on the logarithmic plot, Fig. 4.5 (left), gives the correlation dimension, d_c.

Part of the challenge is that there are simultaneously two unknowns: the correlation dimension and the phase-space dimension in which the fractal dimension should be calculated. For instance, a line has $d_c = 1$ whether it is constructed in 2D space, 3D space, or any higher dimension. The dimension of the construction is called the embedded dimension, d_E, cf. Appendix B.6.8. In this case the dimensionality of the problem is infinite, or $2N$ when the PDEs are solved with a finite difference scheme by discretizing the spatial domain in N segments. However, it may not be necessary to use all $2N$ variables to capture the dynamics of the problem. Therefore, the correlation dimension is first calculated with an embedded dimension of one, which is usually insufficient, and then repeated several times, each time increasing the embedding dimension. At some point the correlation dimension stops changing, giving the minimum embedded dimension.

To create the phase space, the u variable is recorded at d_E equidistant positions in the spatial domain every 250 iterations in the time domain $t = 2.5 \cdot 10^4$ to $6.5 \cdot 10^4$. Fig. 4.5 (left) shows $C(r)$ in logarithmic scale for d_E from one to nine. The successive slopes represent d_C, plotted on Fig. 4.5 (right) as a function of d_E.

The correlation dimension saturates at $d_C \cong 2.8$ with a minimum embedded dimension of $d_E = 6$. This indicates that in principle the dynamics of the system may be modeled without overlapping using only six state variables (Abarbanel 1996).

4.2.5 The Route to Chaos

Each of the two previous sections was devoted to the calculation of a single number, namely, ω_L and the fractal dimension. Here, the emphasis will be the more qualitative evolution from stable to chaotic states. For this study the coefficient C in the KY Eq. (2.99) will be the control parameter, which seems to be the best choice since it was already shown to be directly related to the wave growth speed from the linear dispersion relation, Eq. (2.102). Accordingly, the dynamics of the system are explored varying C starting with negative (stable) values and progressively increasing it to positive (unstable) values. All the tests are performed with the viscosities, boundary and initial conditions, and numerical grid ($N = 512$ and $\Delta t = 0.0002$) used previously in Sect. 4.2. The equations are solved numerically using the scheme developed in Sect. 2.5.5 for a stationary state, which in some cases takes up to $t = 2.5 \cdot 10^4$ as was shown in Fig. 4.4, and then the long-term dynamics are analyzed.

The dynamics are visualized by the trajectory of the state variables at the center of the domain, labeled α_0 and u_0 in the following figures. With this phase-space construction it is easy to see how the variables are approaching the attractor, i.e., the equilibrium position, which corresponds to the constant uniform values $\alpha = 0$ and $u \approx 0.141$. The former is forced by the dissipative sink in Eq. (2.98) and the latter is a consequence of the periodic boundary conditions that ensure the integral conservation of u which is given by $\int_{-\pi}^{\pi} e^{-4x^2} dx = \sqrt{\pi/4}\,\mathrm{Erf}(2\pi) \approx 0.141$ as $N \rightarrow \infty$. In addition, a third dimension is added to the phase space to help unfold some of the dynamics, although an embedded dimension of six precludes unfolding all the chaotic dynamics into a visualizable space in the chaotic region. The difference between u at the boundary and center of the spatial domain, $u_L - u_0$ was found to provide a useful third dimension. This variable can be used to distinguish between 2π- and π-periodic functions. Its equilibrium value is zero due to the uniform equilibrium condition.

When $C < 0$ the solution quickly approaches the equilibrium, which is the expected behavior for a linearly stable hyperbolic system. When C becomes positive, the system remains linearly stable due to the diffusion, i.e., the domain is smaller than the cutoff wavelength. There is a critical value of C when the size of the domain (and therefore the largest possible wavelength) is equal to the cutoff wavelength, $\lambda_0(C^*) = L$. In the present case $L = 2\pi$, $\nu = 0.05$ and using the equilibrium value $\alpha = 0$ gives $C^* = 0.1025$. The decay rate to the equilibrium position slows dramatically as C approaches C^*. This behavior is shown in Fig. 4.6

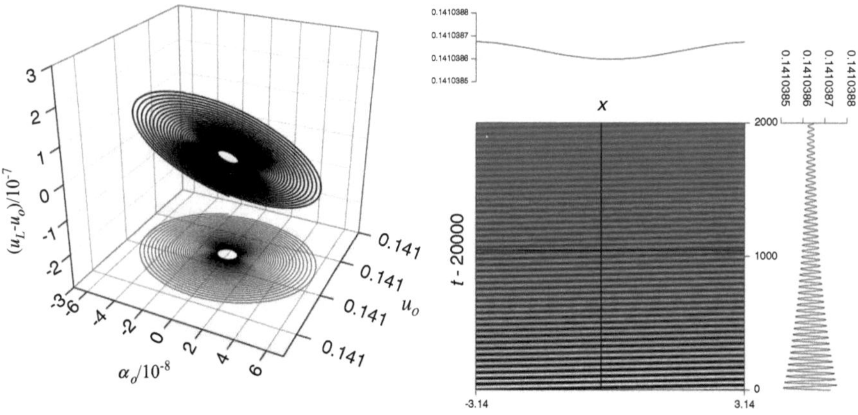

Fig. 4.6 Phase-space and gray scale map of $\boldsymbol{u}(\boldsymbol{x},\boldsymbol{t})$ for $\boldsymbol{C} = 0.102$ showing the asymptotic stability of the system, reprinted from Fullmer et al. (2014), with permission from Elsevier

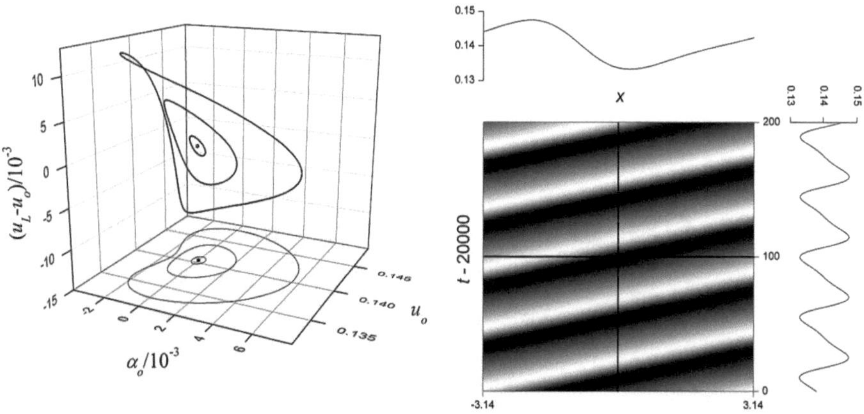

Fig. 4.7 Phase space for $C = 0.102$, 0.1025, 0.103, and 0.105 and gray-scale map of $\boldsymbol{u}(\boldsymbol{x},\boldsymbol{t})$ for $C = 0.103$ showing 2π-periodic limit cycles, increasing in amplitude with increasing C, reprinted from Fullmer et al. (2014), with permission from Elsevier

for $C = 0.102$. The left-hand side of the figure shows the slow spiral of the state variables into the attractor and the right-hand side shows a brief map of the spatiotemporal evolution.

As C increases slightly beyond $C*$, the steady state of the system becomes unstable and a 2π-periodic traveling wave develops. The wave travels at a constant speed which is shown by the parallel lines in the gray-scale contour map of Fig. 4.7 (right), taken after the stationary wave is reached. This behavior corresponds to a limit cycle in phase space. The amplitude of the wave and the limit cycle grow as C increases, as seen on the left of Fig. 4.7. Near $C = 0.104$ the limit cycle becomes deformed and begins to appear heart shaped, as shown on the bottom left of Fig. 4.7 for $C = 0.105$. Beyond $C = 0.107$ the limit cycle becomes significantly more complex, with a curl in the trajectory as can be seen in Fig. 4.8. This corresponds to a steepening wave front and the development of a secondary wave pattern.

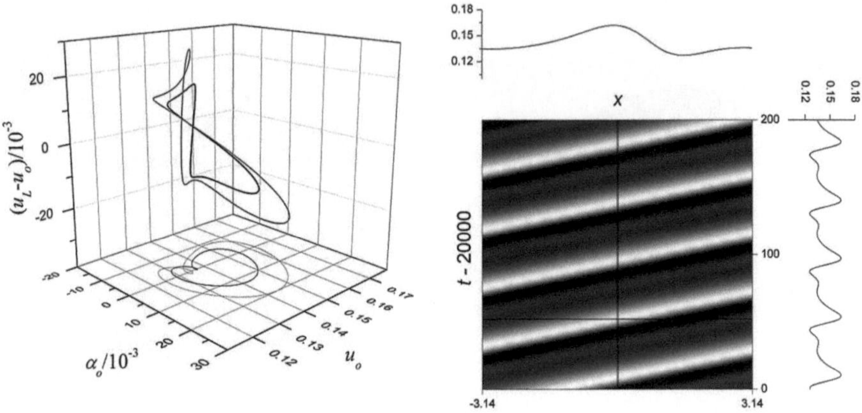

Fig. 4.8 Phase space for $C = 0.107$, 0.109, and 0.111 and gray-scale map of $u(x,t)$ for $C = 0.109$, showing asymmetric 2π-periodic limit cycles, increasing in amplitude with increasing C, reprinted from Fullmer et al. (2014), with permission from Elsevier

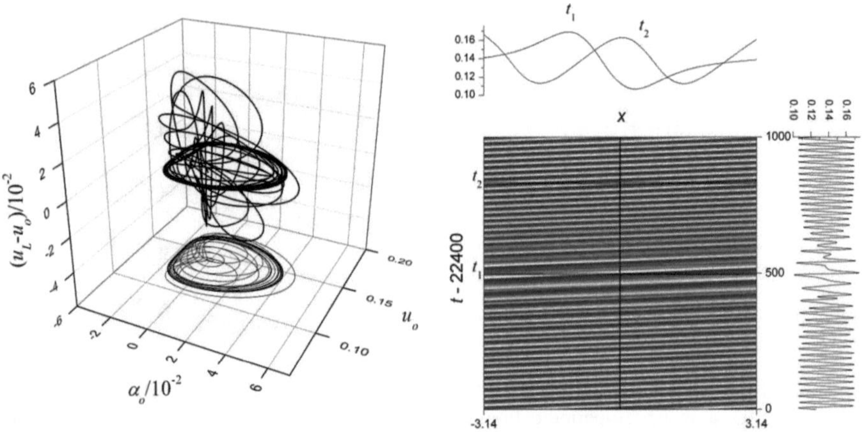

Fig. 4.9 Phase-space and gray-scale map of $u(x,t)$ for $C = 0.115$ showing a burst of brief, intermittent chaos surrounded by π-periodic limit cycles, reprinted from Fullmer et al. (2014), with permission from Elsevier

At $C = 0.112$ the behavior becomes intermittent with relatively long periods of periodic, regular waves interrupted by short bursts of another wave pattern. Intermittency is one of the three basic types of bifurcations leading to chaos, along with quasiperiodic and period doubling (Sprott 2003). In the phase space it is difficult to distinguish between intermittency and chaos for a long time series of data. However, on the right of Fig. 4.9, the brief interruption in the temporal signal is easily recognizable.

This behavior continues up to $C = 0.116$. Between $C = 0.116$ and 0.120 another stable traveling wave appears, with a wavelength of half the spatial domain. This

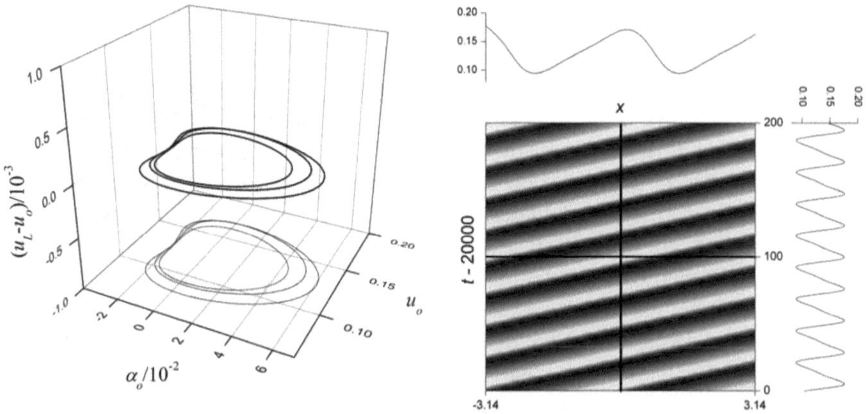

Fig. 4.10 Phase space for $C=0.116, 0.118$, and 0.120 and gray-scale map of $u(x,t)$ for $C=0.120$ showing π-periodic limit cycles, increasing in amplitude with increasing C, reprinted from Fullmer et al. (2014), with permission from Elsevier

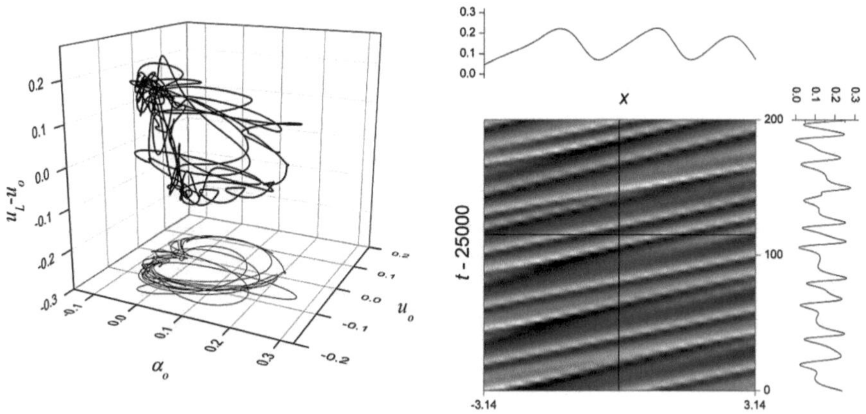

Fig. 4.11 Phase-space and gray-scale map of $u(x,t)$ for $C=0.150$ showing chaotic behavior, reprinted from Fullmer et al. (2014), with permission from Elsevier

feature can be appreciated in Fig. 4.10 which shows that the limit cycle is now flat in the third dimension due to the π-periodic wave which is shown above the contour map.

Finally, after another brief region of intermittency, the system becomes chaotic at $C=0.125$. Figure 4.11 shows the chaotic state for $C=0.150$. Only a brief part of the trajectory in the phase space is plotted, since the evolution is so entangled that it is difficult to visualize it in this representation; we would need six dimensional spaces to disentangle the chaotic trajectory. On the right of Fig. 4.11, the contour map of $u(x,t)$ shows that the regularity of the standing wave is lost, i.e., uniform amplitudes, periods, propagation speeds, etc., and the dynamics has become more complex. This chaotic state increases in complexity as C increases up to the original KY model value of $C=1$.

4.2.6 Numerical Convergence

Now that it has been shown that the KY equations are chaotic, it is possible to consider the stability of the model in light of modern nonlinear statistical theory, widely used for turbulence. In the first place, the nonlinear transport of energy between frequencies implies a new dissipation mechanism, not considered by linear theory, i.e., the formation of material shocks, which leads to Lyapunov stability. Second, the problem of numerical convergence, i.e., how to judge when a grid for the numerical scheme is sufficient, needs to be reassessed with chaos in mind.

A different method than direct comparison is needed to define convergence for chaos vs. a limit cycle since, due to the positive ω_L, any perturbation no matter how small will eventually diverge. Chaos theory provides several possible options, some of which have already been discussed, namely, ω_L and the fractal dimension. These measures are invariants, that is, unlike the solution which varies with time; these measures are insensitive to perturbations of the initial conditions (Abarbanel 1996). Unfortunately calculating these quantities for several grids is computationally expensive. Additionally, since it is hoped that the analysis of the KY equations will serve as guidance to the FFM, calculating these quantities for physical problems may be difficult. Therefore, it appears appropriate to borrow concepts from a well-known physical model that exhibits similar spatiotemporal convergence challenges, i.e., the Navier–Stokes theory of turbulence.

When the Navier–Stokes equations are solved by large eddy simulations, for a sufficiently large Reynolds number, one cannot compare the solution at a particular time as a function of grid size. Instead it is quite typical to assess convergence based on averaged quantities such as the law of the wall (for internal flows), the autocorrelation function or one of the energy spectra. While these may not be so well defined for the nonphysical KY system, a common feature is that they are all averaged measures. Therefore, convergence is assessed by averaging the Fourier transform of $\alpha(x) \rightarrow \alpha(k)$ and $u(x) \rightarrow u(k)$ as shown in Fig. 4.12 (bottom). The first sample is taken at $t = 4$ when the wave pattern fills the domain and 36 additional samples are taken up to $t = 40$.

The averaged amplitude spectra for $C = 0.150$ are shown in Fig. 4.12 (bottom) for successively refined grids from $N = 256$ to $N = 4096$. From here it is up to the user to determine what is "resolved enough." For the present numerical method it seems that the $N = 512$ grid is sufficiently resolved since it correctly captures 99.99 % of the amplitude spectrum, i.e., from the maximum amplitude of 0.1 down to about 0.00001 where the numerical tail introduces significant error. By contrast $N = 256$ only captures about 99 % of the spectrum correctly and higher resolution grids loose much of their potential value to the growing numerical tails. This shows that even though the numerical method may be considered "high-resolution" compared to linear upwind methods, it is still quite coarse in the spectral domain.

It is also revealing to compare the spectra of the limit cycle in Fig. 4.12 (top) with those of chaos. In the former case, the various Fourier components are isolated and do not interact, whereas in the latter case the interaction leads to Lyapunov stability.

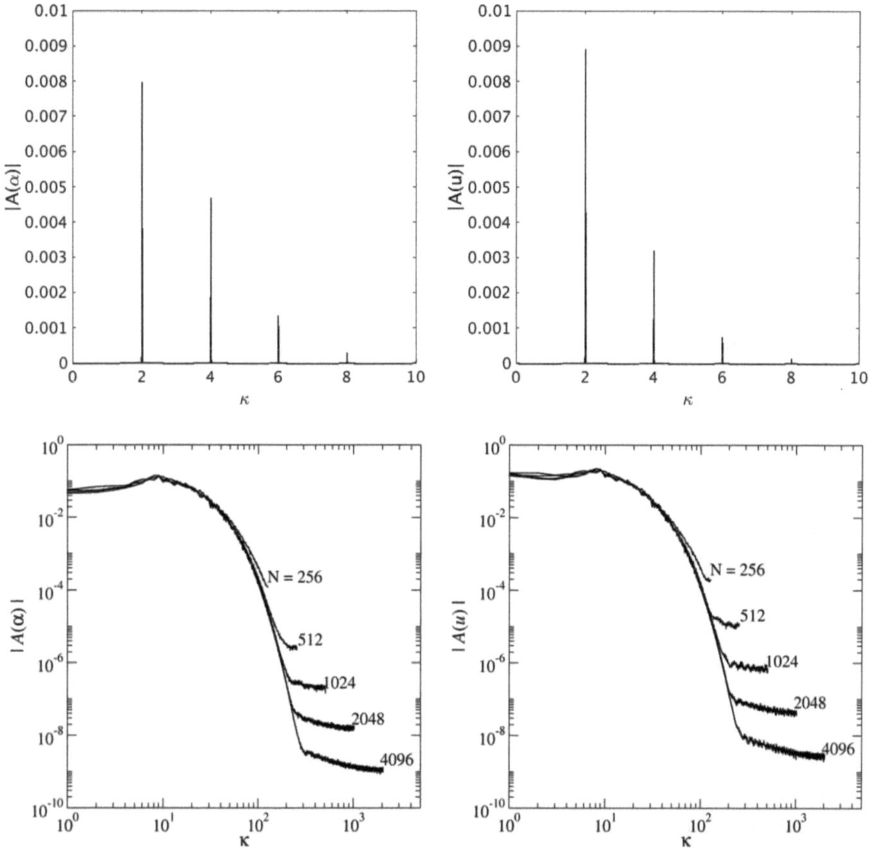

Fig. 4.12 Averaged power spectra of the amplitude of each variable in the time domain for (*top*) $C = 0.120$ and (*bottom*) $C = 0.150$, $t \in [4, 40]$ sampled at a frequency of 1 Hz, for different mesh refinements, reprinted from Fullmer et al. (2014), with permission from Elsevier

At this point it is hard not to draw a comparison between the present KY equations and the Kuramoto–Sivashinsky (KS) equation which was physically derived to model chemical reactions (Hyman and Nicolaenko 1986). In both cases, the chaotic behavior is similar: energy is generated at small wavenumbers where the model is linearly unstable and is transferred to larger wavenumbers through shock-like structures by the nonlinearity until it reaches a sufficiently large wavenumber where the energy is dissipated by viscosity. This is the reason that truncation errors that grow exponentially will cause the same initial condition on two different grids to evolve into two different solutions, rather than just growing exponentially forever causing a numerical excursion.

Lastly, it should be noted that Fig. 4.12 helps explain the different behavior of the divergent MMS cases in Fig. 2.19, where a solution to the KY equations could not be obtained for a resolution less than $N = 256$. This appears to be the result of not having a sufficient amount of high wavenumber nodes to dissipate the energy generated at the large scales, which is not a universal grid resolution requirement to the governing equations, but specific to the numerical method. A different scheme with a better spectral resolution would likely be able to obtain a solution with less mesh refinement. Therefore in all of the cases presented in Fig. 2.19, the underlying KY equations have dominated the MMS and are trying to produce solutions similar to the KY equations, but with a large modified source term. The finer grids are capable of solving this system and reach an asymptotic value for the error when a quasi-steady state is reached. On the other hand, the coarser grids simply do not resolve enough of the spectrum to dissipate the energy that is generated at the large scale, which leads to the blow up.

4.3 Fixed-Flux Model Chaos

4.3.1 Nonlinear Simulations with the FFM

Now that KY chaos has been identified, the objective is to perform FFM simulations to pursue chaos in a similar fashion. There are a couple of encouraging precedents to the nonlinear simulations in this section; Barnea and Taitel (1994) and more recently Picchi and Poesio (2016) have used a model that is very similar to the FFM Eqs. (2.33) and (2.34) to simulate the nonlinear evolution of the waves in a round tube under the kinematic instability. While these results are not chaotic they exhibit distinct nonlinear behavior.

The task ahead turns out to be more complicated compared to the Chap. 3 TFM simulations because, unlike the case of Thorpe, the flow is now fully developed and wavy. Also it is necessary to employ a more elaborate FFM compared to the one in Chap. 2 that is applicable to a density ratio $r_\rho \approx 1$.We proceed to develop a FFM that includes velocity shape factors (Kocamustafaogullari 1985; Picchi et al. 2014), and an interfacial shear model for wavy flow (Andritsos and Hanratty 1987) and a more rigorous derivation of the viscous stresses.

4.3.2 Extension of Thorpe Experiment into Chaos

A water–gasoline experiment in a round tube was performed to extend Thorpe's (1969) observations further into the chaotic regime (Vaidheeswaran et al. 2016). The test section consisted of an acrylic tube of dimensions $D = 0.02\,\text{m}$ and $L = 2.4\,\text{m}$. We were interested in extending the duration of Thorpe's experiment (1969) so the waves could evolve further into chaos, therefore the L/H ratio is two times bigger, where $H = D$ for the tube. The new experiment lasts approximately 10 s vs. 0.350 s reported by Thorpe. We were also interested in reducing the Reynolds numbers, so D_{hyd} is three times smaller than Thorpe's. Thus, the turbulence instabilities are further from the KH interfacial instability.

Following the linear TFM stability results that will be presented in Sect. 4.3.4, which are in accord with the more complete Orr–Sommerfeld stability analysis of Barmak et al. (2016) for fully developed laminar stratified flow, the laminar FFM presented in the following sections should be capable to capture the linear KH unstable behavior of the experimental flow in a similar way to the results presented in Chap. 3 for Thorpe's experiment, considering that the focus of this chapter is the nonlinear chaotic behavior that occurs later.

The tube test section shown in Fig. 4.13, which is closed at both ends, is filled with water and gasoline with densities $\rho_1 = 1000\,\mathrm{kg/m^3}$ and $\rho_2 = 720\,\mathrm{kg/m^3}$. The remaining material properties are $\mu_1 = 0.001\ \mathrm{Pa \cdot s}$, $\mu_2 = 0.0005\,\mathrm{Pa \cdot s}$, and $\sigma = 0.04\,\mathrm{N/m}$. Initially the tube was at rest horizontally to allow the mixture to completely separate and reach a state of stratified equilibrium. Then it was suddenly tilted to a specified small angle. A countercurrent flow pattern evolved as the water rushed down pushing the gasoline up.

A series of photographs of the flow viewed from the side are shown in Fig. 4.14 for an angle of tilt of 3.1°. The grid that appears behind the tube is 5 mm × 5 mm. These photographs may be compared with Thorpe's for an angle of tilt of 4.1°, shown in Fig. 4.13. The first four profiles corresponding to the initial rapid growth of the waves, which has a duration of approximately 320 ms, are comparable to Thorpe's both in shape and time. However, the focus now is on the evolution of the waves which follows the initial growth and lasts approximately 10 s, represented by the last two frames in Fig. 4.14. During this time the waves have stopped growing and the wave contours are no longer periodic and far from linear.

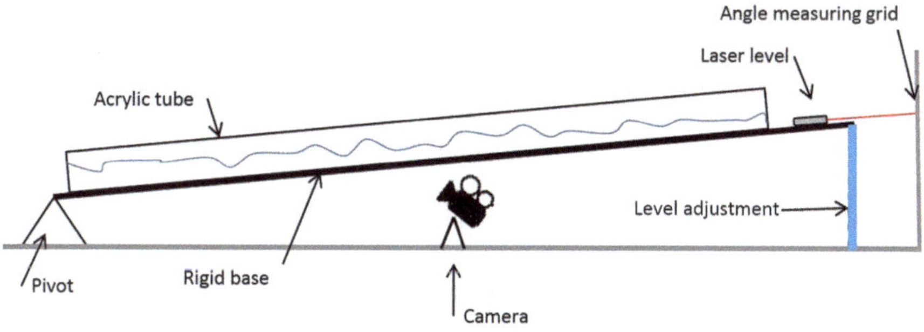

Fig. 4.13 Schematic of new experiment

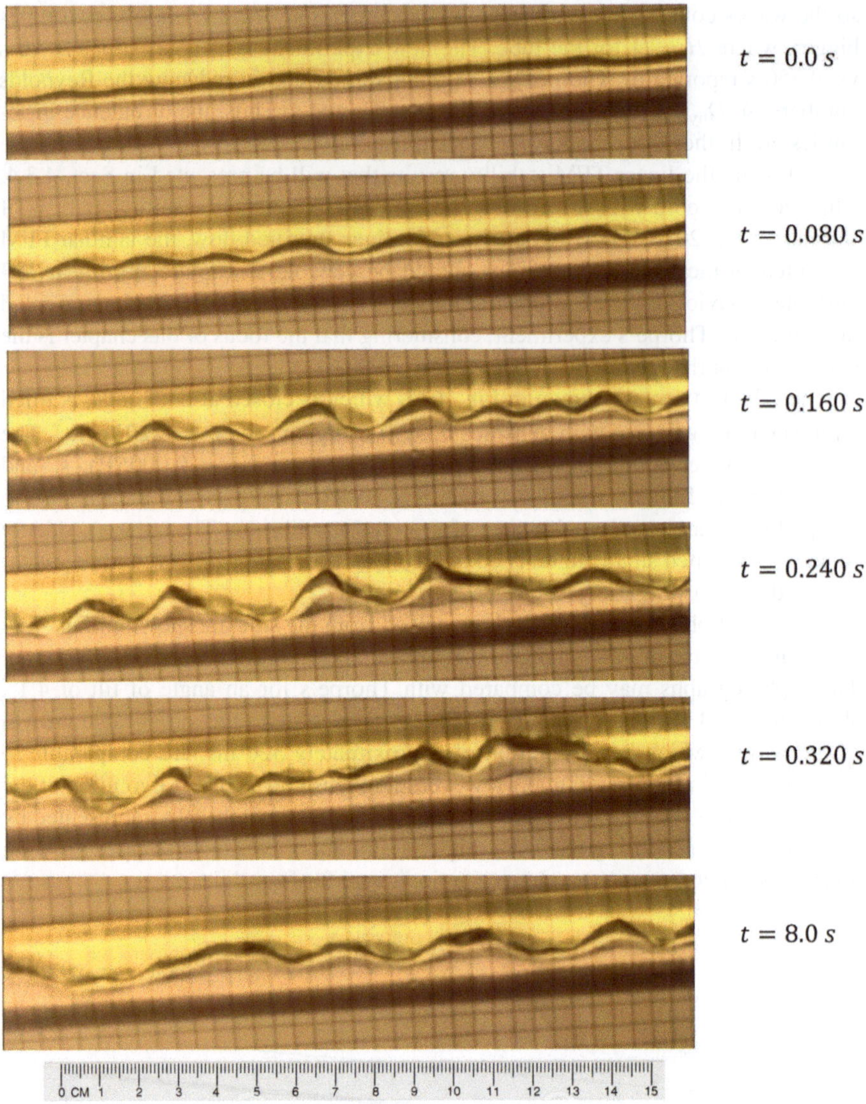

Fig. 4.14 Wave evolution of new experiment where $t = 0$ corresponds to the onset of waves at the interface, reprinted from Vaidheeswaran et al. (2016), with permission from ASME

4.3.3 Fixed-Flux Model for Fully Developed Laminar Flow in a Pipe

The incompressible FFM given by Eqs. (2.33)–(2.35) is not adequate to model the experiment presented in Sect. 4.3.2 for several reasons:

1. $r_\rho \approx 1$, so the approximate Eqs. (2.29) and (2.30) are no longer valid. Equations (2.27) and (2.28) must be used instead.
2. Covariance terms (related to the momentum flux distribution parameters or velocity profile shape factors) are now needed to simulate fully developed laminar two-phase flow, in contrast to the case of Thorpe (1969) for the initial growth of the waves where the flow was approximately plug flow.
3. The interfacial shear correlation for fully developed waves differs from that for a smooth interface.
4. The geometry is now cylindrical, not rectangular.

Unfortunately these considerations add a significant amount of complexity to the model but they may be incorporated in a straightforward way. Following the treatment of covariance or shape factors by Kocamustafaogullari (1985) the incompressible momentum equations now become

$$\begin{aligned}\rho_1\frac{D_1u_1}{Dt}+\rho_1\frac{\partial}{\partial x}\alpha_1\mathrm{cov}\left[{u_1}^2\right]=&-\frac{\partial p_{2i}}{\partial x}+\rho_1 g_y D\frac{\partial h/D}{\partial x}+\sigma D\frac{\partial^3 h/D}{\partial x^3}+\rho_1 g_x\\&-\frac{1}{\frac{\alpha_1 A}{P_{w1}}}\frac{f_1}{2}\rho_1 u_1^2+\frac{1}{\frac{\alpha_1 A}{P_i}}\frac{f_i}{2}\rho_2|u_R|(u_2-u_1)\\&+\frac{\rho_1}{\alpha_1}\frac{\partial}{\partial x}\left(\alpha_1\nu_1\frac{\partial u_1}{\partial x}\right)\end{aligned}\tag{4.6}$$

$$\begin{aligned}\rho_2\frac{D_2u_2}{Dt}+\rho_2\frac{\partial}{\partial x}\alpha_2\mathrm{cov}\left[{u_2}^2\right]=&-\frac{\partial p_{2i}}{\partial x}+\rho_2 g_y D\frac{\partial h/D}{\partial x}+\rho_2 g_x-\frac{1}{\frac{\alpha_2 A}{P_{w2}}}\frac{f_2}{2}\rho_2 u_2^2\\&-\frac{1}{\frac{\alpha_2 A}{P_i}}\frac{f_i}{2}\rho_2|u_R|(u_2-u_1)\\&+\frac{\rho_2}{\alpha_2}\frac{\partial}{\partial x}\left(\alpha_2\nu_2\frac{\partial u_2}{\partial x}\right)\end{aligned}\tag{4.7}$$

where, using the area averaging notation of Kocamustafaogullari (1985),

$$\mathrm{cov}\left[{u_k}^2\right]=\langle {u_k}^2\rangle-\langle u_k\rangle^2.\tag{4.8}$$

Assuming that each phase flows through an equivalent circular duct, rather than a semicircular one, and that the flow is laminar results in

$$\mathrm{cov}\left[{u_k}^2\right]=\frac{1}{3}\langle u_k\rangle^2\tag{4.9}$$

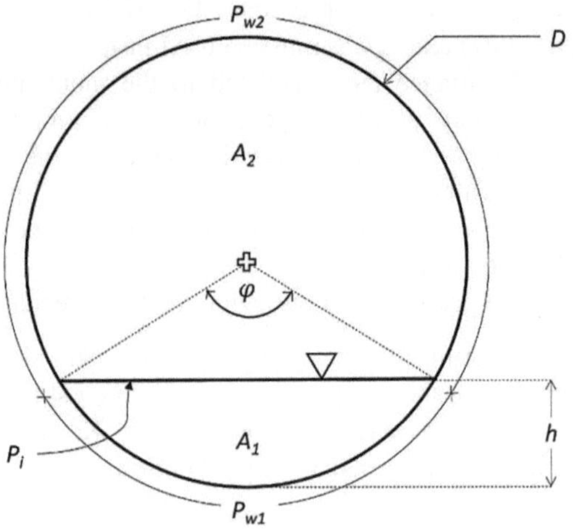

Fig. 4.15 Cross section of stratified flow in pipe

Furthermore, the pipe two-phase flow geometry, shown in Fig. 4.15, leads to a modification of the definition of the interfacial level with respect to a rectangular channel. Linearizing the basic relations: $\varphi = 2\ \mathrm{Arcos}\left(1 - 2\frac{h}{D}\right)$ and $\alpha_1 = 1 - \frac{1}{2}(\varphi - \sin\varphi)$, around $\alpha_1 = 1/2$ one obtains $\varphi \cong \pi(1/2 + \alpha_1)$. Then,

$$\frac{h}{D} = \frac{1}{2}(1 - \cos\varphi) \cong \frac{1}{2}\left(1 - \cos\left[\frac{\pi}{2}\left(\frac{1}{2} + \alpha_1\right)\right]\right) \tag{4.10}$$

and,

$$\frac{\partial\, h/D}{\partial x} \cong \frac{\pi}{4}\sin\left[\frac{\pi}{2}\left(\frac{1}{2} + \alpha_1\right)\right]\frac{\partial\,\alpha_1}{\partial x}. \tag{4.11}$$

The present simulations of fully developed chaotic wavy flow are different from those of Chap. 3, where the main concern was the initial development of the waves from a smooth interface. Now the target of interest is the sustained chaotic state. The most uncertain aspect of this case is the characterization of the interfacial force of the fully developed waves because the highly nonlinear flow structures around the interface are not well known. The state of the art for a 1D formulation is the correlation of Andritsos and Hanratty (1987) for air–water flows:

$$\frac{f_{\mathrm{i}}}{f_2} = 1 + 15\sqrt{\frac{h}{D}}\left[\frac{(u_2 - u_1)}{(u_2 - u_1)_{\mathrm{c}}} - 1\right] \tag{4.12}$$

where $(u_2 - u_1)_{\mathrm{c}}$ is the relative velocity corresponding to the critical condition $C = 0$ in Eq. (4.15). Notice that the correlation does not include the sheltering effect given by Eq. (2.64), among other things.

The FFM conservation equations may now be derived starting from Eqs. (4.6) and (4.7) with these assumptions using new assumptions following the same procedure performed in Sect. 2.2.1 with the TFM momentum equations for flow between parallel plates. The FFM continuity equation is still given by Eq. (2.33):

$$\boxed{\frac{\partial \alpha_1}{\partial t} + u_1 \frac{\partial \alpha_1}{\partial x} + \alpha_1 \frac{\partial u_1}{\partial x} = 0.} \tag{4.13}$$

The momentum equation including the velocity shape factors is,

$$\boxed{\begin{array}{l}\frac{\partial u_1}{\partial t} + B_{22}\frac{\partial u_1}{\partial x} - \frac{1-\alpha_1}{1-\alpha_1+\alpha_1 r_\rho} C \frac{\partial \alpha_1}{\partial x} = \frac{1-\alpha_1}{1-\alpha_1+\alpha_1 r_\rho} \\ \left[\frac{\sigma D}{2\,\rho_1}\frac{\partial^3}{\partial x^3}\left(1 - \cos\left[\frac{\pi}{2}\left(\frac{1}{2}+\alpha_1\right)\right]\right) + F_{\text{visc}} + F\right],\end{array}} \tag{4.14}$$

where

$$\boxed{C = \left[\left(\frac{1}{1-\alpha_1} + \frac{1}{3}\right) r_\rho (u_2 - u_1)^2 - \frac{1}{3}(1 + r_\rho) u_1{}^2 - \frac{\pi}{4}\sin\left[\frac{\pi}{2}\left(\frac{1}{2} + \alpha_1\right)\right](1 - r_\rho) g_y D\right]} \tag{4.15}$$

and

$$\boxed{B_{22} = \frac{1-\alpha_1}{1 - \alpha_1 + \alpha_1 r_\rho}\left[u_1 + r_\rho \frac{\alpha_1}{1-\alpha_1}(2u_2 - u_1) + \frac{2}{3}\frac{\alpha_1}{1-\alpha_1}(1 - \alpha_1 - \alpha_1 r_\rho) u_1\right]} \tag{4.16}$$

This model is more complete than the FFM of Chap. 2 because the assumption $r_\rho \to 0$ has been removed, and so it is now equivalent to the full incompressible TFM of Chap. 3.

The laminar flow assumption also leads to the definition of the Fanning friction factors $f_k = 16/Re_k$ where $D_{hk} = 4\frac{\alpha_k A}{P_{wk}+P_i}$. Then the algebraic force terms are as follows:

$$\begin{array}{l} F = (1 - r) g_x - \frac{1}{\frac{\alpha_1 A}{P_{w1}}}\frac{16}{\frac{D_{h1}}{\nu_1}} u_1 + \frac{1}{\frac{\alpha_2 A}{P_{w2}}}\frac{16}{\frac{D_{h2}}{\nu_2}} r_\rho u_2 + \left(\frac{1}{\alpha_1} + \frac{1}{\alpha_2}\right)\frac{1}{\frac{A}{P_i}}\frac{16}{\frac{D_{h2}}{\nu_2}} \\ \left(1 + 15\sqrt{\frac{h}{D}}\left[\frac{(u_2 - u_1)}{(u_2 - u_1)_c} \quad 1\right]\right) r_\rho u_2. \end{array} \tag{4.17}$$

Finally it can be shown that the combined viscous force, assuming the effective viscosity is the same for both phases, is

$$F_{\text{visc}} = \nu_T\left[\left(\frac{1}{\alpha_1} + \frac{r_\rho}{\alpha_2}\right)\frac{\partial}{\partial x}\alpha_1\frac{\partial u_1}{\partial x} + \frac{r_\rho}{\alpha_2}\frac{\partial}{\partial x}\frac{u_1}{\alpha_2}\frac{\partial \alpha_1}{\partial x}\right]. \tag{4.18}$$

In addition to Eqs. (4.13)–(4.18), the void condition, Eq. (2.17), and the fixed-flux condition for the experiment, $j = \alpha_1 u_1 + \alpha_2 u_2 = 0$, are needed for closure.

The effective viscosity, which is the same for both phases, is specified using the same model developed in Sect. 3.6.2 for the pseudo-turbulence of the eddies generated by the waves,

$$\nu_T = 0.0015\, D|u_2 - u_1|. \tag{4.19}$$

Fullmer et al. (2011) then proposed a 1D Reynolds stress term that produces the same amount of energy dissipation as the Reynolds stresses obtained with a multidimensional CFD simulation. It is constituted in terms of the turbulent viscosity and the covariance between the CFD and the 1D strain rate tensors. For the case of fully developed flow, the asymptotic covariance obtained was approximately 7, which is multiplied to Eq. (4.19) yielding an effective viscosity $\nu_T = 0.00004\,\mathrm{m}^2/\mathrm{s}$ for the new experiment.

4.3.4 Kelvin–Helmholtz Instability

We start by comparing the 1D FFM linear stability analysis with the more complete analysis using two Orr–Sommerfeld equations by Barmak et al. (2016). The liquid–liquid flow regime map shown in Fig. 4.16 is similar to those obtained in Sect. 2.4.2 for gas–liquid stratified flow. The conditions are very similar to the present experiment: water–oil flow with a kinematic viscosity ratio $\frac{\nu_1}{\nu_2} = m = 2$, a density ratio

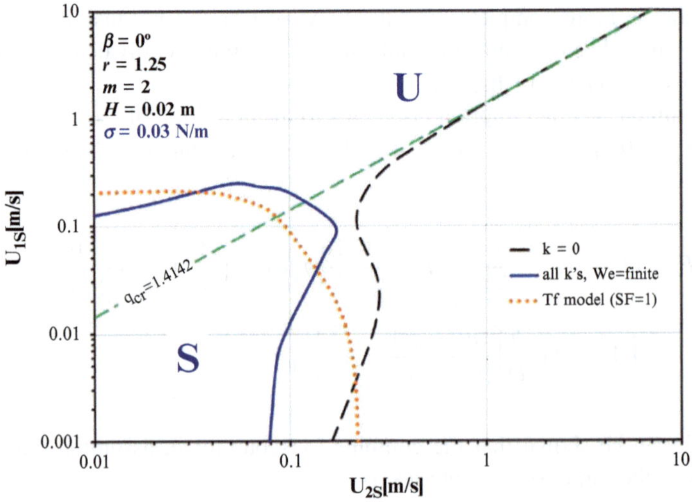

Fig. 4.16 Stability analysis of Barmak et al. (2016) comparing Orr–Sommerfeld analysis in *blue* and TFM in brown dots, reprinted from Barmak et al. (2016), with permission from AIP Publishing

$r_\rho = \frac{1}{r} = 0.8$, and surface tension $\sigma = 0.03\,\mathrm{N/m}$. The parallel plate channel dimension is $H = 0.02\,\mathrm{m}$. The stability analysis of Barmak et al. (2016) considered all the wavelengths. The brown boundary of Fig. 4.6 refers to the long wave linear stability boundary of the TFM neglecting shape factors, the memory terms in the shear stress closures, and the corrections for the shear stresses. The blue one is the Orr–Sommerfeld stability boundary of the exact multidimensional Navier-Stokes solution. Barmak et al. (2016) observed that, for those conditions which are quite similar to our experiment, the TFM long wave stability boundary without the corrections is close to the Orr–Sommerfeld stability boundary considering all the effects.

We consider next the FFM dispersion relation with $F = 0$ since we are interested in the dynamics of the KH instability. Equations (4.13) and (4.14) may be written as:

$$\boldsymbol{A}\frac{\partial}{\partial t}\underline{\phi} + \boldsymbol{B}\frac{\partial}{\partial x}\underline{\phi} + \boldsymbol{D}\frac{\partial^2}{\partial x^2}\underline{\phi} + \boldsymbol{E}'\frac{\partial^3}{\partial x^3}\underline{\phi} = 0, \tag{4.20}$$

where,

$$\boldsymbol{B} = \begin{bmatrix} u_1 & \alpha_1 \\ -\dfrac{1-\alpha_1}{1-\alpha_1+\alpha_1 r_\rho}C & B_{22} \end{bmatrix}, \tag{4.21}$$

$$\boldsymbol{D} \cong \begin{bmatrix} 0 & 0 \\ 0 & -\dfrac{1-\alpha_1}{1-\alpha_1+\alpha_1 r_\rho}\nu_T \end{bmatrix}, \tag{4.22}$$

$$\boldsymbol{E}' = \begin{bmatrix} 0 & 0 \\ -\dfrac{1-\alpha_1}{1-\alpha_1+\alpha_1 r_\rho}\dfrac{\sigma D}{\rho_1} & 0 \end{bmatrix}. \tag{4.23}$$

where C and B_{22} do not include shape factors, i.e., the potential flow assumption during the initial growth of the waves (Thorpe 1969) as opposed to fully developed laminar flow later on.

Following the procedure of Sect. 2.4.1 the dispersion analysis, including kinematic viscosity and surface tension, is performed with the experiment conditions: $D = 0.02\,\mathrm{m}$, $\alpha_1 = 0.542$, $j = 0$, $u_1 = 0.12\,\mathrm{m/s}$, $r_\rho = 0.72$, $\nu_T = 40\,\mathrm{mm^2/s}$, and $\frac{\sigma H}{\rho_1} = 0.8 \times 10^{-6}\,\mathrm{m^4/s^2}$. The results are shown in Fig. 4.17. The first significant outcome is the fast wave growth rate for all cases under the KH instability.

The basic, or Euler, 1D TFM dispersion relation is obtained from the solution to Eq. (2.39) with $\nu = 0$ and $\sigma = 0$. Beyond the KH limit this model is linearly ill-posed, i.e., growth rates increase unboundedly as the wavelength shrinks to zero. The zero wavelength growth rate is infinite for any relative velocity except for the trivial case of homogeneous flow, i.e., zero relative velocity. This is the well-known ill-posed TFM condition (Gidaspow 1974). Adding kinematic viscosity makes the

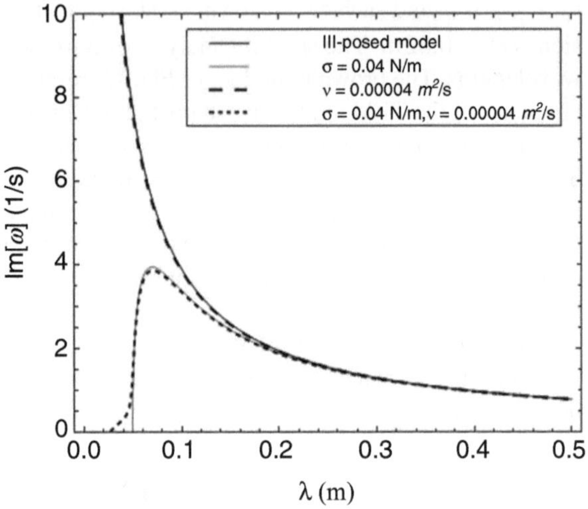

Fig. 4.17 Dispersion relation for FFM, reprinted from Vaidheeswaran et al. (2016), with permission from ASME

model well-posed but the growth rate is still very high at zero wavelength, $\omega_i = \frac{\alpha_1 C}{\nu} = 1000\ \mathrm{s}^{-1}$, which is practically ill-posed.

On the other hand, Fig. 4.17 shows that surface tension makes the model well-posed with the advantage that the growth rate is reduced to a value that is comparable to the observed initial growth. The cutoff wavelength corresponding to the surface tension of water–gasoline, i.e., $\sigma = 0.04$ N/m, is approximately 50 mm. Furthermore, the most dangerous wavelength is approximately 60 mm which is about three times the length of the waves shown in Fig. 4.14 for $t = 0.08$ s.

The figure also shows that the effect of viscosity on the linear stability is negligible compared to surface tension. Surface tension stabilization (Ramshaw and Trapp 1978) is the earliest published demonstration that the TFM may be rendered well-posed for unstable flow by including appropriate short wavelength physics. Nevertheless, even when the model is well-posed, there is still strong exponential wave growth which presents a challenge to practical applications of the 1D TFM. In the next section, a nonlinear simulation that overcomes this difficulty is presented. It will be shown that viscosity plays the key role in a mechanism of nonlinear stability.

4.3.5 *Nonlinear Simulations*

The current nonlinear simulations with the FFM developed in Sect. 4.3.3 were performed by Vaidheeswaran et al. (2016). Simulations with a similar model were also performed by Lopez de Bertodano et al. (2016). The primary objective of the simulations is to understand how the linearly unstable but well-posed FFM behaves in the long term when the waves have become fully developed. The corresponding 1D model is solved with the second-order method described in Sect. 2.5.5, also used in Sect. 4.3 for the KY equations. It should be reiterated that the model is not optimized for the initial growth of the waves.

The typical initial nonlinear behavior of the FFM exhibits Burgers-like steepening of the waves (Whitham 1974). Then, as previously shown in Sects. 4.2 and 4.3 for the KY equations, the waves stop growing due to the nonlinear interaction of viscosity and the steep wave fronts.

Simulations were run for a period of 2000 s, more than two orders of magnitude longer than the duration of the experiment. Periodic boundary conditions were applied in order to emulate the conditions in the center of the pipe in an infinite domain. Other than that the geometrical domain is the same as the experiment, i.e., the length is $L = 2.4\,\text{m}$ and the diameter is $D = 0.02\,\text{m}$. The inclination angle of the simulations is $\theta = 2.4°$, to obtain similar results to $\theta = 3.1°$ for the experiment. It was found that starting the simulations from the kinematic condition, i.e., $F = 0$ in Eq. (4.17), plus an initial perturbation in the void fraction, minimizes unnecessary initial transients. The initial conditions are $\alpha(x,0) = \alpha_{10} + \delta\left[\exp\left(\frac{x-0.8}{0.01}\right)^2 - \exp\left(\frac{x-1.6}{0.01}\right)^2\right]$ where $\alpha_{10} = 0.542$, and $\delta = 0.02$. The domain was initialized with a uniform velocity, $u_1 = -0.1\,\text{m/s}$ calculated from the kinematic condition, $F = 0$. The numerical simulations were performed for three grid sizes, $\Delta x = 1, 0.5$ and $0.025\,\text{mm}$, and the corresponding time step sizes are $\Delta t = 0.2, 0.1$ and $0.05\,\text{ms}$.

The comparisons of the numerical simulation with the experiment are shown in Figs. 4.18 and 4.19. Figure 4.18 compares a snap shot of the void fraction profile with a photograph of the interphase taken during the experiment. The shapes of the waves are similar even though they can never be exactly the same if they are chaotic. However, it should be noted that the spatial scales are different, i.e., the simulated wavelength is approximately twice the one observed experimentally.

Unfortunately the short time of the experiment, i.e., approximately 10 s, makes it difficult to obtain sufficient statistics for a conclusive comparison in the time domain. At any rate, Fig. 4.19 attempts a comparison of the time series. Again the chaotic time profiles are similar but the time scales are again different by a factor of 2.

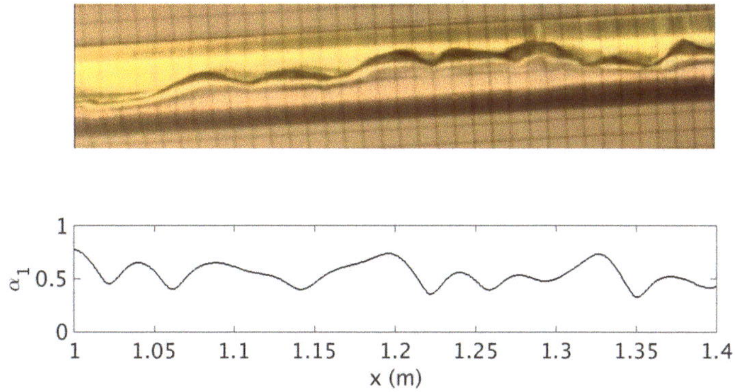

Fig. 4.18 Comparison of waveform of experiment with FFM simulation, reprinted from Vaidheeswaran et al. (2016), with permission from ASME

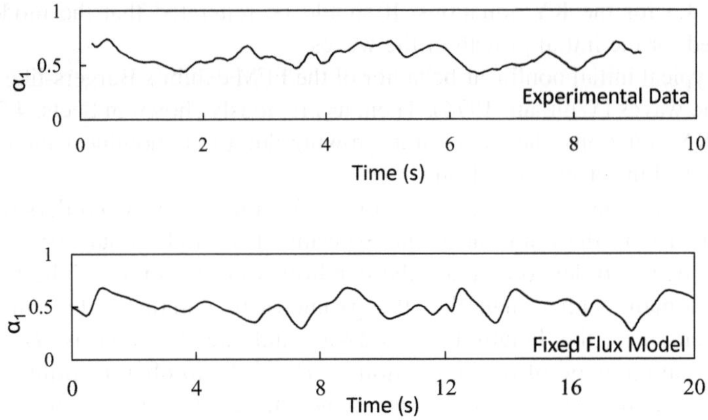

Fig. 4.19 Comparison of time series of experiment with FFM simulation

There are several reasons that can account for the discrepancies in the inclination angle and wavelengths, namely: (a) the limitations of the 1D model; (b) uncertainties in the closure laws, in particular the interfacial force, Eq. (4.12), has a significant impact on the profiles; (c) the periodic boundary conditions are an infinite domain approximation of the finite length test section. While these uncertainties are not trivial and require further investigation, it is remarkable that such a simple model captures the chaotic wave dynamics at all.

4.3.6 Lyapunov Exponent

The largest Lyapunov exponent, ω_L, was calculated for the simulations described in the previous section, following the procedure described in Sect. 4.2.2. The initial condition for the simulations is perturbed as follows,

$$\alpha_1(x,0) = \alpha_{10} + (\delta + \delta_{01})e^{\left(\frac{x-0.8}{0.01}\right)^2} - (\delta + \delta_{02})e^{\left(\frac{x-1.6}{0.01}\right)^2}. \tag{4.24}$$

Following Sprott (2003) the magnitude of perturbation is chosen to be $\delta_{01} = \delta_{02} = \pm 10^{-8}$. Four different perturbations of the void fraction initial condition, Eq. (4.24), are simulated, corresponding to the choice signs of $(\delta_{01}, \delta_{02})$, namely, $(+,+)$, $(+,-)$, $(-,+)$, and $(-,-)$.

The difference between each of the perturbed solutions with the unperturbed case is treated as an error measured by the L_2 norm given by Eq. (2.115). A positive ω_L indicates that the solutions are diverging and hence that the system is chaotic. The magnitude of ω_L is an estimation of the rate at which the predictability of the system is lost. Figure 4.20 shows the L_2-norm deviation of each perturbed solutions with slightly perturbed initial conditions compared to the initially perturbed states

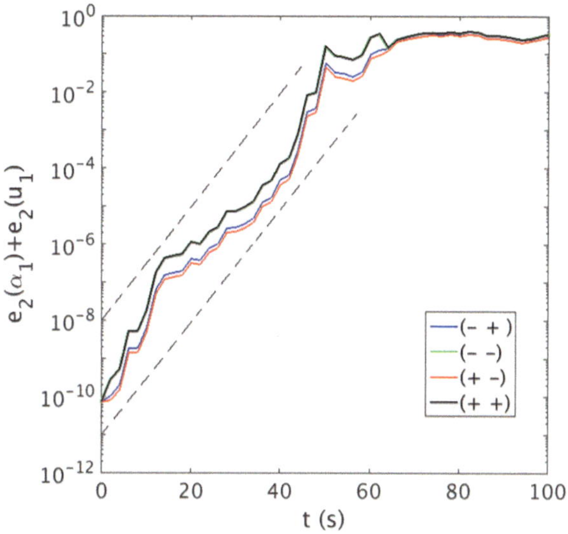

Fig. 4.20 The divergence of four solutions with slightly perturbed initial conditions for $\theta = 2.4°$, reprinted from Vaidheeswaran et al. (2016), with permission from ASME

with respect to the unperturbed case. The corresponding estimation of ω_L resulted in a positive value of $\omega_L \cong 0.27\,\mathrm{s}^{-1}$. This may be compared with the maximum linear growth rate from Fig. 3.2, $\omega_{i,\max} \cong 4\,\mathrm{s}^{-1}$, which is more than one order of magnitude greater, i.e., $\omega_L \ll \omega_{i,\max}$. This indicates that the chaotic wavy instability is significantly gentler than the initial instability of the smooth interface. Furthermore, the asymptotic behavior beyond 60 s shown in Fig. 4.20 indicates that chaos allows the model to become bounded. Thus, this nonlinear analysis illustrates the Lyapunov stability of the FFM.

4.3.7 Numerical Convergence

Chaos is also confirmed by the averaged amplitude void spectra shown in Fig. 4.21, which are used to check the convergence of the simulations with respect to the mesh resolution, following the argument already proposed in Sect. 4.2.6 and illustrated in Fig. 4.12. For this analysis the domain was 1 m long and the simulations ran for 2000 s after the waves become chaotic.

It is clear from the Figure that the simulations converge in a statistical sense as the mesh is refined. This is a demonstration that in principle the unstable FFM converges, as it should, provided it is well-posed, i.e., nonelliptic, and has a short wavelength dissipation mechanism. One important consequence of these results concerning the nature of the TFM is that it does capture the unstable nonlinear behavior of local material waves only when it ceases to be hyperbolic. Therefore, while it may be desirable to have a hyperbolic model for some engineering applications, there is no fundamental mathematical or physical requirement to do so. Furthermore, if the simulation of the local instabilities is of interest a hyperbolic model may be undesirable.

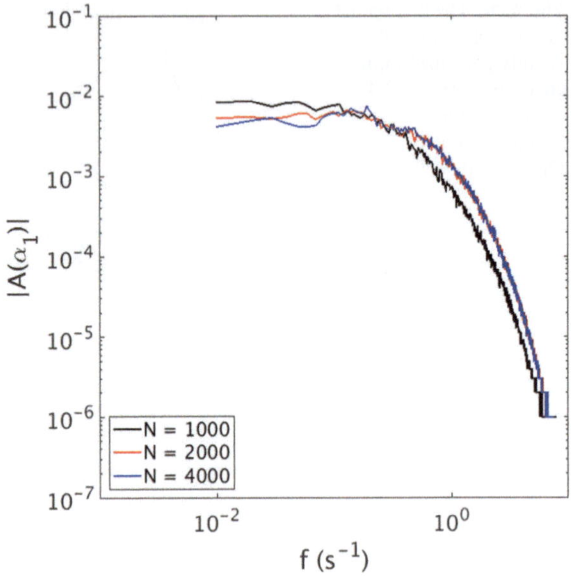

Fig. 4.21 Fourier transform of chaos, $\theta = 2.4°$

4.3.8 *Fractal Dimension*

Another informative test of chaos is the fractal dimension presented in Sect. 4.3.2. The slope of $C(r)$ vs. r, which gives the correlation dimension d_C, is shown in Fig. 4.22. As was discussed in Sect. 4.3.2 there are simultaneously two unknowns: the fractal correlation dimension d_C of the trajectory in phase space and the dimension of the phase space into which the attractor is embedded, d_E. The curves of $C(r)$ shown in Fig. 4.22 (left) correspond to the number of points in a radius-r ball of a succession of dimensions d_E. The maximum log–log slope of each curve, d_C, is plotted on Fig. 4.22 (right) as a function of d_E. It can be seen that in this case the correlation dimension saturates at $d_C \approx 7.35$ requiring a minimum embedded dimension of $d_E = 18$. Comparing with the KY equations, which have a minimum embedded dimension of 6 (cf. Fig. 4.5), the stratified-flow FFM needs in principle at least 18 modes to reflect the nonlinear dynamics. Such type of behavior is considered a high-dimensional chaos.

4.4 Summary and Discussion

This chapter presented a detailed nonlinear analysis based on simulations of the equations of Kreiss and Yström (2002) and the FFM. In Chap. 2 it was shown that the linear dynamics of the KY equations are similar to the FFM and, like the TFM, they are conditionally ill-posed as an IBVP, i.e., the characteristics of the system are complex, but the model becomes parabolic–hyperbolic and well-posed when viscosity is included.

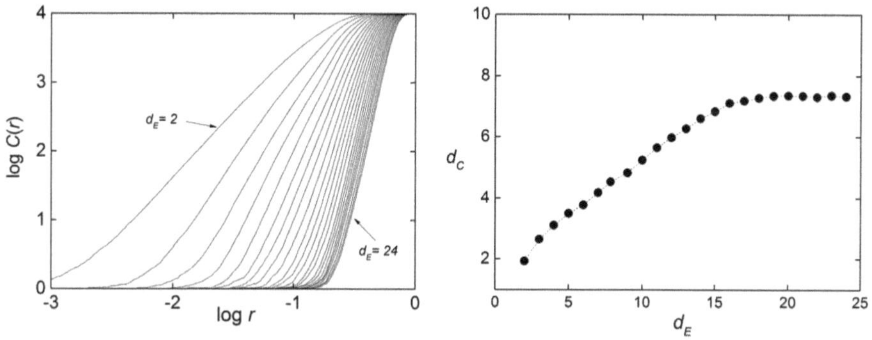

Fig. 4.22 Calculation of the embedded dimension (*right*) and the correlation dimension as a function of the embedded dimension (*right*) for the chaotic oscillation with an inclination angle $\theta = 2.4°$

The second-order finite-difference numerical method developed in Sect. 2.5.5 was used to solve the KY equations. These simulations were compared to the KY simulations performed with a spectral method. The calculated solutions matched each other up to $t = 4$. However, for longer times, $t = 40$, the two solutions were different. First it was assumed that the grid was too coarse and further refinement was required. However, upon four successive refinements, it was observed that the solutions did not converge, either with the KY results or with each other. This convergence problem was further demonstrated by using a single grid ($N = 512$) and imposing four different small perturbations to the initial conditions. It was found that each slightly perturbed solution diverged from the unperturbed solution. Such extreme sensitivity to initial conditions indicated that those KY solutions are chaotic. It was then determined that the solutions are bounded, i.e., Lyapunov stable, and ω_L was calculated in the long term when the solution reached a stationary state around the equilibrium condition. The calculations showed that ω_L was positive, confirming that the equations are indeed chaotic. The strange attractor was further characterized by calculating the fractal correlation dimension. The minimum embedded dimension necessary to unfold the dynamics of the chaos was shown to be only 6.

Once chaos was quantified the route from steady state to chaos was investigated using the C coefficient as the control variable. For small, positive values of C the solution asymptotically approached the equilibrium steady-state solution. As C approached a critical value C^*, the rate of decay toward the steady state slowed until a limit cycle behavior emerged, with a wavelength equal to the size of the computational domain, i.e., L-periodic. Further increasing C increased the amplitude of the limit cycle and the asymmetry of the waveform. Eventually intermittency appeared: stable $L/2$-periodic waves punctuated by short bursts of chaos-like behavior. The frequency of the intermittency decreased until a stable $L/2$-periodic limit cycle was reached. A similar pattern repeated itself until chaotic behavior was finally reached around $C = 0.125$. In essence then, the regime transition from stable to wavy chaotic flow is described by the transition of a well-posed model from a stable attractor to a strange attractor.

Then the issue of convergence was reexamined, and stochastic means to quantify the effect of the grid were sought, which did not rely on comparing specific solutions at specific times. The wavenumber (Fourier) spectra of the two solution variables, averaged from times $t = 4$–40, were used instead for comparison. While the different grids produce different solutions, the averaged wavenumber spectra over the time range converge to an invariant spectrum, i.e., independent of the initial conditions. This allows a statistical assessment of convergence.

After the KY equations, the FFM was solved numerically in a periodic domain. The simulations were set to match the conditions of a new experiment, based on the experiment performed by Thorpe (1969). The new experiment focuses on the chaotic behavior of the waves past their initial growth, which was not reported by Thorpe. The FFM nonlinear simulations produced a wavy, highly dynamic pattern. Chaos was confirmed with two quantitative measures. First, a positive ω_L was calculated. It turned out to be one order of magnitude smaller than the linear counterpart making the model more docile. The Lyapunov analysis demonstrates that diverging trajectories eventually became bounded into a strange attractor. Second, the embedded dimension of the attractor obtained from the fractal dimension analysis was 18, suggesting that the FFM leads to high-dimensional chaos in contrast to the KY equations.

Convergence in the stochastic sense was demonstrated from the continuous Fourier spectra obtained from the simulations. A remarkable observation regarding these spectra is that they look similar to the Kolmogorov turbulence spectrum, which is consistent with the high embedded dimension. However, the similarity cannot be taken too far because the spectra represent different phenomena, i.e., the interfacial behavior of a 1D FFM vs. the behavior of turbulent 3D eddies. Nevertheless, an analogous scenario may be suggested, namely, in as much as the mechanism of eddy stretching and turning combined with viscosity ensures Lyapunov stability of the Navier–Stokes equations, the mechanism of interfacial wave steepening combined with viscosity appears to bound the dynamics of the FFM, ultimately stabilizing it in the Lyapunov sense.

In closing, the TFM stability analysis has been extended far beyond linear theory via the FFM. Nonlinear theory describes growth saturation and chaos. Of course this was possible only because the specific FFM used for the simulations is well-posed. Ultimately, dwelling only on the ill-posed question is not enough and it is desirable to get past it and look at the nonlinear dynamics. The newly encountered FFM chaotic behavior differs significantly from the well-known linear theory which only determines whether the unstable TFM blows up instantaneously (ill-posed) or exponentially (well-posed) and which is valid for a very short interval of time. On the other hand non-linear stability analysis determines whether the behavior is bounded in the long term and may lead to a better understanding of two-phase flow dynamics, e.g., some flow regime transitions may be analyzed as a transition of a well-posed model between two attractors.

References

Abarbanel, H. D. I. (1996). *Analysis of observed chaotic data*. New York: Springer.

Andritsos, N., & Hanratty, T. J. (1987). Interfacial instabilities for horizontal gas–liquid flows in pipelines. *International Journal of Multiphase Flow, 13*, 583–603.

Arai, M. (1980). Characteristics and stability analyses for two-phase flow equation systems with viscous terms. *Nuclear Science and Engineering, 74*, 77–83.

Barmak, I., Gelfgat, A., Ullmann, A., Brauner, N., & Vitoshkin, H. (2016). Stability of stratified two-phase flows in horizontal channels. *Physics of Fluids, 28*, 044101.

Barnea, D., & Taitel, Y. (1994). Interfacial and structural stability of separated flow. *International Journal of Multiphase Flow, 20*, 387–414.

Lopez de Bertodano, M. A., Fullmer, W.D., & Clausse, A. (2016). One-dimensional two-fluid model for wavy flow beyond the Kelvin-Helmholtz instability: Limit cycles and chaos. *Nuclear Engineering and Design*. Retrieved from, http://authors.elsevier.com/sd/article/S0029549316301716.

Fullmer, W. D., Lopez de Bertodano, M. A., & Ransom, V. H. (2011). The Kelvin–Helmholtz instability: Comparisons of one- and two-dimensional simulations. In *14th International Topical Meeting on Nuclear Reactor Thermal Hydraulics (NURETH-14)*, Toronto.

Fullmer, W. D., Lopez de Bertodano, M. A., & Clausse, A. (2014). Analysis of stability, verification and chaos with the Kreiss–Yström equations. *Applied Mathematics and Computation, 248*, 28–46.

Grassberger, P., & Procaccia, I. (1983). Characterization of strange attractors. *Physical Review Letters, 50*(5), 346–349.

Gidaspow, D. (1974) Round Table Discussion (RT-1-2): Modeling of Two-Phase Flow, Proc. 5th Int. Heat Transfer Conf., Tokyo, Japan, September 3–7.

Hyman, L. M., & Nicolaenko, B. (1986). The Kuramoto-Sivashinsky equation: A bridge between PDE'S and dynamical systems. *Physica D Nonlinear Phenomena, 18*, 113–126.

Keyfitz, B. L., Sever, M., & Zhang, F. (2004). Viscous singular shock structure for a non-hyperbolic Two Fluid model. *Nonlinearity, 17*, 1731–1747.

Kocamustafaogullari, G. (1985). Two-fluid modeling in analyzing the interfacial stability of liquid film flows. *International Journal of Multiphase Flows, 11*, 63–89.

Kreiss, H.-O., & Yström, J. (2002). Parabolic problems which are ill-posed in the zero dissipation limit. *Mathematical and Computer Modelling, 35*, 1271–1295.

Picchi, D., Correra, S., & Poesio, P. (2014). Flow pattern transition, pressure gradient, hold-up predictions in gas/non-Newtonian power-law fluid stratified flow. *International Journal of Multiphase Flow, 63*, 105–115.

Picchi, D., & Poesio, P. (2016). A unified model to predict flow pattern transitions in horizontal and slightly inclined two-phase gas/shear-thinning fluid pipe flows. *International Journal of Multiphase Flow, 84*, 279–291.

Ramshaw, J. D., Trapp, J. A. (1978). Characteristics, stability and short wavelength phenomena in two-phase flow equation systems. *Nuclear Science and Engineering, 66*, 93–102.

Richardson, L. F. (1926). Atmospheric Diffusion Shown on a Distance-Neighbour Graph. *Proceedings of the Royal Society of London. Series A, 110*, 709–737.

Sprott, J. C. (2003). *Chaos and time series analysis*. Oxford, UK: Oxford University Press.

Thorpe, J. A. (1969). Experiments on the instability of stratified shear flow: Immiscible fluids. *Journal of Fluid Mechanics, 39*, 25–48.

Vaidheeswaran, A., Fullmer, W. D., Chetty, K., Marino, R. G., & Lopez de Bertodano, M. (2016). Stabiility analysis of chaotic wavy stratified fluid-fluid flow with the 1D fixed-flux two-fluid model. In *Proceedings of ASME 2016 HT/FEDSM/ICNMM*, Washington, DC, USA, July 10–14.

Whitham, G. B. (1974). *Linear and nonlinear waves*. New York: Wiley.

Part II
Vertical Bubbly Flow

Chapter 5
Fixed-Flux Model

Abstract Following the procedures developed in Chap. 2, the Fixed-Flux Model (FFM) is applied here to the stability of vertical bubbly flow. The virtual mass force and the interfacial pressure are introduced to obtain a conditionally well-posed TFM. Then a collision-induced pressure is considered and we adopt the interfacial collision force derived from the Enskog kinetic equation by Alajbegovic et al. (Chemical Engineering Communications 174: 85–133, 1999). When this force is incorporated into the FFM it removes the Kelvin–Helmholtz-type instability completely, resulting in an unconditionally well-posed model. These mechanisms are also associated with the acoustic and material wave speeds so they have an effect on the fidelity of the model. Therefore, while the pursuit of a complete TFM may be impractical, it is at least possible to obtain a well-posed model with the correct wave speeds. Finally, the nonlinear behavior of the well-posed FFM for bubbly flow, first analyzed by Park et al. (International Journal of Multiphase Flow 24: 1205–1244, 1998), is investigated with numerical simulations for stable flow and for kinematically unstable waves for conditions similar to a Guinness draught beer.

5.1 Introduction

Vertical bubbly flow cover a wide range of conditions from frothy columns to a single bubble rising in a liquid pool. Bubbly flows are never steady since the bubble size and distribution are continuously changing so a steady flow is an idealization. Bubbly flows are significant to many energy technologies because of boiling heat transfer. Even though we will not consider phase change or change in the bubble size, in this chapter we will address the SWT and KH fluid dynamic instabilities with a physically well-posed FFM.

The linear stability theory of the well-posed bubbly flow TFM was first formulated in terms of two forces: the interfacial pressure of Stuhmiller (1977) and the virtual mass force. These were added together to obtain the TFM of Pauchon and Banerjee (1986). Following the procedure of Chap. 2, a two-equation FFM for bubbly flow is derived that produces identical results to those of Pauchon and Banerjee (1986) and is consistent with the more complete model of Haley et al. (1991). This model results in a critical void fraction beyond which the equations are ill-posed. It is demonstrated that this restriction is removed by an

M.L. de Bertodano et al., *Two-Fluid Model Stability, Simulation and Chaos*,
DOI 10.1007/978-3-319-44968-5_5

elastic collision force that is derived from the collision kernel of the Enskog kinetic equation, an extension of Boltzmann's equation for nondilute gases (Garzó and Santos 2003). This is common practice for particle flows where it is referred to as the "particle pressure." Finally, even though the two-equation FFM for bubbly flow is not identical to SWT, the dispersion analysis shows that the kinematic instability condition of Whitham (1974) is still valid.

A nonlinear analysis of the stable TFM for vertical bubbly flow was first performed by Park et al. (1998). Here the nonlinear analysis is extended to nonlinear simulations of stable and kinematically unstable flows. Material shock waves are obtained as predicted by Park. A limit cycle is also obtained for the case of waves in Guinness (Robinson et al. 2008). Thus, the methodology already developed in the previous chapters to derive a physically well-posed TFM for unstable horizontal stratified flow conditions is also demonstrated for the vertical bubbly flow regime.

5.2 Compressible Two-Fluid Model

5.2.1 Compressible Model Equations

Following the procedure of Sect. 2.2.1 Chap. 2, a very simplified bubbly flow TFM for compressible bubbly flow is considered first, containing only drag and virtual mass forces, M_1^{D} and M_1^{VM}, to analyze the effect of the virtual mass force on the speed of isentropic sound waves. Thus the TFM equations are:

$$\frac{D_1\alpha_1}{Dt}+\alpha_1\frac{\partial u_1}{\partial x}+\frac{\alpha_1}{\rho_1 c_1^2}\frac{D_1 p}{Dt}=0 \tag{5.1}$$

$$-\frac{D_2\alpha_1}{Dt}+\alpha_2\frac{\partial u_2}{\partial x}+\frac{\alpha_2}{\rho_2 c_2^2}\frac{D_2 p}{Dt}=0 \tag{5.2}$$

$$\alpha_1\rho_1\frac{D_1 u_1}{Dt}=-\alpha_1\frac{\partial p}{\partial x}-\rho_1 g+M_1^{\mathrm{D}}+M_1^{\mathrm{VM}} \tag{5.3}$$

$$\alpha_2\rho_2\frac{D_2 u_2}{Dt}=-\alpha_2\frac{\partial p}{\partial x}-\rho_2 g-M_1^{\mathrm{D}}-M_1^{\mathrm{VM}} \tag{5.4}$$

where it is assumed that $p_2=p_1=p$ and M_1^{VM} is the interfacial virtual mass force which is described in the next section. M_1^{D} is the drag force described later in Sect. 5.4.4. A constitutive model for the drag is not required at this point because algebraic terms do not enter into the characteristic analysis.

5.2.2 Virtual Mass Force

The virtual mass force accounts for the transient drag force acting on the bubbles when their acceleration differs from that of the surrounding liquid. It accounts for the additional inertia that the bubbles need to displace the surrounding liquid in order to accelerate. It is significant for transient gas-liquid bubbly flows because it is proportional to the density of the continuous phase and therefore becomes more important than the inertia of the bubbles themselves. The general form of the virtual mass force given by Drew and Lahey (1987) is,

$$M_1^{\mathrm{VM}} = -M_2^{\mathrm{VM}} = \alpha_2\rho_1 C_{\mathrm{VM}}\left[\left(\frac{\partial u_1}{\partial t} + u_1\frac{\partial u_1}{\partial x}\right) - \left(\frac{\partial u_2}{\partial t} + u_2\frac{\partial u_2}{\partial x}\right)\right]. \tag{5.5}$$

Other forms of the virtual mass force can be found in the literature; however, Eq. (5.5) is by far the most common found in multiphase CFD codes. A model for the virtual mass force coefficient, C_{VM}, was derived by Zuber (1964) using potential flow theory. Here, we consider the dilute limit so that $C_{\mathrm{VM}} = 0.5$. The characteristic equation, $\mathrm{Det}[\boldsymbol{B} - c\boldsymbol{A}] = 0$, is similar to that of Sect. 2.2.2 but the matrices now include virtual mass instead of hydrostatic terms:

$$\boldsymbol{A} = \begin{bmatrix} 1 & 0 & 0 & \dfrac{\alpha_2}{c_2^2\rho_2} \\ -1 & 0 & 0 & \dfrac{\alpha_1}{c_1^2\rho_1} \\ 0 & \alpha_2(C_{\mathrm{VM}}\rho_1 + \rho_2) & -\alpha_2 C_{\mathrm{VM}}\rho_1 & 0 \\ 0 & -\alpha_2 C_{\mathrm{VM}}\rho_1 & \alpha_1\rho_1 + \alpha_2 C_{\mathrm{VM}}\rho_1 & 0 \end{bmatrix} \tag{5.6}$$

$$\boldsymbol{B} = \begin{bmatrix} u_2 & \alpha & 0 & \dfrac{\alpha_2 u_2}{c_2^2\rho_2} \\ -u_1 & 0 & \alpha_1 & \dfrac{\alpha_1 u_1}{c_1^2\rho_1} \\ 0 & \alpha_2(C_{\mathrm{VM}}\rho_1 + \rho_2)u_2 & -\alpha_2 C_{\mathrm{VM}}\rho_1 u_1 & \alpha_2 \\ 0 & -\alpha_2 C_{\mathrm{VM}}\rho_1 u_2 & [\alpha_1 + \alpha_2 C_{\mathrm{VM}}]\rho_1 u_1 & \alpha_1 \end{bmatrix} \tag{5.7}$$

For the special case of homogeneous flow, $u_1 = u_2 = u$, the characteristics of this simple TFM are $c_{1-4} = u,\ u,\ u + c_{2\phi},\ u - c_{2\phi}$, where the first two characteristics are the material wave speeds and the last two characteristics are the acoustic wave speeds

$$c_{2\phi} = \sqrt{\frac{\rho_2(C_{\mathrm{VM}}\rho_1 + \alpha_1\widetilde{\rho})}{\left(\frac{\alpha_2\rho_1}{c_1^2} + \frac{\alpha_1\rho_2}{c_2^2}\right)(C_{\mathrm{VM}}\overline{\rho} + \alpha_1\rho_2)}}, \tag{5.8}$$

where $\widetilde{\rho} = \alpha_2\rho_1 + \alpha_1\rho_2$ and $\overline{\rho} = \alpha_1\rho_1 + \alpha_2\rho_2$. In turn, for $u_1 \neq u_2$ the material characteristics are imaginary so this model is ill-posed. The energy equations

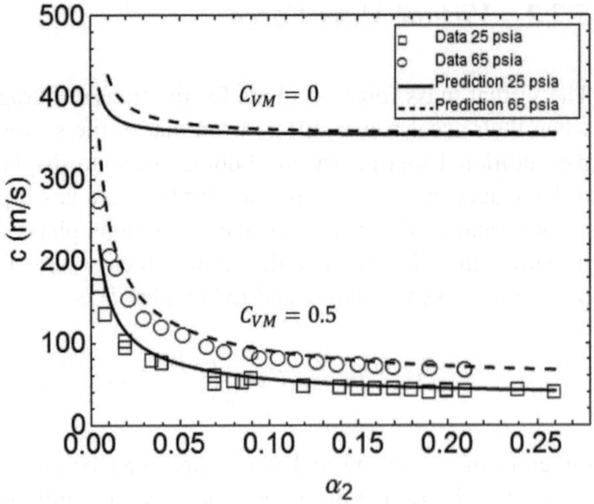

Fig. 5.1 Two-phase sound speed predictions comparison with water–steam data of Henry et al. (1971)

simply add two more real characteristics, $c_{5,6} = u_1,\ u_2$, which do not change the stability of the TFM.

It is well known that the virtual mass force has a significant effect on the two-phase sound speed (Watanabe et al. 1990; Watanabe and Kukita 1992). Figure 5.1 shows the dependence of the sound speed, Eq. (5.8), on the dispersed-phase concentration. The sound speed predictions with virtual mass are in good agreement with the experimental data of Henry et al. (1971).

5.3 Incompressible Two-Fluid Model

The adiabatic incompressible one-dimensional TFM of Pauchon and Banerjee (1986) for vertical bubbly flow between two parallel plates is used as the starting point. The main difference with the model in the previous section is the addition of the interfacial pressure difference to the liquid momentum equation:

$$\frac{D_1\alpha_1}{Dt} + \alpha_1\frac{\partial u_1}{\partial x} = 0 \tag{5.9}$$

$$-\frac{D_2\alpha_1}{Dt} + \alpha_2\frac{\partial u_2}{\partial x} = 0 \tag{5.10}$$

$$\rho_1\frac{D_1u_1}{Dt} = -\frac{\partial p_{1\mathrm{i}}}{\partial x} - \rho_1 g + \frac{1}{\alpha_1}\left[M_1^{\mathrm{D}} + M_1^{\mathrm{VM}} - (p_{1\mathrm{i}} - p_1)\frac{\partial \alpha_1}{\partial x}\right] - \frac{f_1}{H}\rho_1|u_1|u_1 \tag{5.11}$$

$$\rho_2\frac{D_2u_2}{Dt} = -\frac{\partial p_{1\mathrm{i}}}{\partial x} - \rho_2 g - \frac{1}{\alpha_2}\left(M_1^{\mathrm{D}} + M_1^{\mathrm{VM}}\right) \tag{5.12}$$

where it is assumed that $p_2 = p_{1i}$. The wall friction has been included in Eq. (5.11) for completeness, but it won't be used in the rest of the chapter. The incompressibility condition implies that the acoustic speed $c_{2\phi} = \pm\infty$. An important aspect is the definitions of the virtual mass force already addressed and the interfacial pressure p_{1i} which makes the model conditionally well-posed as shall be demonstrated. The material wave characteristic speeds will be analyzed using the fixed flux assumption that reduces the model to two equations, as in Chap. 2. However, we must first introduce a model for the interfacial pressure. A model for the collision force, which turns the TFM unconditionally well-posed, will be introduced in Sect. 5.4.2.

5.3.1 Interfacial Pressure

Stuhmiller (1977) proposed an interfacial pressure obtained from the integrated difference between the interfacial and the bulk pressures of the liquid due to the relative motion of the bubbles. This has a significant effect on the model stability and it is mathematically similar to the hydrostatic pressure for stratified flows, i.e., as the magnitude of this force increases so does the region of stability. The model has a physical interpretation for bubbly flow but has been applied artificially to other flow regimes in some industrial codes to achieve unconditional hyperbolicity, see e.g. Bestion (1990). For the present case, the interfacial pressure is given by

$$p_{1i} - p_1 = -C_p \rho_1 (u_1 - u_2)^2, \tag{5.13}$$

where it may be shown that $C_p = 1/4$ for dilute spheres in potential flow. The corresponding value of $C_{VM} = 1/2$ completes the TFM presented in Eqs. (5.9)–(5.12).

5.3.2 Fixed-Flux Model Derivation

The incompressible TFM presented in the previous sections is now used as the starting point to obtain a FFM following the derivation by Lopez de Bertodano et al. (2013). In Sect. 5.4 it will be shown how the interfacial pressure force produces a stabilization of the bubbly flow equations that is mathematically similar to the hydrostatic force of horizontal stratified flow. The effect of virtual mass, which is also significant for incompressible bubbly flow stability, is also considered and makes the model somewhat more complicated.

We now proceed to reduce the four-equation TFM to a two-PDE model similar to SWT following the procedure of Sect. 2.3.2. The first PDE is the sum of the two continuity Eqs. (5.9) and (5.10) written in conservative form,

$$\frac{\partial}{\partial t}(\rho_1\alpha_1 + \rho_2\alpha_2) + \frac{\partial}{\partial x}(\rho_1\alpha_1 u_1 + \rho_2\alpha_2 u_2) = 0. \tag{5.14}$$

The difference of the momentum Eqs. (5.11) and (5.12) allows the elimination of the pressure terms,

$$\begin{aligned}&\left(1+\frac{C_{\mathrm{VM}}}{\alpha_1}\right)\frac{D_1}{Dt}\rho_1 u_1-\left(1+\frac{C_{\mathrm{VM}}}{r\alpha_1}\right)\frac{D_2}{Dt}\rho_2 u_2+\frac{\partial}{\partial x}C_{\mathrm{p}}\rho_1|u_2-u_1|(u_2-u_1)\\&+\frac{C_{\mathrm{p}}}{\alpha_1}\rho_1(u_2-u_1)^2\frac{\partial\alpha_1}{\partial x}=-(\rho_1-\rho_2)g\\&-\frac{2}{\alpha_1 H}\frac{f_1}{2}\rho_1|u_1|u_1+\left(\frac{1}{\alpha_1}+\frac{1}{\alpha_2}\right)M^{\mathrm{D}}.\end{aligned} \tag{5.15}$$

The two other equations for closure are again the void fraction condition,

$$\alpha_1+\alpha_2=1, \tag{5.16}$$

and the volumetric flux condition,

$$\alpha_1 u_1+\alpha_2 u_2=j, \tag{5.17}$$

where j is the total volume flux. As before, j is in general a function of time but we assume that it is a constant, i.e., the fixed flux approximation. We now recast the two PDEs in matrix form,

$$\frac{d}{dt}\underline{\psi}+\frac{d}{dx}\underline{\varphi}=\underline{\varsigma}, \tag{5.18}$$

such that

$$\underline{\psi}=\begin{bmatrix}\rho_1\alpha_1+\rho_2\alpha_2\\(1+\beta)\rho_1 u_1-\left(1+\frac{\beta}{r}\right)\rho_2 u_2\end{bmatrix} \tag{5.19}$$

and

$$\underline{\varphi}=\begin{bmatrix}\rho_1\alpha_1 u_1+\rho_2\alpha_2 u_2\\\frac{1}{2}(1+\beta)\rho_1 u_1^2-\frac{1}{2}\left(1+\frac{\beta}{r_\rho}\right)\rho_2 u_2^2+C_{\mathrm{p}}\rho_1(u_2-u_1)^2+\gamma\alpha_1\end{bmatrix} \tag{5.20}$$

where $\gamma=\frac{C_{\mathrm{p}}}{\alpha_1}\rho_1(u_2-u_1)^2, \beta=C_{\mathrm{VM}}/\alpha_1, r_\rho=\rho_2/\rho_1$ and the source term is given by

$$\underline{\varsigma}=\begin{bmatrix}0\\-(\rho_1-\rho_2)g-\frac{2}{\alpha_1 H}\frac{f_1}{2}\rho_1|u_1|u_1+M^{\mathrm{D}}\end{bmatrix}. \tag{5.21}$$

The next step is to convert this system into primitive variables, $\underline{\phi} = [\alpha_1, u_1]^T$, so that:

$$A\frac{\partial}{\partial t}\underline{\phi} + B\frac{\partial}{\partial x}\underline{\phi} = \underline{F}\,, \tag{5.22}$$

where $\boldsymbol{A} = \boldsymbol{I}$. To simplify the analysis γ and β are arbitrarily treated as constants in the following transformation of variables with the heuristic justification that the final result remains indistinguishable from the full TFM analysis of Pauchon and Banerjee (1986). Following the procedure in Section 2.3.2 the components of matrix $\boldsymbol{B}$ are obtained by linear algebra analysis and expanded in Taylor series of the density ratio $r_\rho = \rho_2/\rho_1$. Again it is assumed that $r_\rho \ll 1$, which is satisfactory for air–water flow experiments at atmospheric conditions, and the terms of order r_ρ are neglected. Thus, the simplified matrix $\boldsymbol{B}$ is obtained:

$$\boldsymbol{B} \cong \begin{bmatrix} u_1 & \alpha_1 \\ \dfrac{(1+\alpha_1)C_\mathrm{p} - C_\mathrm{VM}}{C_\mathrm{VM} + \alpha_1 - \alpha_1^2}(u_2 - u_1)^2 & u_1 + \dfrac{2\alpha_1\left(C_\mathrm{VM} - C_\mathrm{p}\right)}{C_\mathrm{VM} + \alpha_1 - \alpha_1^2}(u_2 - u_1) \end{bmatrix} \tag{5.23}$$

and the source term becomes

$$\underline{F} \cong \frac{\alpha_1\alpha_2}{C_\mathrm{VM} + \alpha_1\alpha_2}\begin{bmatrix} 0 \\ -g - \dfrac{2}{\alpha_1 H}\dfrac{f_1}{2}|u_1|u_1 + \left(\dfrac{1}{\alpha_1} + \dfrac{1}{\alpha_2}\right)\dfrac{M^\mathrm{D}}{\rho_1} \end{bmatrix} \tag{5.24}$$

Now the TFM may be written in the familiar shallow-water theory form:

$$\boxed{\frac{\partial \alpha_1}{\partial t} + u_1\frac{\partial \alpha_1}{\partial x} + \alpha_1\frac{\partial u_1}{\partial x} = 0,} \tag{5.25}$$

$$\boxed{\begin{gathered}\frac{\partial u_1}{\partial t} + \left[u_1 + \frac{2\alpha_1\left(C_\mathrm{VM} - C_\mathrm{p}\right)}{C_\mathrm{VM} + \alpha_1\alpha_2}(u_2 - u_1)\right]\frac{\partial u_1}{\partial x} + C\frac{\partial \alpha_1}{\partial x} \\ = \frac{\alpha_1\alpha_2}{C_\mathrm{VM} + \alpha_1\alpha_2}\left[-g - \frac{1}{\alpha_1 H}\frac{f_1}{2}|u_1|u_1 + \left(\frac{1}{\alpha_1} + \frac{1}{\alpha_2}\right)\frac{M^\mathrm{D}}{\rho_1}\right],\end{gathered}} \tag{5.26}$$

where

$$\boxed{C = \frac{(1+\alpha_1)C_\mathrm{p} - C_\mathrm{VM}}{C_\mathrm{VM} + \alpha_1\alpha_2}(u_2 - u_1)^2.} \tag{5.27}$$

This two-equation model is mathematically more complex than1D SWT, Eqs. (2.33) and (2.34), because the convective term of the momentum equation has an additional term.

5.4 Linear Stability

We now perform the linear stability analysis of the 1D TFM for bubbly flow following the procedure of Chap. 2 and the analysis of Lopez de Bertodano et al. (2013). We start with the analysis of Pauchon and Banerjee (1986) which demonstrates that the TFM is conditionally KH stable, and well-posed, due to the interfacial pressure difference. The collision force is introduced next and it is demonstrated that the model becomes unconditionally well-posed. However, the kinematic instability, which was shown to be well-posed in Chap. 2, remains under certain conditions. A multidimensional instability analysis was not considered in Chap. 2 nor in this one, which may result in different instabilities. One multidimensional instability, i.e., the plume instability, is considered at the end of Chap. 9.

5.4.1 *Characteristic Analysis*

The characteristic equation, $\mathrm{Det}[\boldsymbol{B} - c\boldsymbol{A}] = 0$, is now solved for the characteristic velocities, c. The result is:

$$c^*_{1,2} = \frac{\alpha_1}{C_{\mathrm{VM}} + \alpha_1 - \alpha_1^2}\left(C_{\mathrm{VM}} - C_{\mathrm{p}}\right)\left(1 \pm \sqrt{1 + \frac{C_{\mathrm{p}}\left[C_{\mathrm{VM}}(1+\alpha_1) + \alpha_1 - \alpha_1^3\right] - C_{\mathrm{VM}}\left(C_{\mathrm{VM}} + \alpha_1 - \alpha_1^2\right)}{\alpha_1\left(C_{\mathrm{VM}} - C_{\mathrm{p}}\right)^2}}\right) \tag{5.28}$$

where $c^* = \frac{c-u_1}{u_2-u_1}$. Since $0 \le \alpha_1 \le 1$ the conditional stability criterion derived from the expression for the square root, i.e., stable for some range of α_2, is simply $C_{\mathrm{p}} \ge 0$ and the full stability criterion is $\frac{C_{\mathrm{p}}\left[C_{\mathrm{VM}}(1+\alpha_1)+\alpha_1-\alpha_1^3\right]-C_{\mathrm{VM}}\left(C_{\mathrm{VM}}+\alpha_1-\alpha_1^2\right)}{\alpha_1\left(C_{\mathrm{VM}}-C_{\mathrm{p}}\right)^2}$. For example, if

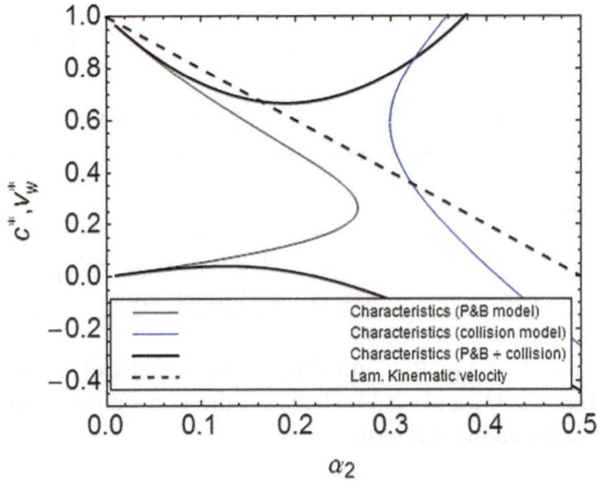

Fig. 5.2 Material wave speeds and kinematic speed for Stokes bubble

$C_p = \frac{1}{4}$ and $C_{VM} = \frac{1}{2}$ the model is stable for $\alpha_2 < 0.26$ as shown in Fig. 5.2. The upper and lower branches of the curve represent the two characteristic speeds. This is the well-known conditional stability result of Pauchon and Banerjee (1986) reproduced here with the FFM. However, stable bubbly flows are observed in nature at higher void fractions. One way to overcome this shortcoming is to include more physics. The effect of a collision force, that makes the TFM unconditionally stable, will be analyzed next.

5.4.2 Collision Force

The surface tension force in bubbly flow plays a stabilizing role that is similar to stratified flow; however, it is more difficult to model directly. Instead, a very simple assumption is made that surface tension keeps the bubbles spherical and elastic so that bubbles are able to undergo collisions without coalescence or breakup. This assumption may not be applicable to industrial flows of "large" distorted bubbles but is surprising relevant for microscale bubbles—such as the nitrogen bubbles in a draught beer. We recommend interested readers view the videos Alexander and Zare (2004) that show the motion of Guinness bubbles under magnification. An interfacial collision force model was derived from the Enskog equation by Alajbegovic et al. (1999),

$$\boldsymbol{M}^{\text{coll}} = -\nabla \cdot \left[(\rho_2 + C_{\text{VM}}\rho_1)\chi(\alpha_2)\alpha_2^2 \left(2\overline{\boldsymbol{u}_2'\boldsymbol{u}_2'} + \overline{\boldsymbol{u}_2' \cdot \boldsymbol{u}_2'}\boldsymbol{I} \right) \right], \tag{5.29}$$

where $\overline{\boldsymbol{u}_2'\boldsymbol{u}_2'}$ is the normal turbulent stress tensor of the bubbles and $\chi(\alpha_2)$ is the pair correlation function at contact, also known as the radial distribution function (RDF), which accounts for finite bubble number density effects (Garzó and Santos 2003). Equation (5.29) represents a dispersed-phase pressure which is proportional to the fluctuating kinetic energy of the dispersed phase. The relationship between a hydrodynamic pressure and the fluctuations at the microscale (i.e., temperature) is analogous to the ideal gas law from the kinetic theory of gases. It is well known that the TFM may be hyperbolized by adding interfacial forces that contain void gradient terms (Bestion 1990) and the collision force fits the description. The particle flow community uses the collision force extensively, i.e., the "particle pressure," to stabilize the TFM (Fullmer and Hrenya 2017).

Considering potential flow around an ensemble of spheres (Drew and Passman 1999), the liquid phase bubble-induced turbulence may be expressed as:

$$\overline{\boldsymbol{u}_1'\boldsymbol{u}_1'} = \begin{pmatrix} 4/5 & 0 & 0 \\ 0 & 3/5 & 0 \\ 0 & 0 & 3/5 \end{pmatrix} \alpha_2 \frac{1}{2} C_{\text{VM}} |\boldsymbol{u}_2 - \boldsymbol{u}_1|^2. \tag{5.30}$$

Furthermore, if the bubbles are in turbulent equilibrium with the liquid, it may be shown following a procedure similar to that of sect. 9.3.6 that

$$\overline{u_2'u_2'} = \frac{1}{1+\tau_b/\tau_e}\overline{u_1'u_1'}, \tag{5.31}$$

where τ_e is the time constant of the bubble-induced eddies and τ_b is the time constant related to the bubble inertia. Inserting Eqs. (5.30) and (5.31) into (5.29) the collision force becomes:

$$\boldsymbol{M}^{\text{coll}} = -\nabla\cdot\left[\begin{pmatrix} 3.6 & 0 & 0 \\ 0 & 3.2 & 0 \\ 0 & 0 & 3.2 \end{pmatrix}\frac{1}{1+\dfrac{\tau_e}{\tau_b}}(\rho_2 + C_{\text{VM}}\rho_1)\chi(\alpha_2)\alpha_2^3\frac{1}{2}C_{\text{VM}}|\boldsymbol{u}_2 - \boldsymbol{u}_1|^2\right]. \tag{5.32}$$

The radial distribution function (RDF) at contact, χ, is a monotonically increasing function of the dispersed phase concentration, here α_2. One of the most widely used models for the RDF was proposed by Carnahan and Starling (1969) and is given by:

$$\chi(\alpha_2) = \frac{2-\alpha_2}{2(1-\alpha_2)^3}. \tag{5.33}$$

Theoretically the RDF should approach infinity at maximum packing, which occurs at approximately 0.64 for a randomly assembly of spherical particles. The RDF model of Ma and Ahmadi (1988) can be used to account for the effect of maximum packing. The maximum packing density of bubbles may be significantly closer to unity though, (i.e., a foam) due to the deformability of the bubbles (Ishii and Hibiki 2006). However, such a state certainly violates the assumptions of binary and instantaneous collisions which were used to derive Eq. (5.29). Therefore, we choose to use the simpler RDF of Eq. (5.33) and forego results with concentrations $\alpha_2 > 0.6$. Finally, combining Eqs. (5.32) and (5.33) and replacing $\alpha_2 = 1 - \alpha_1$, the axial component of the collision force is obtained:

$$M_x^{\text{coll}} = -\frac{\partial}{\partial x}\left[\frac{1.8}{1+\frac{\tau_e}{\tau_b}}\frac{(1+\alpha_1)(1-\alpha_1)^3}{2\alpha_1^3}C_{\text{VM}}(\rho_2 + C_{\text{VM}}\rho_1)(u_2-u_1)^2\right]. \tag{5.34}$$

Inserting the momentum transfer term due to collisions into the TFM, using the fixed flux assumption, Eq. (5.17), and letting the density ratio $r_\rho \ll 1$, the momentum Eq. (5.26) gets modified as follows,

$$\boxed{\begin{aligned}&\frac{\partial u_1}{\partial t} + \left[u_1 + \frac{2\alpha_1\left(C_{\text{VM}} - C_p + \dfrac{1.8}{1+\tau_e/\tau_b}\dfrac{(1+\alpha_1)(1-\alpha_1)^2 C_{\text{VM}}^2}{2\,\alpha_1^3}\right)}{C_{\text{VM}}+\alpha_1\alpha_2}(u_2-u_1)\right] \\ &\frac{\partial u_1}{\partial x} + C'\frac{\partial \alpha_1}{\partial x} = \frac{\alpha_1\alpha_2}{C_{\text{VM}}+\alpha_1\alpha_2}\left[-g - \frac{2}{\alpha_1 H}\frac{f_1}{2}|u_1|u_1 + \left(\frac{1}{\alpha_1}+\frac{1}{\alpha_2}\right)\frac{M^{\text{D}}}{\rho_1}\right],\end{aligned}} \tag{5.35}$$

where,

$$C' = C + \frac{1.8}{1 + \tau_e/\tau_b} C_{VM}^2 \frac{(3 - \alpha_1^2)(1 - \alpha_1)^2}{2\,\alpha_1^3(C_{VM} + \alpha_1\alpha_2)} (u_2 - u_1)^2. \tag{5.36}$$

and C as given previously by Eq. (2.27).

Figure 5.2 shows the characteristics including the collision force for conditions similar to the nitrogen bubbles, $d_B = 120\,\mu m$, in a glass of Guinness draught beer (Robinson et al. 2008). For Stokes flow with $\nu_1 = 2 \times 10^{-6}\,m^2/s$ the time constant of the bubbles and bubble-induced eddies are $\tau_b = \frac{1}{18} C_{VM} \frac{d_B^2}{\nu_1}$ and $\tau_e \cong \frac{d_B}{u_2 - u_1}$, respectively, where $u_2 - u_1 = \frac{1}{18} g \frac{d_B^2}{\nu_1}$. The time constant ratio is then $\tau_e/\tau_b \cong 0.01 \ll 1$. In the end, it is the effect of the drastic increase in the RDF that makes the model unconditionally well-posed.

It is now possible to compare the collision force of bubbly flow with the hydrostatic force of near horizontal wavy flow mathematically because both mechanisms are represented by a first-order void gradient. The main difference is that in this case the term makes the model stable at all relative velocities, i.e., $\mathrm{Det}[\boldsymbol{B} - c\boldsymbol{A}] = 0$ means that the KH instability is precluded. Thus, the case of small spherical bubbles is always dynamically stable. However, the system may still be kinematically unstable which will be explored in the following section.

5.4.3 Dispersion Relation: Kinematic Instability

The dispersion relation for Eqs. (5.25) and (5.35) is obtained from

$$\mathrm{Det}[-i\omega \boldsymbol{A} + ik\boldsymbol{B} - \boldsymbol{F}'] = 0, \tag{5.37}$$

where

$$\boldsymbol{F}' = \frac{\partial F}{\partial \underline{\phi}} = \begin{bmatrix} 0 & 0 \\ F_\alpha & F_u \end{bmatrix}. \tag{5.38}$$

The resulting equation, which is more complicated for bubbly flow than for stratified flow, is,

$$\omega = uk - bk - i\frac{F_u}{2} \pm \sqrt{\left(i\frac{F_u}{2}\right)^2 + iF_u(b + v_w)k + a(bk)^2}, \tag{5.39}$$

where $a = \frac{C_p\left[C_{VM}(1+\alpha_1)+\alpha_1-\alpha_1^3\right]-C_{VM}\left(C_{VM}+\alpha_1-\alpha_1^2\right)}{\alpha_1\left(C_{VM}-C_p\right)^2}$ and $b = \frac{\alpha_1}{C_{VM}+\alpha_1-\alpha_1^2}\left(C_{VM} - C_p\right)$ $(u_2 - u_1)$ for the model of Pauchon and Banerjee (1986). This dispersion relation may be rewritten as

$$\omega = uk - bk - i\frac{F_u}{2} \pm \sqrt{\left(\sqrt{1+a}\, bk + i\frac{F_u}{2}\right)^2 + iF_u\left[(1-\sqrt{1+a})b + v_{\mathrm{w}}\right]k}. \tag{5.40}$$

Finally, it can be shown from this equation that the kinematic stability condition is the same as that of SWT (Whitham 1974), i.e., $v_{\mathrm{w}} \le c$.

5.4.4 *Drag Force*

5.4.4.1 Laminar Regime (Stokes Flow)

We now consider the laminar interfacial drag force of the TFM. For Stokes flow the well-known correlation is

$$M^{\mathrm{D}} = \frac{18\,\nu}{d_{\mathrm{B}}^2}\rho_1\alpha_2(u_2 - u_1) \tag{5.41}$$

Dividing the momentum equation by $g\rho_1$, the RHS of Eq. (5.26) may be cast in nondimensional form as follows,

$$\frac{F}{g} = \frac{\alpha_1\alpha_2}{C_{\mathrm{VM}} + \alpha_1\alpha_2}\left[-1 + \left(\frac{1}{\alpha_1} + \frac{1}{\alpha_2}\right)4.5\alpha_2\left(\frac{u_2}{u_0} - \frac{u_1}{u_0}\right)\right] \tag{5.42}$$

where $u_0 = \frac{gd_{\mathrm{B}}^2}{4\nu}$. For the condition $j = 0$ the velocities obtained for the kinematic condition, $F = 0$, become

$$\frac{u_1}{u_0} = -\frac{1}{4.5}\alpha_2(1-\alpha_2), \quad \frac{u_2}{u_0} = \frac{1}{4.5}(1-\alpha_2) \tag{5.43}$$

Then the resulting kinematic wave speed obtained from Eq. (2.50) is

$$v_{\mathrm{w}}^* = \frac{v_{\mathrm{w}}}{u_0} = \frac{1}{4.5}\frac{\alpha_1\left[C_{\mathrm{VM}}(-1 + 2\,\alpha_1) + \alpha_1\left(-1 + 3\,\alpha_1 - 2\,\alpha_1^2\right)\right]}{C_{\mathrm{VM}} + \alpha_1\left(1-\alpha_1\right)} \tag{5.44}$$

For $C_{\mathrm{p}} = \frac{1}{4}$ and $C_{\mathrm{VM}} = \frac{1}{2}$ the kinematic instability condition ($v_{\mathrm{w}} > c$) for the model including the collision force is satisfied for $\alpha_2 < 0.17$, as can be observed in Fig. 5.2.

Figure 5.3 shows the wave growth speed, $v_{\mathrm{i}}^* = \frac{\omega_{\mathrm{i}}/k}{u_0}$, as a function of the wavelength, $\lambda^* = \frac{2\,\pi/k}{u_0/\tau_{\mathrm{b}}}$, obtained from the dispersion relation for the positive characteristic in the range $\alpha_2 = 0.1 - 0.2$. This is consistent with the results

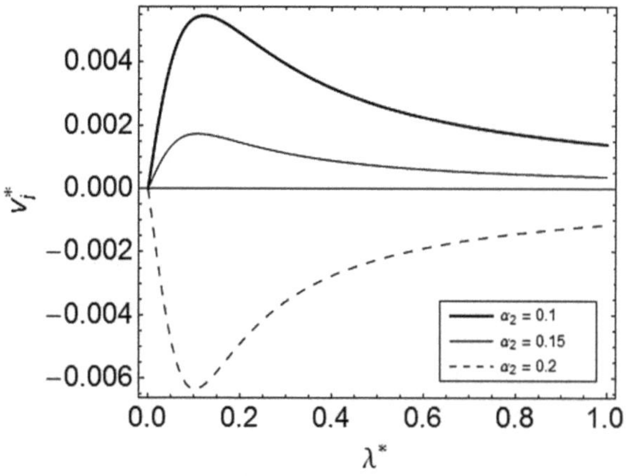

Fig. 5.3 Dispersion relation for bubbly flow in the Stokes regime

shown in Fig. 5.2, i.e., the waves grow for $\alpha_2 < 0.17$. Furthermore, the growth rate, ω_i, for zero wavelength is zero so the model is well-posed. This was also shown in Chap. 2 for the kinematic instability for stratified flow.

5.4.4.2 Turbulent Regime (Distorted Bubbles)

We now consider the turbulent interfacial drag force,

$$M^{\mathrm{D}} = \frac{3}{4}\frac{C_{\mathrm{D}}}{d_{\mathrm{B}}}\rho_1\alpha_2|u_2 - u_1|(u_2 - u_1) \tag{5.45}$$

but we leave the definition of the distorted bubble drag coefficient for later. At this point we make the momentum equation dimensionless in terms of the drag coefficient. Following the method of the previous section, the RHS of the momentum equation becomes,

$$\frac{F}{g} = \frac{\alpha_1\alpha_2}{C_{\mathrm{VM}} + \alpha_1\alpha_2}\left[-1 + \left(\frac{1}{\alpha_1} + \frac{1}{\alpha_2}\right)\alpha_2\left(\frac{u_2}{u_0} - \frac{u_1}{u_0}\right)^2\right], \tag{5.46}$$

where $u_0 = \sqrt{\frac{4}{3}\frac{gd_{\mathrm{B}}}{C_{\mathrm{D}}}}$. For the condition $j = 0$ the velocities obtained for the kinematic condition, $F = 0$, become

$$\frac{u_1}{u_0} = -\alpha_2\sqrt{1-\alpha_2}, \quad \frac{u_2}{u_0} = \sqrt{1-\alpha_2} \tag{5.47}$$

and the kinematic wave speed is

$$v_{\mathrm{w}}^* = \frac{v_{\mathrm{w}}}{u_0} = \frac{C_{\mathrm{VM}}(-1 + 3\,\alpha_1 - \alpha_1^2) - \alpha_1^2 + 4\alpha_1^3 - 3\alpha_1^4}{2\sqrt{1-\alpha_1}[C_{\mathrm{VM}} + (1-\alpha_1)\alpha_1]}. \tag{5.48}$$

Figure 5.4 shows the resulting characteristics for $C_p = 1/4$ and $C_{VM} = 1/2$. The flow is kinematically unstable over a considerable range, i.e., the condition $v_w > c$ is satisfied up to 30 % void fraction, so this model for turbulent-bubbly flow is well-posed but kinematically unstable. However, these results should be taken with a grain of salt, since the kinematic instability is not observed in typical turbulent flow air–water experiments. In the first place, the model is more incomplete than others, e.g., Haley et al. (1991), and does not include turbulent mechanisms like the turbulent diffusion force. In the second place, the values of the coefficients for distorted bubbles in turbulent flow differ significantly from those for a perfect sphere in potential flow. For example, using the arbitrary but not unreasonable values for typical 3–5 mm bubbles of $C_p = 1$ and $C_{VM} = 2$ in Fig 5.4 (right) the kinematic instability is practically removed.

The collision model is built on the assumption that the collisions between the bubbles are elastic. However, collisions of distorted bubbles, e.g., 3–5 mm air bubbles in water at standard temperature and pressure, are far from elastic. Figure 5.5 shows a comparison of the characteristic speed for $C_p = 1/4$ and $C_{VM} = 1/2$ with the data of

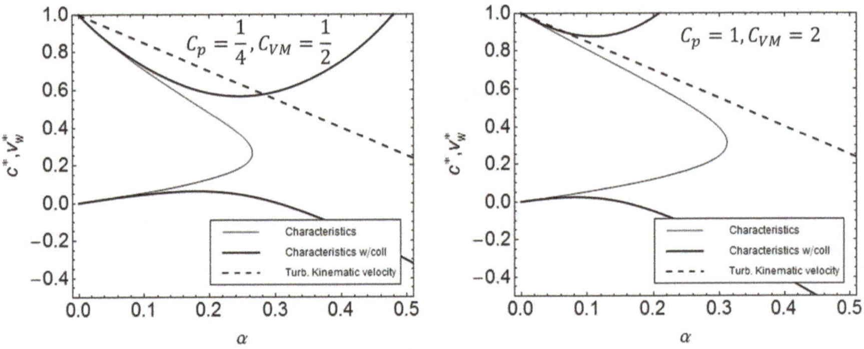

Fig. 5.4 Material wave speeds and kinematic speed for turbulent flow

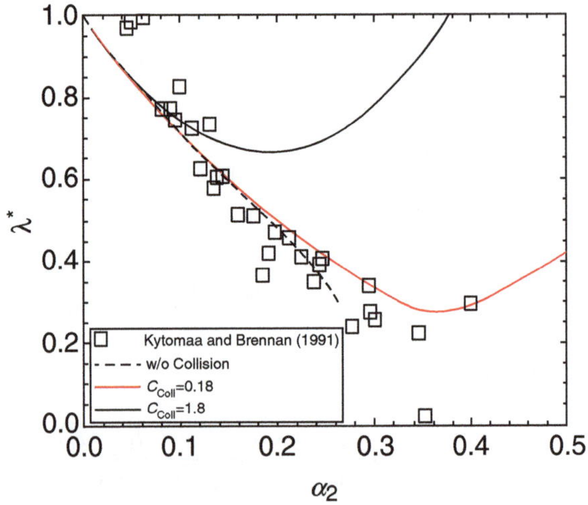

Fig. 5.5 Comparison of the bubbly flow characteristics with the data of Kytomaa and Brennen (1991)

Kytomaa and Brennen (1991). The change in the collision coefficient, from $C_{\text{coll}} = \frac{1.8}{1+\tau_e/\tau_b} \cong 1.8$ to $C_{\text{coll}} = 0.18$, accounts for the departure from elasticity and other inaccuracies in the model assumptions. It is worth noting that even under such a drastic change the model remains well-posed.

5.5 Nonlinear Simulations

5.5.1 Stable Wave Evolution

In this section, the FFM developed in the previous sections is used to show the nonlinear stability effect of adding interfacial pressure and a collision mechanism to the formulation. A viscous term is added ad hoc to the momentum equation, as was done in Chap. 2, for completeness:

$$\frac{\partial \alpha_1}{\partial t} + u_1 \frac{\partial \alpha_1}{\partial x} + \alpha_1 \frac{\partial u_1}{\partial x} = 0, \tag{5.49}$$

$$\frac{\partial u_1}{\partial t} + \left[u_1 + \frac{2\alpha_1 \left(C_{\text{VM}} - C_{\text{p}} + \frac{1.8}{1+\tau_e/\tau_b} \frac{(1+\alpha_1)(1-\alpha_1)^2 C_{\text{VM}}^2}{2\,\alpha_1^3} \right)}{C_{\text{VM}} + \alpha_1\alpha_2} (u_2 - u_1) \right] \frac{\partial u_1}{\partial x} + C' \frac{\partial \alpha_1}{\partial x} = \frac{\alpha_1\alpha_2}{C_{\text{VM}} + \alpha_1\alpha_2} \left[-g - \frac{2}{\alpha_1 H} \frac{f_1}{2} |u_1| u_1 + \left(\frac{1}{\alpha_1} + \frac{1}{\alpha_2} \right) \frac{M^{\text{D}}}{\rho_1} \right] + \nu \frac{\partial^2 u_1}{\partial x^2}, \tag{5.50}$$

where M^{D} is given by Ishii and Chawla (1979), Eq (9–7), and

$$C' = \frac{(1+\alpha_1)C_{\text{p}} - C_{\text{VM}}}{C_{\text{VM}} + \alpha_1\alpha_2} (u_2 - u_1)^2 + \frac{1.8}{1+\tau_e/\tau_b} C_{\text{VM}}^2 \frac{(3-\alpha_1^2)(1-\alpha_1)^2}{2\,\alpha_1^3 (C_{\text{VM}} + \alpha_1\alpha_2)} (u_2 - u_1)^2. \tag{5.51}$$

The second-order numerical scheme of Sect. 2.5.5 is used to minimize the effect of numerical damping. A simple traveling void wave *Gedankenexperiment* ("thought experiment") is chosen to demonstrate the need for stabilizing of the ill-posed problem. The initial void fraction distribution is a Gaussian wave superimposed with a high frequency sine wave given by,

$$\alpha_2 = \alpha_{2,0} + \delta_1 \exp\left[-\left(\frac{2\pi}{\lambda_1} \right)^2 (x - x_0)^2 \right] + \delta_2 \sin\left(\frac{2\pi x}{\lambda_2} \right), \tag{5.52}$$

where $\delta_1 = 0.1, \delta_2 = 0.005, \lambda_1 = 0.1$, and $\lambda_2 = 0.01$. Setting $\alpha_{2,0} = 0.25$ the model is kinematically stable according to Fig. 5.4 and, therefore, viscosity is not required for boundedness since there is no wave growth. Nevertheless, for the present

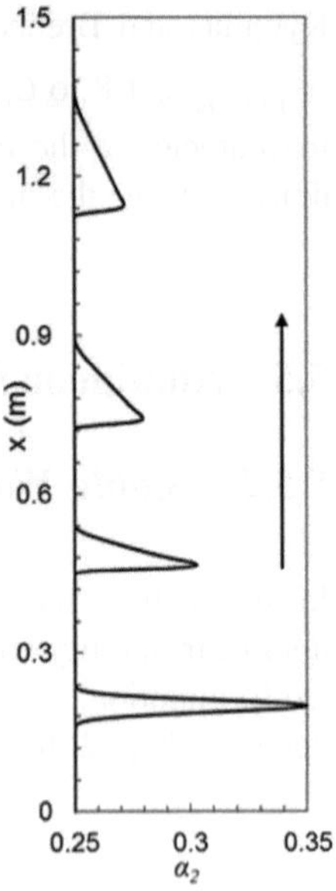

Fig. 5.6 Evolution of upward moving void wave simulation from Gaussian initial condition

calculations the kinematic viscosity of water is used, $\nu = 10^{-6}\,\text{m}^2/\text{s}$. The length of the domain is 1 m. and the number of grid points chosen for the simulations are 500, 1000, and 2000. The time step size is chosen such that $Co = (u_2 + v_w)\Delta t/\Delta x = 0.04$. Figure 5.6 shows four instances of the evolution of the void wave from Gaussian to triangular. These results converged without the need for artificial or numerical regularizations.

The nonlinear shape of the wave is comprised of a shock at the rear and an expansion wave at the front. This is opposite to the case of Burgers' equation. An explanation of this behavior may be obtained more clearly with the drift-flux void propagation equation which is derived in Sect. 6.2 and described in Appendix B.3.2. Consider the wave propagation Eq. (6.10):

$$\frac{\partial \alpha_2}{\partial t} + (u_2 + v_w)\frac{\partial \alpha_2}{\partial x} = 0, \tag{5.53}$$

where the void wave propagation velocity is given by

$$u_2 + v_w = C_0 j + V_{gj} + \alpha_2 \frac{dV_{gj}}{d\alpha_2}. \tag{5.54}$$

and the drift velocity may be obtained from the simplified relation:

$$V_{gj} = (1 - \alpha_2)u_R. \tag{5.55}$$

Using Eq. (6.3) to define the relative velocity, $u_R = \sqrt{\dfrac{g}{\left(\dfrac{1}{\alpha_1} + \dfrac{1}{\alpha_2}\right)\dfrac{3}{4}\dfrac{C_D}{d_B}\alpha_2}}$, with the drag coefficient of Ishii and Chawla (1979), Eq. (9.7), for distorted bubbles it can easily be shown that $u_2 + v_w$ increases with decreasing α_2. Therefore, the low concentration at the leading foot of the initial Gaussian wave travels faster than the rest of the wave behind it, causing the leading expansion fan. Conversely, the initial downstream tail quickly catches up to the rest of the waveform ahead of it, creating the shock-like structure at the trailing end.

Figure 5.7 shows high frequency numerical oscillations obtained with the model without considering bubble–bubble collisions, i.e., ill-posed. For $N = 500$ short wavelengths are prevented and the solution is similar to the well-posed FFM, i.e.,

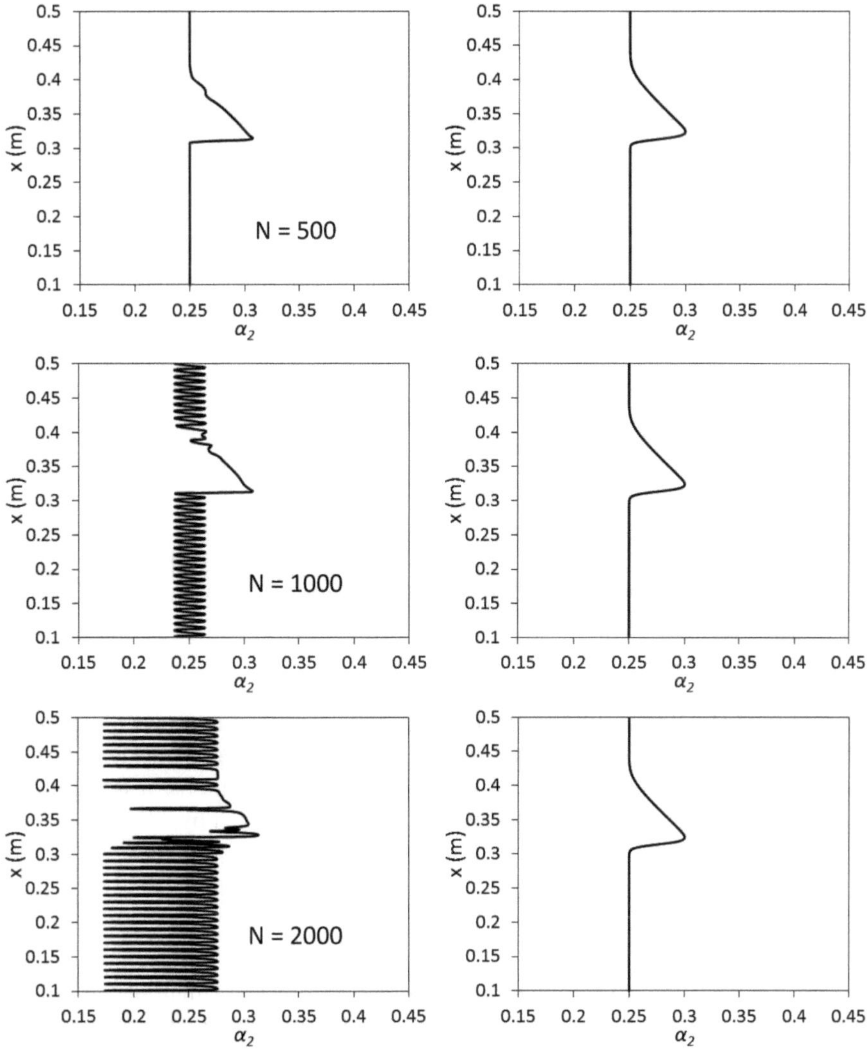

Fig. 5.7 Comparison of simulations of upward moving nonlinear material wave with ill-posed (*left*) and well-posed FFM (*right*)

including interfacial pressure and collision forces. However when the grid is refined, the ill-posed FFM develops high frequency oscillations which eventually dominate the solution, similar to those shown in Fig. 2.23.

5.5.2 Kinematically Unstable Waves in Guinness

We have seen in Sect. 5.4.2 that the Stokes bubbly flow FFM may become kinematically unstable for a range of void fractions. Robinson et al. (2008) performed an interesting analysis and experiments of waves in Guinness and reported a kinematic instability. The diameter of the Guinness bubbles was estimated to be approximately 0.12 mm, which corresponds to the Stokes regime in Sect. 5.4.4, i.e., $Re_b < 1$. Using the length and time scales,

$$l_o = \frac{u_o^2}{g}, \quad t_o = \frac{l_o}{u_o} = \frac{u_o^2}{g}, \tag{5.56}$$

Equations (5.35) and (5.41) can be nondimensionalized and combined to give:

$$\frac{\partial u_1}{\partial t} + \left[u_1 + \frac{2\alpha_1 \left(C_{VM} - C_p + \frac{0.9}{1+\tau_e/\tau_b} \frac{(1+\alpha_1)(1-\alpha_1)^2 C_{VM}^2}{\alpha_1^3} \right)}{C_{VM} + \alpha_1\alpha_2} (u_2 - u_1) \right]$$
$$\frac{\partial u_1}{\partial x} + C' \frac{\partial \alpha_1}{\partial x} = \frac{\alpha_1\alpha_2}{C_{VM}+\alpha_1\alpha_2} \left[-1 + 4.5\frac{u_1}{\alpha_1\alpha_2} + \frac{\nu}{\alpha_1}\frac{\partial}{\partial x}\left(\alpha_1 \frac{\partial u_1}{\partial x}\right) \right] + \nu \frac{\partial^2 u_1}{\partial x^2} \tag{5.57}$$

where C' is defined by Eq. (5.51). The last term on the RHS was introduced by Robinson et al. (2008) with $\nu = 0.05$ to bound the growth of the waves. Following the simulations of Lopez de Bertodano et al. (2013) the domain considered here is ten nondimensional units long with periodic boundary conditions. It is initialized with perturbed liquid void fraction and velocity profiles given by,

$$\alpha_1 = \alpha_{1,0} + \sum_{i=1}^{5} \delta_0 e^{-k^2\left(x - x_{0,i}\right)^2}, \quad u_1 = u_{1,0} + \sum_{i=1}^{5} \delta_0 e^{-k^2\left(x - x_{0,i}\right)^2}, \tag{5.58}$$

where $k = 2\pi$ is the wave number corresponding to a wave length of 1 unit. The values of $x_{0,i}$ are 1, 3, 5, 7, and 9 dimensionless units along the length of the domain. The amplitude of the perturbation, δ_0, is chosen to be 10^{-4}. The values of $\alpha_{1,0}$ used are 0.7 (kinematically stable), and 0.9 (kinematically unstable), and the initial velocity, $u_{1,0}$, is 0 for both cases. Furthermore, $N = 2000$ and the time step size is

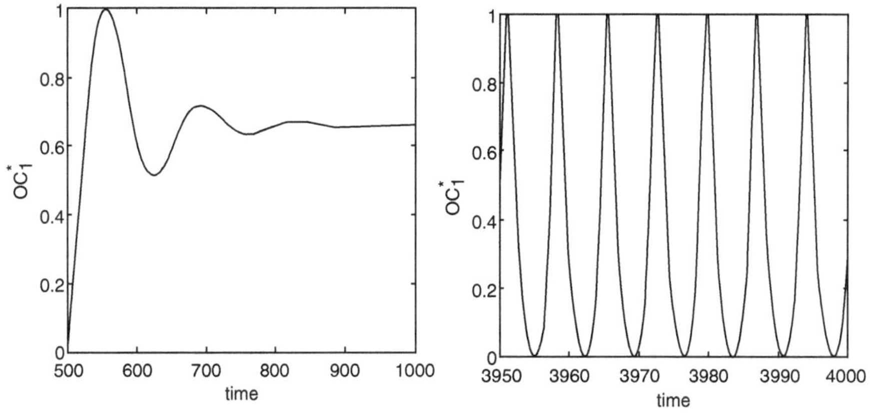

Fig. 5.8 Time evolution of void fraction for stable and unstable cases, reprinted from Lopez de Bertodano et al. (2013), with permission from Begell House

given by $\Delta t/_{\Delta x} = 0.2$. The calculations are run up to a nondimensional time of $t = 4000$. The results of the simulations are in accordance with the linear stability theory. For values of $\alpha_{2,0}$ greater than 0.17, the flow is stable as shown in Fig. 5.2. The independent variables are normalized as

$$\phi^* = \frac{\phi - \phi_{\min}}{\phi_{\max} - \phi_{\min}}. \tag{5.59}$$

where $\phi = [u_1, \alpha_1]^T$.

Figure 5.8 (left) shows the temporal evolution of α_1^* when $\alpha_{1,0} = 0.7$ at the center of the channel. The case is well-posed because of the collision term. It can be seen that the amplitude of the perturbation decreases until it reaches steady state. On the other hand when $\alpha_{1,0} = 0.9$ the flow becomes kinematically unstable, as predicted by the linear stability analysis, and the solution reaches a periodic oscillation as seen in Fig. 5.8 (right). A useful way to visualize the long-term development is through the phase-space plot which provides insight into the nonlinear stability and boundedness of the model. In the present case, the time series data of the liquid velocities obtained from the simulations at $x = 5.0(\text{i.e.}, v_1^*)$, and $x = 5.5(\text{i.e.}, v_2^*)$, are used. For the stable case, Fig. 5.9 (left), it can be seen that the solution decays rapidly to the point attractor. In turn a limit cycle is obtained for the kinematically unstable case, Fig. 5.9 (right). The case is well-posed because of the interfacial pressure. A convergence study was performed with $N = 500$, 1000, 2000, and 4000 grid points keeping $\Delta t/_{\Delta x} = 0.2$. Figure 5.8 (right) shows how the limit cycle converges as the grid is refined. Hence, the well-posed model results in convergence. Figure 5.10 shows the discrete Fourier components of the limit cycle which are comparable with the KY limit cycle spectra shown in Fig. 4.12.

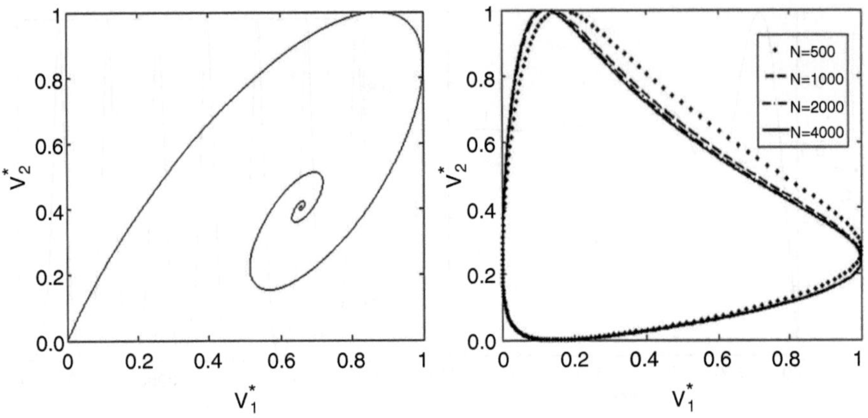

Fig. 5.9 Phase diagram for stable and unstable cases, reprinted from Lopez de Bertodano et al. (2013), with permission from Begell House

Fig. 5.10 Fourier spectrum of limit cycle shown in Fig. 5.9

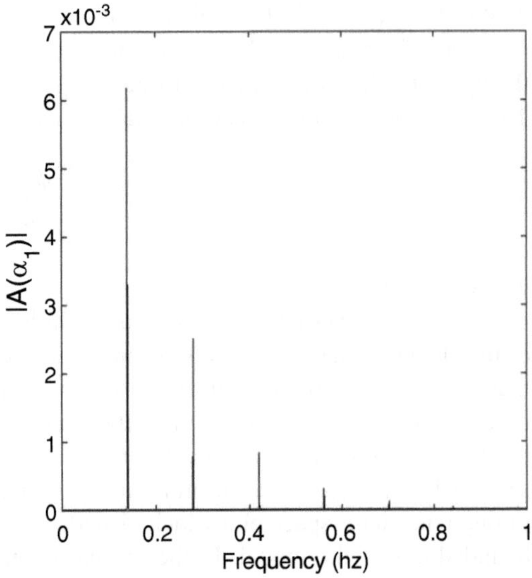

The wavelength is consistent with the results of Robinson et al. (2008), two orders of magnitude below the experimental observations. Furthermore, the viscosity is too low. Benilov et al. (2013) have shown that the shape of the container is an important aspect of the problem, i.e., pint-shaped glasses generate a shear layer near the wall, so that the instability may be multidimensional.

Nevertheless, the simulation with the 1D model is useful for TFM verification. Furthermore, it is relevant that chaos could not be obtained for bubbly flow under the kinematic instability. However, the clustering instability of particle flows, which is also a kinematic instability (Batchelor 1988), appears to be chaotic in the TFM CFD simulations of Fullmer and Hrenya (2016).

5.6 Summary and Discussion

The FFM linear stability analysis yields the well-known result of Pauchon and Banerjee (1986) for a bubbly flow TFM that includes the virtual mass and interfacial pressure forces, i.e., the model is conditionally well-posed. A new analysis that incorporates an elastic bubble collision force makes the model unconditionally well-posed, i.e., for all void fractions. This force is derived from kinetic theory as is common practice in the particle flow community where it is called the "particle pressure." The elastic collision mechanism results from the surface tension which allows small bubbles to collide with one another without coalescing while remaining nearly spherical. The incorporation of surface tension for bubbly flow is therefore not explicit as in the case of separated flow in Chap. 2. The fundamental differences in the stabilization mechanisms for vertical bubbly and horizontal stratified flows illustrate that the physical mechanisms used to obtain a well-posed TFM must be pursued independently for each specific flow regime. Furthermore, the physical mechanisms should render the correct material wave speed, so that a well-posed TFM with accurate wave propagation characteristics may be obtained.

Nonlinear simulations were performed for a stable solitary wave and for the SWT kinematic instability. Stable simulations with the well-posed FFM produced Burgers-like waves, reversed with respect to the flow direction. The dispersion analysis shows that the kinematic SWT instability of Whitham (1974) is still there, even though the equations are more complicated. Furthermore, it was demonstrated that the SWT instability simulations with the second-order numerical scheme converge, since the viscous FFM is well-posed. The main difference between the simulations of this chapter with those of the previous chapter is that chaos is not observed. However, chaos cannot be discarded for dispersed flows since the clustering instability of particles, which is also a kinematic instability (Wallis 1969; Batchelor 1988; Fullmer and Hrenya 2017), appears to be chaotic when simulated with the 3D TFM (see e.g., Agrawal et al. 2001 or Fullmer and Hrenya 2016 and references therein).

References

Agrawal, K., Loezos, P. N., Syamlal, M., & Sundaresan, S. (2001). The role of meso-scale structures in rapid gas–solid flows. *Journal of Fluid Mechanics, 445*, 151–185.

Alajbegovic, A., Drew, D. A., & Lahey, R. T., Jr. (1999). An analysis of phase distribution and turbulence in dispersed particle/liquid flows. *Chemical Engineering Communications, 174*, 85–133.

Alexander, C. A., & Zare, R. N. (2004). *Do bubbles in Guinness go down?* Retrieved from http://www.stanford.edu/group/Zarelab/guinness/

Batchelor, G. K. (1988). A new theory of the instability of a uniform fluidized-bed. *Journal of Fluid Mechanics, 193*, 75–110.

Benilov, E. S., Cummins, C. P. and Lee, W. T. (2013). Why do bubbles in Guinness sink? *American Journal of Physics, 81*(2), 88. http://arxiv.org/pdf/1205.5233v1.pdf

Bestion, D. (1990). The physical closure laws in the CATHARE code. *Nuclear Engineering and Design, 124*, 229–245.

Carnahan, N. F., & Starling, K. E. (1969). Equations of state for non-attracting rigid spheres. *Journal of Chemical Physics, 51*, 635–636.

Drew, D. A., & Lahey, R. T., Jr. (1987). The virtual mass and lift force on a sphere in rotating and straining inviscid fluid. *International Journal of Multiphase Flow, 13*(1), 113–121.

Drew, D. A., & Passman, S. L. (1999). *Theory of multicomponent fluids*. Springer, Berlin: Applied Mathematical Sciences.

Fullmer, W. D., & Hrenya, C. M. (2016). Quantitative assessment of fine-grid kinetic-theory-based predictions of mean-slip in unbounded fluidization. *AIChE Journal, 62*(1), 11–17.

Fullmer, W. D., & Hrenya, C. M. (2017). The clustering instability in rapid granular and gas-solid flows. *Annual Review of Fluid Mechanics, 49*, 485–510.

Garzó, V., & Santos, A. (2003). *Kinetic theory of gases in shear flows*. Dordrecht, the Netherlands: Springer.

Haley, T. C., Drew, D. A., & Lahey, R. T. (1991). An analysis of the eigenvalues of bubbly two-phase flows. *Chemical Engineering Communications, 106*, 93–117.

Henry, R. E., Grolmes, M. A., & Fauske, H. K. (1971) *Pressure-pulse propagation in two-phase one- and two-component mixtures* (Technical Report ANL-7792). Argonne National Laboratory.

Ishii, M., & Chawla, T. C. (1979). *Local drag laws in dispersed two-phase flow*. Argonne: ANL.

Ishii, M., & Hibiki, T. (2006). *Thermo-fluid dynamics of two-phase flow*. Berlin: Springer.

Kytomaa, H. K. & Brennen, C. E. (1991). Small Amplitude Kinematic Wave Propagation in Two-component Media, *International Journal of Multiphase Flow, 17*(1), 13–26.

Lopez de Bertodano, M. A., Fullmer W., Vaidheeswaran, A. (2013). One-Dimensional Two-Equation Two-Fluid Model Stability. *Multiphase Science and Technology, 25*(2), 133–167.

Ma, D., & Ahmadi, G. (1988). A kinetic-model for rapid granular flows of nearly elastic particles including interstitial fluid effects. *Powder Technology, 56*(3), 191–207.

Park, S. W., Drew, D. A., & Lahey, R. T., Jr. (1998). The analysis of void wave propagation in adiabatic monodispersed bubbly two-phase flows using an ensemble-averaged two fluid model. *International Journal of Multiphase Flow, 24*, 1205–1244.

Pauchon, C., & Banerjee, S. (1986). Interphase momentum interaction effects in the averaged multifield model. *International Journal of Multiphase Flow, 12*, 559–573.

Robinson, M., Fowler, A. C., Alexander, A. J., & O'Brien, S. B. (2008). Waves in Guinness. *Physics of Fluids, 20*, 067101.

Stuhmiller, J. H. (1977). The influence of interfacial pressure forces on the character of two-phase flow model equations. *International Journal of Multiphase Flow, 3*, 551–560.

Wallis, G. B. (1969). *One-dimensional two-phase flow*. New York: McGraw-Hill.

Watanabe, T., Hirano, M., Tanabe, F., & Kamo, H. (1990). The effect of the virtual mass force term on the numerical stability and efficiency of system calculations. *Nuclear Engineering and Design, 120*, 181. doi:10.1016/0029-5493(90)90371-4.

Watanabe, T., & Kukita, Y. (1992). The effect of the virtual mass term on the stability of the two-fluid model against perturbations. *Nuclear Engineering and Design, 135*, 327. doi:10.1016/0029-5493(92)90200-F.

Whitham, G. B. (1974). *Linear and nonlinear waves*. New York: Wiley.

Zuber, N. (1964). On the dispersed two-phase flow in the laminar flow regime. *Chemical Engineering Science, 19*, 897. doi:10.1016/0009-2509(64)85067-3.

Chapter 6
Drift-Flux Model

Abstract This chapter addresses the stable and linearly unstable Drift-Flux Model (DFM). The well-known Drift-Flux wave propagation equation is derived applying the kinematic condition to the FFM of Chap. 5. This removes the SWT and KH instabilities but preserves the material wave speed and nonlinear evolution of the waves, allowing the analysis of several stable problems of engineering interest, e.g., level swell, drainage, and the propagation of material discontinuities. Then the fixed flux approximation is removed and the DFM mixture momentum equation of Ishii and Hibiki (Thermo-fluid dynamics of two-phase flow, Springer, 2006) that incorporates the drift-flux assumption, i.e., the kinematic equilibrium condition, is introduced. This represents the counterpart to the fixed flux assumption of earlier chapters because it fixes the relative velocity whereas the total flux, j, is now allowed to fluctuate. To illustrate what this difference implies, the DFM is applied to the linear analysis of two global instabilities for boiling channels: the flow excursion and the density wave instability.

The DFM is the optimal TFM approximation to analyze global material wave instabilities, if the flow regime is stable, precisely because it precludes the local instabilities specifically addressed by the FFM. Thus, FFM and DFM are natural counterparts that render a broad picture of TFM stability. This wide stability spectrum is one reason why the TFM is so versatile for engineering two-phase flow applications.

6.1 Introduction

Historically the Drift-Flux Model (DFM) may be considered the direct predecessor of the TFM and it exhibits the long material wave instability aspect of it, cf. Fig. 1.1. Although a complete and unified set of governing equations for the DFM was not presented until Ishii's seminal work typically referenced for its development of the TFM (Ishii 1975, Ch. 10), the model was developed by Zuber's group at General Electric beginning in the early 1960s (Wallis 2013). Constitutive relations for the drift velocity began appearing in the open literature by the mid to late 1960s (Zuber and Findlay 1965; Zuber and Staub 1967). While Zuber is widely considered the father of the DFM, such an approach, particularly for dispersed two-phase flow, was widely used throughout the multiphase flow community for decades, e.g. (Wallis

M.L. de Bertodano et al., *Two-Fluid Model Stability, Simulation and Chaos*,
DOI 10.1007/978-3-319-44968-5_6

1969; Hewitt 1982), and continues to be relied upon by nuclear reactor safety codes (ISL 2003).

Hierarchically the DFM may be considered a simplified version of the TFM, as it contains only three fluid dynamic equations. Eqs. (6.41)–(6.43). The essence of the DFM is that only a mixture velocity is actively solved, i.e., one of the unknown variables of the system. The phasic velocities, specifically the dispersed phase velocity, are determined via closure algebraic relations relating the relative or drift velocity between the phasic velocity and the center-of-volume velocity, i.e., the total volumetric flux (Ishii 1977). While simplified compared to the TFM, the DFM still exhibits the long-wavelength material instability, cf. Fig. 1.1.

The DFM may also be further reduced to the well-known void propagation equation which accurately predicts the void fraction waves that govern the dynamics of two-phase flow system instabilities. First, the kinematic condition is applied, together with the fixed flux condition, to turn the TFM momentum equations into an algebraic relation which is inserted into the gas continuity equation. The kinematic condition removes the SWT and KH instabilities, which were the focus of the previous chapters, making the void propagation equation unconditionally stable. Thus, a robust and stable one-equation DFM, that preserves the material wave speed and the nonlinear evolution of the wave profiles, is obtained. This model is applicable to the analysis of several problems of practical interest, e.g., drainage, level swell, the propagation of material discontinuities, etc.

Second, the dynamic DFM is obtained by removing the fixed flux condition and replacing the two momentum equations of the TFM with a mixture momentum equation, while preserving the equilibrium kinematic condition that fixes the relative velocity. Thus, the DFM acquires the capability to perform global instability analysis, while the troublesome local KH and SWT instabilities of the FFM are precluded. Since these instabilities are associated with flow-regime changes, in a stability sense the DFM is the counterpart of the FFM and it is optimal for the analysis of system instabilities where it is known a priori that the flow regime is stable. This completes the picture of TFM material wave stability since it isolates the global from the local material instabilities. Of course, this is the result of using the mixture momentum equation with the kinematic equilibrium condition, which should be contrasted with the FFM relative-velocity momentum equation with fixed-flux condition, i.e., the TFM momentum equations are added for the DFM and subtracted for the FFM.

It is of historical interest that the dynamic DFM was formulated and solved (Ishii 1971) before the full TFM was rigorously derived. We adopt the drift-flux momentum equation of Ishii and Hibiki (2006). The well-known density wave (DW) instability and the flow excursion, i.e., the static Ledinegg instability (1938), may be analyzed within this mathematical framework. While these instabilities do not comprise the full spectrum of two-phase flow global instabilities, they are very important in engineering boiler design and represent other instabilities related to other phase change phenomena such as flashing and condensation. More complicated physics like thermal nonequilibrium, etc., may be also analyzed with the DFM but are beyond the scope of this chapter. For the interested reader, there are several excellent reviews of two-phase flow global instabilities available in the literature, e.g., Ruspini et al. (2014).

The integral of the momentum equation along the boiling channel must be performed because the DW instability depends on the interaction of the subcooled and two-phase regions. The density wave oscillations are characterized by the pressure loss at the inlet of a heated channel and the time it takes for a fluid particle to reach the exit of the channel and thus modify the density and the associated two-phase pressure loss there. Integral systems are well known in linear control theory and it is no coincidence that the mathematical approach is the corresponding Laplace Transform, which has the capability to handle time delays.

There is a considerable body of literature for the integral momentum equation stability analysis using the Homogeneous Equilibrium Model. For example, Wallis and Heasley (1961) obtained the solution to the Lagrangian homogeneous void propagation equation and formulated the momentum integral in terms of it. On the other hand, the linear analysis of the density wave instability using the integral DFM was not performed in its entirety until the work of Ishii (1971). Ishii's method consists of three steps. First the continuity and energy equations were decoupled from the momentum equation to obtain the Lagrangian void propagation equation that was integrated along the characteristic up to an arbitrary position along the channel. Then the momentum equation was integrated along the entire test section to obtain a dynamic ordinary differential-integral equation. Finally, using linear perturbation analysis and the Laplace Transform, Ishii obtained a transfer function with time delays, i.e., exponential terms in the frequency domain, that he solved using the D-partition method.

We shall employ the equivalent formulation of Achard et al. (1985) instead of Ishii's, because the presentation of the integral-differential momentum equation with time delays is clearer. This equation is then linearized by the same technique used in Chap. 2 to obtain a linear perturbation equation for the global material stability. The difference with Chap. 2 is that instead of using Fourier analysis to obtain the dispersion relation it is now more convenient to use the Laplace Transform to obtain the transfer function. While the mathematics are more elaborate, the basic idea is the same and we will preserve the notation of Fourier analysis, i.e., $i\omega$ instead of s, for consistency. Finally, the D-partition technique renders the roots of the transfer function with multiple time delays. This results in stability curves in terms of the subcooling and phase change dimensionless numbers identified by Ishii (1971). A similar map is also obtained for the flow excursion. It should not be surprising that the results obtained with Achard's equations agree well with Ishii's.

6.2 Void Propagation Equation

The stable drift-flux void propagation equation is a single-equation model that may be derived from Eq. (5.25) and from the kinematic condition, i.e., setting $F = 0$ in Eq. (5.26). This approach is a further simplification of the FFM that removes the possibility of the SWT and KH instabilities. The kinematic condition may be written as

$$-g - \frac{1}{\alpha_1 H}\frac{f_1}{2}\rho_1|u_1|u_1 + \left(\frac{1}{\alpha_1} + \frac{1}{\alpha_2}\right)\frac{3}{4}\frac{C_D}{d_B}\alpha_2|u_2 - u_1|(u_2 - u_1) = 0. \tag{6.1}$$

The equation is further simplified by neglecting the wall friction term, that is:

$$-g+\left(\frac{1}{\alpha_1}+\frac{1}{\alpha_2}\right)\frac{3}{4}\frac{C_D}{d_B}\alpha_2|u_2-u_1|(u_2-u_1)=0 \tag{6.2}$$

which may be solved for $u_R=u_2-u_1$:

$$u_R=\sqrt{\frac{g}{\left(\frac{1}{\alpha_1}+\frac{1}{\alpha_2}\right)\frac{3}{4}\frac{C_D}{d_B}\alpha_2}}. \tag{6.3}$$

We now introduce mass transfer between the phases, because the DFM has been most successful in analyses of transient boiling and flashing systems. The TFM continuity equations with mass transfer between the phases, Γ, are as follows:

$$\frac{\partial\alpha_1}{\partial t}+\frac{\partial\alpha_1 u_1}{\partial x}=-\frac{\Gamma}{\rho_1}, \tag{6.4}$$

$$\frac{\partial\alpha_2}{\partial t}+\frac{\partial\alpha_2 u_2}{\partial x}=\frac{\Gamma}{\rho_2}. \tag{6.5}$$

The phase transfer rate Γ will be considered constant to simplify the analysis, corresponding to uniform heat flux and/or uniform flashing. This restriction, of course, may be removed by considering the TFM energy equations, with considerable more mathematical complication.

The objective is to now obtain the drift-flux void propagation equation with phase change starting with the gas continuity equation (6.5). The one-dimensional drift-flux algebraic model defined by Zuber and Findlay (1965) is

$$\langle j_2\rangle=C_0\langle\alpha_2\rangle\langle j\rangle+\langle\alpha_2\rangle V_{gj}, \tag{6.6}$$

where the operator $\langle\,\rangle$ represents the cross sectional area average. $C_0=\frac{\langle j_2\rangle}{\langle\alpha_2\rangle\langle j\rangle}$ and $V_{gj}=\frac{\langle\alpha_2\,(u_2-j)\rangle}{\langle\alpha_2\rangle}$ are called the distribution parameter and the drift velocity, respectively. Since all quantities defined so far for the one-dimensional models are cross sectional area averages, the averaging operator may be dropped and the gas velocity may be defined as

$$u_2=\frac{j_2}{\alpha_2}=C_0 j+V_{gj}, \tag{6.7}$$

where the volumetric flux is given by

$$j=\alpha_1 u_1+\alpha_2 u_2. \tag{6.8}$$

The drift velocity, V_{gj}, and the distribution parameter, C_0, are described and correlated in detail in several well-known texts, e.g., Wallis (1969) and Ishii and Hibiki (2006). They turn out to be straightforward to measure experimentally for a wide range of vertical gas–liquid flow regimes including bubbly, cap bubbly, slug, churn, and frothy flows. The DFM correlates all these data surprisingly well so it is better validated than most other two-phase flow models. Due to this fact, the DFM has received wide industrial acceptance. Inserting Eq. (6.7) into Eq. (6.5) we may obtain the drift-flux void propagation equation,

$$\frac{\partial \alpha_2}{\partial t} + \frac{\partial \alpha_2 (C_0 j + V_{gj})}{\partial x} = \frac{\Gamma}{\rho_2}, \tag{6.9}$$

or, in nonconservative form,

$$\frac{\partial \alpha_2}{\partial t} + \left(C_0 j + V_{gj} + \alpha_2 \frac{dV_{gj}}{d\alpha_2} \right) \frac{\partial \alpha_2}{\partial x} + \alpha_2 \frac{\partial j}{\partial x} = \frac{\Gamma}{\rho_2}. \tag{6.10}$$

For the stable DFM, the volumetric flux, j, is either constant for adiabatic flow or is a prescribed linear function of x for constant Γ, as shall be demonstrated in the next section. Hence, this model is amenable to analytic solution. Furthermore, the dispersion relation is the well-known void-propagation speed, which is always real,

$$\frac{\omega}{k} = C_0 j + V_{gj} + \alpha_2 \frac{dV_{gj}}{d\alpha_2}. \tag{6.11}$$

Thus, the TFM is reduced to a stable first-order one-way wave equation that preserves the nonlinear material wave behavior, i.e., material shocks and expansion waves. The kinematic condition precludes any local instability, resulting in a practical model for engineering analysis. One-way wave equations may be solved analytically by the method of characteristics, a procedure that will be thoroughly employed in this and the following chapter.

6.3 Applications of Void Propagation Equation

6.3.1 Level Swell

The problem of level swell in a vertical tank, which is significant in nuclear reactor safety, is a wave propagation problem. The geometry may be considered to be a vertical pipe closed at the inlet and partially filled with saturated water. The problem begins when the pipe undergoes a sudden and steady depressurization.

The solution of the linearized void propagation equation for this problem, i.e., $\frac{dV_{gj}}{d\alpha_2} = 0$, was solved by Wulff (1985) using the method of characteristics. First, adding the TFM continuity Eqs. (6.4) and (6.5) we obtain the drift-flux continuity equation

$$\frac{\partial j}{\partial x} = \left(\frac{\rho_1 - \rho_2}{\rho_1 \rho_2}\right) \Gamma. \tag{6.12}$$

Assuming uniform and constant mass transfer rate Γ this equation may be integrated so that the total flux now increases linearly with elevation as

$$j = j_0 + \left(\frac{\rho_1 - \rho_2}{\rho_1 \rho_2}\right) \Gamma x, \tag{6.13}$$

This important equation will also be used for the dynamic DFM in the Sect. 6.4 to derive the integral momentum equation. To simplify the problem, it is assumed that $\frac{dV_{gj}}{d\alpha_2} = 0$, $j_0 = 0$ and $C_0 = 1$, so the void propagation equation becomes

$$\frac{\partial \alpha_2}{\partial t} + (j + V_{gj}) \frac{\partial \alpha_2}{\partial x} = \frac{\Gamma}{\rho_2} \left(1 - \frac{\rho_1 - \rho_2}{\rho_1} \alpha_2\right). \tag{6.14}$$

Defining,

$$t^* = \frac{\Gamma}{\rho_2} t, \quad x^* = \frac{x}{x_0}, \quad \beta = \frac{\rho_1 - \rho_2}{\rho_1}, \quad V_{gj}^* = \frac{\rho_2 V_{gj}}{x_0 \Gamma}, \tag{6.15}$$

where x_0 is the initial surface level, the void propagation equation may be expressed in dimensionless form as

$$\frac{\partial \alpha_2}{\partial t^*} + \left(\beta x^* + V_{gj}^*\right) \frac{\partial \alpha_2}{\partial x^*} = 1 - \beta \alpha_2 \tag{6.16}$$

with a corresponding characteristic equation,

$$\frac{dx^*}{dt^*} = \beta x^* + V_{gj}^*. \tag{6.17}$$

The void distribution is given by two solutions corresponding to the two regions divided by the main characteristic $x^* = \frac{V_{gj}^*}{\beta}\left(e^{\beta t^*} - 1\right)$, as shown in Fig. 6.1. On the left, the void propagation equation corresponds to the region where the wave that starts at the bottom of the tank at $t^* = 0$ has not arrived yet to x, so $\frac{\partial \alpha_2}{\partial x^*} = 0$. The void fraction in this region is therefore uniform and given by:

$$\alpha_2 = \frac{1}{\beta}\left(1 - e^{-\beta t^*}\right). \tag{6.18}$$

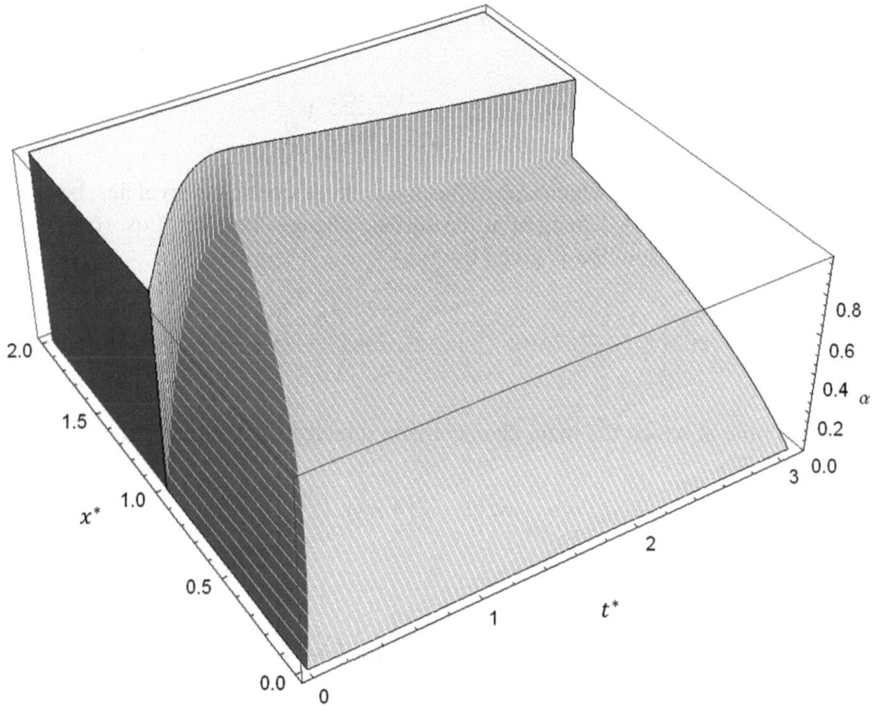

Fig. 6.1 Void propagation equation solution for level swell

To the right of the dividing characteristic, the wave has passed and the void distribution depends only on the position, i.e., $\frac{\partial \alpha_2}{\partial t^*} = 0$, and

$$\alpha_2 = \frac{x^*}{V^*_{gj} + \beta x^*}. \tag{6.19}$$

The equation for the level is obtained separately from a vapor balance at the surface interface, that is:

$$\alpha_{2i}\left(u_{2\,int} - \frac{dx_i}{dt}\right) = j_{int} - \frac{dx_i}{dt}, \tag{6.20}$$

where the velocity of the bubbles arriving to the interface is:

$$u_{2int} = j_{int} + V_{gj} \tag{6.21}$$

The subindex int from now on refers to the interface.

The differential equation of the surface level is obtained combining Eqs. (6.20) and (6.21), which leads to:

$$\frac{dx_{int}}{dt} = j_{int} - \frac{1 - \alpha_{2int}}{\alpha_{2int}} V_{gj}. \tag{6.22}$$

Combining Eq. (6.22) with Eq. (6.13) the final nondimensional characteristic equation is obtained:

$$\frac{dx^*_{\text{int}}}{dt^*} - \beta x^*_{\text{int}} = -\frac{1-\alpha_{2\text{int}}}{\alpha_{2\text{int}}} V^*_{\text{gj}}. \tag{6.23}$$

The solution depends on which side of the main characteristic the level lies, because the equation for the void fraction at the surface changes between Eqs. (6.18) and (6.19). Initially the solution is given by

$$x^*_{\text{int}} = \mathrm{e}^{\beta t^*}\left[1 + \frac{V^*_{\text{gj}}}{\beta}\left(1 - \mathrm{e}^{-\beta t^*}\right) + V^*_{\text{gj}} \log\left(\frac{\beta - 1 + \mathrm{e}^{-\beta t^*}}{\beta}\right)\right] \tag{6.24}$$

until the time at which the main characteristic intersects the surface, given by:

$$t^*_{\text{c}} = -\frac{1}{\beta}\log\left[1 - \beta\left(1 - \mathrm{e}^{-1/V^*_{\text{gj}}}\right)\right], \tag{6.25}$$

which corresponds to the level,

$$x^*_{\text{int c}} = \mathrm{e}^{\beta t^*_{\text{c}}}\left[1 + \frac{V^*_{\text{gj}}}{\beta}\left(1 - \mathrm{e}^{-\beta t^*_{\text{c}}}\right) + V^*_{\text{gj}} \log\left(\frac{\beta - 1 + \mathrm{e}^{-\beta t^*_{\text{c}}}}{\beta}\right)\right]. \tag{6.26}$$

After that, the level is given by the implicit equation,

$$t^*_{\text{int}} = 1 + \frac{1}{\beta}\log\left[\frac{V^*_{\text{gj}} + x^*_{\text{int}}}{V^*_{\text{gj}} + x^*_{\text{intc}}}\right] - \frac{1}{1-\beta}\log\left[\frac{x^*_{\text{int}}\left(V^*_{\text{gj}} + x^*_{\text{intc}}\right)}{x^*_{\text{intc}}\left(V^*_{\text{gj}} + x^*_{\text{int}}\right)}\right]. \tag{6.27}$$

Figure 6.1 shows the result of the model for water–steam at $p = 5\,\text{MPa}$, $\rho_1 = 780$ kg/m^3, $\rho_2 = 25\,\text{kg/m}^3$, $\Gamma = 4\,\text{kg/m}^3\text{s}$, $V_{\text{gj}} = 0.5\,\text{m/s}$ and the initial surface level at $x_0 = 3.125$ m.

6.3.2 *Drainage*

Another example which demonstrates the versatility of the DFM is the case of the drainage of a frothy column (Wallis 1969). Consider the experiment of Miles et al. (1943). The geometry is a pipe closed at both ends initially filled with a uniform frothy mixture. In this case the column is 0.367 m. tall and the initial void fraction is $\alpha_{20} = 0.977$. This problem was solved with an analytic-graphical method by Wallis (1969). Unfortunately the most accurate DFM correlations are usually nonlinear which makes the problem more difficult to solve analytically than the previous level swell example, so here it is solved instead by numerical simulation. Wallis recommends the drift-flux velocity correlation $V_{\text{gj}} = 0.139\alpha_1^{1.8}$ m/s.

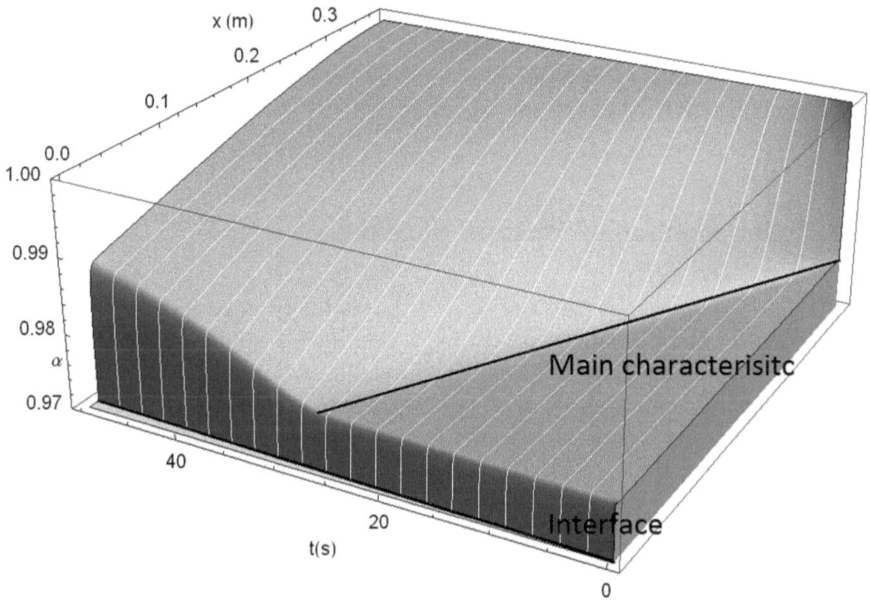

Fig. 6.2 Void propagation equation solution for drainage of frothy column of Miles et al. (1943)

A part of the solution may be obtained analytically by the method of characteristics and it is instructive to compare it to a full numerical solution, obtained with *Mathematica*, shown in Fig. 6.2. For this case there is no vapor source so the void propagation Eq. (6.10) is applied without the phase change source. In addition, $j = 0$ and it is assumed that $C_0 = 1$.

The rarefaction waves propagate from the top of the column downward toward the interface, opposite to the previous case, as nonlinear expansion waves illustrated in Fig. 6.2. The froth which is 0.367 m. tall drains over a period of more than 100 s while a small amount of water collects at the bottom of the column with an interface between the liquid and the froth which reaches approximately 10 mm at the end of the transient. The first expansion wave reaches the bottom of the column when the main characteristic given by the equation,

$$\frac{dx}{dt} = V_{\text{gj0}} + \alpha_{20}\frac{dV_{\text{gj0}}}{d\alpha_2} = 0.139\left[(1-\alpha_{20})^{1.8} + 1.8\,\alpha_{20}(1-\alpha_{20})^{0.8}\right]\text{m/s}, \quad (6.28)$$

meets the liquid–frothy interface at the bottom of the column. This interface moves during this period with velocity

$$\frac{dx_{\text{i}}}{dt} = V_{\text{gj0}} = 0.139\,(1-\alpha_{20})^{1.8}\,\text{m/s}. \quad (6.29)$$

After the main characteristic meets with the liquid–froth interface the void fraction above the interface begins to increase as shown in Fig. 6.2. Eventually the frothy column drains completely. In the case of the numerical solution shown in Fig. 6.2, the interface position is solved implicitly by the void propagation equation. This can be done because V_{gj} is properly defined on both sides of the interface, so the interface equation need not be solved explicitly like in the previous example.

6.3.3 Propagation of Material Shocks

In this section the propagation of a void discontinuity is considered, using the experiment of Bernier (1982) on drainage in a bubble column as base case. In Bernier's experiment a bubble column 2.17 m tall is initially supplied with bubbles through a sparger imposing an initial void fraction along the column of $\alpha_{20} = 0.22$. The air flow at the sparger is then reduced and a transient ensues where the void fraction decreases to $\alpha_2 = 0.1$. This is illustrated by the numerical solution of the void propagation given by Eq. (6.10), without the phase change source, shown in Fig. 6.3. To obtain the solution the volumetric flux was derived from Eq. (6.7) with $j_2 = j$ and $C_0 = 1$, yielding:

$$j = \frac{\alpha}{1-\alpha} V_{gj}. \tag{6.30}$$

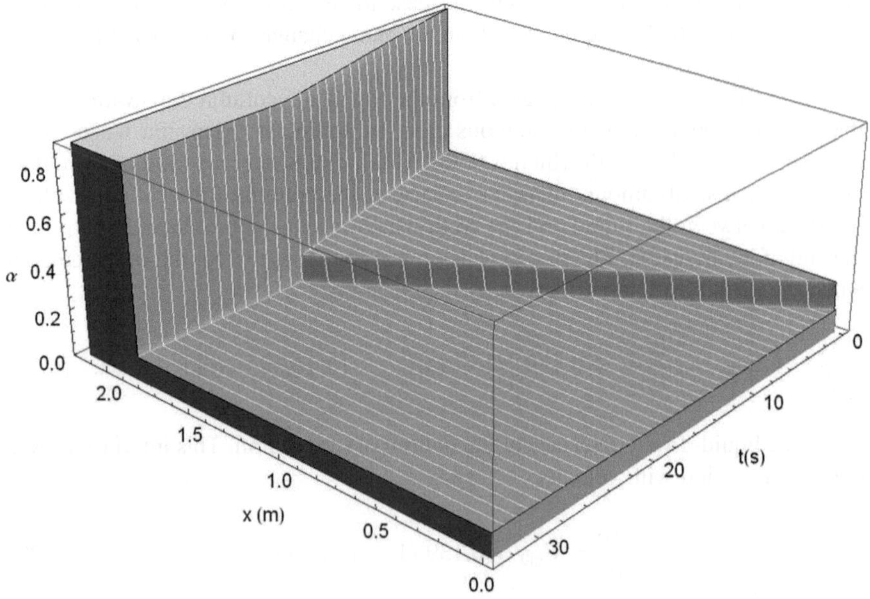

Fig. 6.3 Void propagation equation solution for void discontinuity experiment of Bernier (1982)

The Drift-Flux velocity correlation proposed by Ishii (1977) for distorted bubbles was used for this problem, that is:

$$V_{\mathrm{gj}} = V_{\mathrm{gj}}^{\mathrm{ZF}} \alpha_1^{1.75}, \tag{6.31}$$

where the Zuber and Findlay (1965) drift-flux velocity is given by

$$V_{\mathrm{gj}}^{\mathrm{ZF}} = 1.41 \left[\frac{\sigma g (\rho_1 - \rho_2)}{\rho_1^2} \right]^{1/4} \tag{6.32}$$

At atmospheric conditions this becomes $V_{\mathrm{gj}} = 0.162\, \alpha_1^{1.75}\, \mathrm{m/s}$. It is remarkable how similar the drift velocity correlations of this section and the previous one are, considering the difference in the flow regimes. Figure 6.3 shows the void shock that propagates upward until it reaches the surface and the column settles at a new lower steady void fraction. The velocity of propagation of the shock may be obtained in similar fashion to the shock for Burgers' equation obtained in Appendix B.3.1, namely:

$$\frac{dx}{dt} = \frac{(j + V_{\mathrm{gj}})\alpha_{\mathrm{R}} - (j + V_{\mathrm{gj}})\alpha_{\mathrm{L}}}{\alpha_{\mathrm{R}} - \alpha_{\mathrm{L}}}. \tag{6.33}$$

The water volume in the column is held constant and therefore needs to drain to fill in the reduced amount of bubbles. Thus, the water level decreases from 2.17 to 1.88 m. The time required for the shock to propagate to the surface according to Eq. (6.33) is 19.8 s which is very close to the numerical result. In fact this problem may be analytically solved with Eq. (6.33).

6.4 Dynamic Drift-Flux Model

Up to now the transients we dealt with were all driven by the material transport, that is, the kinematics of the void fraction waves. In this section, the dynamic effects introduced by the momentum equation will be included in the analysis, leading to richer features, albeit mathematically more complicated. The present analysis will be limited to boiling channel flows, which have concentrated most of the related theoretical and experimental development.

6.4.1 Mixture Momentum Equation

The continuity and the void propagation equations, Eqs. (6.12) and (6.10), remain valid because the kinematic condition is still maintained for the dynamic DFM to

Fig. 6.4 Diagram of a boiling channel

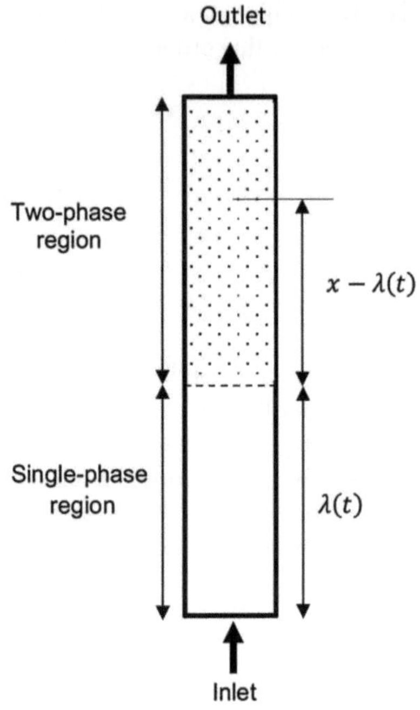

avoid the KH instability. The mass transfer source, defined by the thermodynamic equilibrium condition for the uniformly heated channel shown in Fig. 6.4, becomes $\Gamma=\frac{q_0''P_H}{A_{xs}h_{12}}$. In turn, the fixed flux assumption will be removed and therefore a momentum equation will be needed to determine the total flux j. For simplicity, the analysis will be restricted to a case with constant V_{gj} and $C_o=1$. Also, the channel has constant friction f uniformly distributed along the run and a concentrated friction K_i at the inlet.

The corresponding mixture momentum equation for the two-phase region in terms of the center of mass velocity was derived by Ishii and Hibiki (2006),

$$-\frac{\partial p}{\partial x}=\rho_m\left(\frac{\partial v_m}{\partial t}+v_m\frac{\partial v_m}{\partial x}\right)+\frac{\partial}{\partial x}\left(\frac{\alpha}{1-\alpha}\frac{\rho_1\rho_2}{\rho_m}V_{gj}^2\right)+\frac{f\rho_m v_m^2}{2D_H}+\rho_m g, \quad (6.34)$$

where

$$\rho_m=(1-\alpha)\rho_1+\alpha\rho_2, \quad v_m=\frac{(1-\alpha)\rho_1u_1+\alpha\rho_2u_2}{\rho_m}=j+\left(1-\frac{\rho_1}{\rho_m}\right)V_{gj}. \quad (6.35)$$

Let us now obtain a set of corresponding nondimensional equations, which will reduce the amount of constant parameters and yield the natural dimensionless groups. For convenience we will preserve the same nomenclature for the dimensional and the dimensionless variables since from this point on there will be no more dimensional ones. Achard et al. (1981) the following convenient timescale,

$$t_0 \triangleq \frac{\rho_1 A_{xs}}{q_0'' P_{\mathrm{H}}}(h_1 - h_{\mathrm{i}}), \tag{6.36}$$

which is the time needed for a fluid particle entering the channel with enthalpy h_{i} to reach the saturation enthalpy h_1, and a velocity scale,

$$v_0 \triangleq \frac{L_{\mathrm{H}}}{t_0}, \tag{6.37}$$

corresponding to the steady-state inlet velocity when the steady-state boiling boundary is exactly the channel length L_{H}. Then the dimensionless variables are $x = \frac{x}{L_{\mathrm{H}}}$, $t = \frac{t}{t_0}$, $v_{\mathrm{m}} = \frac{v_{\mathrm{m}}}{v_0}$, $j = \frac{j}{v_0}$, $V_{\mathrm{gj}} = \frac{V_{\mathrm{gj}}}{v_0}$. Furthermore, the dimensionless density and pressure are defined as $\rho = \frac{\rho}{\rho_1}$ and $p = \frac{p}{\rho_1 v_0^2}$. Thus, the following dimensionless numbers appear naturally,

$$N_{\mathrm{SUB}} = \frac{(\rho_1 - \rho_2)(h_1 - h_{\mathrm{i}})}{\rho_2 h_{12}} \text{(Subcooling Number)}, \tag{6.38}$$

$$Fr \triangleq \frac{v_0^2}{g L_{\mathrm{H}}} \text{(Froude Number)}, \tag{6.39}$$

$$\Lambda = \frac{f L_{\mathrm{H}}}{2 D_{\mathrm{H}}} \text{(Friction Number)}. \tag{6.40}$$

Furthermore, even though j is the variable that is solved, Ishii (1971) defined the Phase Change Number instead to present his results,

$$N_{\mathrm{PCH}} = \frac{N_{\mathrm{SUB}}}{\bar{j}} \text{(Phase Change Number)}.$$

where $\bar{j}$ is the steady-state dimensionless total flux. Then the dimensionless dynamic DFM may be written as the mixture continuity equation, the void propagation equation, and the mixture momentum equation:

$$\boxed{\frac{\partial j}{\partial x} = N_{\mathrm{SUB}},} \tag{6.41}$$

$$\boxed{\frac{\partial \alpha}{\partial t} + (j + V_{\mathrm{gj}}) \frac{\partial \alpha}{\partial x} + N_{\mathrm{SUB}} \alpha = \frac{\rho_1}{\rho_1 - \rho_2} N_{\mathrm{SUB}},} \tag{6.42}$$

$$-\frac{\partial p}{\partial x} = \rho_m \left(\frac{\partial v_m}{\partial t} + v_m \frac{\partial v_m}{\partial x} + \Lambda v_m^2 + Fr^{-1} \right) + \frac{\partial}{\partial x} \left(\frac{\alpha}{1-\alpha} \frac{\rho_2}{\rho_m} V_{gj}^2 \right). \tag{6.43}$$

The characteristic speed of these equations is $c = j + V_{gj}$. The dispersion relation is still given by Eq. (6.11) which shows that the model is locally stable, i.e., hyperbolic. This may be explained by the simple fact that the void propagation equation is a one-way wave equation that imposes null wave growth along a characteristic for the adiabatic case and uniform wave growth for the case with mass transfer. So, essentially, the reason for local stability is that the KH and SWT instabilities have been "surgically" removed from the TFM by the kinematic condition. However, the DFM is not always stable in a global or integral sense, as we shall see in Sects. 6.5 and 6.6, and this is its greatest virtue.

6.4.2 *Integral Momentum Equation*

It was already shown in Sect. 6.3.1 that Eqs. (6.41) and (6.42) may be integrated analytically along a characteristic to obtain the flux along the channel. We will now insert the resulting flux and void fraction into Eq. (6.43) and integrate it. However, before the integration is carried out it is first necessary to define the system and the thermodynamic process. The components of the system shown in Fig. 6.4 are kept to the minimum for the purpose of this chapter, but more components may be added as needed by the engineering requirements, e.g., Lahey and Moody (1977).

The system consists of a uniformly heated channel with a subcooled liquid flow region at the inlet and a two-phase flow region thereon. The procedure is to integrate the momentum equation in these two regions and add them together under the constraint of a constant total pressure drop. The subcooled length $\lambda(t)$ is the key variable in the analysis. It is defined by the position at time t of a fluid particle that entered the channel at time $t-1$ (notice that the dimensionless subcooled residence time equals unity). Regarding that in the subcooled region the flux j is uniform (i.e., $\frac{\partial j}{\partial x} = 0$), then we can write:

$$\lambda(t) = \int_{t-1}^{t} j_i(t')dt', \tag{6.44}$$

where j_i is the flux at the inlet. Note that by differentiating Eq. (6.44) we obtain:

$$\frac{d\lambda}{dt} = \frac{d}{dt}\int_{t-1}^{t} j_i(t')dt' = j_i(t) - j_i(t-1). \tag{6.45}$$

The momentum integral equation over the single-phase region is straightforward, for the flux is uniform and the density is constant, that is:

$$\Delta p_{1\phi} = \int_{x=0}^{x=\lambda(t)} \left(-\frac{\partial p}{\partial x_{1\phi}}\right) dx = \lambda(t)\left[\frac{dj_{\mathrm{i}}}{dt} + \Lambda j_{\mathrm{i}}^2 + Fr^{-1}\right] + K_{\mathrm{i}} j_{\mathrm{i}}^2. \tag{6.46}$$

The momentum integral equation over the two-phase region is more complicated because the mixture density and flux depends on space. The flux in the two-phase region may be obtained by integrating Eq. (6.41),

$$j(x,t) = j_{\mathrm{i}}(t) + N_{\mathrm{SUB}}[x - \lambda(t)]. \tag{6.47}$$

Substituting Eq. (6.47) in the characteristic equation, $\frac{dx}{dt} = j + V_{\mathrm{gj}}$, gives:

$$\frac{dx}{dt} = j_{\mathrm{i}}(t) + N_{\mathrm{SUB}}[x - \lambda(t)] + V_{\mathrm{gj}},$$

which combined with Eq. (6.45) gives:

$$\frac{d(x-\lambda)}{dt} + N_{\mathrm{SUB}}(x-\lambda) = j_{\mathrm{i}}(t-1) + V_{\mathrm{gj}}.$$

Multiplying by the integrating factor and rearranging terms results in:

$$\frac{d(x-\lambda)\mathrm{e}^{-N_{\mathrm{SUB}}t}}{dt} = \mathrm{e}^{-N_{\mathrm{SUB}}t}\left[j_{\mathrm{i}}(t-1) + V_{\mathrm{gj}}\right],$$

which gives after integrating,

$$x - \lambda(t) = \mathrm{e}^{N_{\mathrm{SUB}}t}\int_{t_1}^{t} \mathrm{e}^{-N_{\mathrm{SUB}}t''}\left[j_{\mathrm{i}}\left(t''-1\right) + V_{\mathrm{gj}}\right] dt'',$$

where t_1 is the time at which a fluid element that is at position x in the two-phase region at time t had passed through the boiling boundary. Defining now $t' = t - t''$, we obtain the following remarkable relation that includes the time delay in the boiling region:

$$x - \lambda(t) = \int_{0}^{t-t_1} \mathrm{e}^{N_{\mathrm{SUB}}t'}\left[j_{\mathrm{i}}(t-1-t') + V_{\mathrm{gj}}\right] dt'. \tag{6.48}$$

Note that $(t - t_1)$ is the residence time that a fluid element located in x at time t has spent in the boiling region. The total two-phase residence time, $\tau(t)$, of a particle leaving the channel (i.e., $x = 1$) at time t is then defined by,

$$1 = \lambda(t) + \int_{0}^{\tau(t)} \mathrm{e}^{N_{\mathrm{SUB}}t'}\left[j_{\mathrm{i}}(t-1-t') + V_{\mathrm{gj}}\right] dt'. \tag{6.49}$$

Next, the void propagation equation (6.42) may be written in terms of the material derivative,

$$\frac{D\alpha}{Dt} + N_{\mathrm{SUB}}\alpha = \frac{\rho_1}{\rho_1 - \rho_2} N_{\mathrm{SUB}}, \tag{6.50}$$

which may be integrated in the two-phase region to obtain the void fraction along a characteristic, resulting

$$\alpha(t) = \frac{\rho_1}{\rho_1 - \rho_2}\left[1 - \mathrm{e}^{N_{\mathrm{SUB}}(t-t_1)}\right]. \tag{6.51}$$

Thus, the dimensionless mixture density can be easily obtained,

$$\rho_{\mathrm{m}} = \frac{(1-\alpha)\rho_1 + \alpha\,\rho_2}{\rho_1} = \mathrm{e}^{-N_{\mathrm{SUB}}(t-t_1)}. \tag{6.52}$$

Finally, combining Eqs. (6.35), (6.47), and (6.52), the mixture velocity is given by

$$v_{\mathrm{m}} = j_{\mathrm{i}}(t) + N_{\mathrm{SUB}}[x - \lambda(t)] + V_{\mathrm{gj}}\left[1 - \mathrm{e}^{N_{\mathrm{SUB}}(t-t_1)}\right]. \tag{6.53}$$

With these expressions we are now ready to undertake the two-phase momentum equation by integrating Eq. (6.34) in the boiling region:

$$\Delta p_{2\phi} = \int_{\lambda(t)}^{1}\left[\rho_{\mathrm{m}}\left(\frac{\partial v_{\mathrm{m}}}{\partial t} + v_{\mathrm{m}}\frac{\partial v_{\mathrm{m}}}{\partial x} + \Lambda v_{\mathrm{m}}^2 + Fr^{-1}\right) + \frac{\partial}{\partial x}\left(\frac{\alpha}{1-\alpha}\frac{\rho_2}{\rho_{\mathrm{m}}}V_{\mathrm{gj}}^2\right)\right]dx. \tag{6.54}$$

Equation (6.54) can be simplified by changing the spatial integration variable to time by means of Eq. (6.48). The result is:

$$\Delta p_{2\phi} = \int_0^{\tau(t)} \rho_{\mathrm{m}}\left(\frac{\partial v_{\mathrm{m}}}{\partial t} + v_{\mathrm{m}}\frac{\partial v_{\mathrm{m}}}{\partial x} + \Lambda v_{\mathrm{m}}^2 + Fr^{-1}\right)\mathrm{e}^{N_{\mathrm{SUB}}t'}\left[j_{\mathrm{i}}(t-t'-1) + V_{\mathrm{gj}}\right]dt'.$$

Now, using Eqs. (6.45), (6.52), and (6.53) to express ρ_{m}, v_{m}, and λ in terms of j_{i} and t' we derive with the following nonlinear integral-differential expression in the variable j_{i} for the total channel pressure drop:

$$\begin{aligned}\Delta p = {} & K_{\mathrm{i}}j_{\mathrm{i}}(t)^2 + [\lambda(t) + I_{\mathrm{o}}]\left[\Lambda j_{\mathrm{i}}(t)^2 + Fr^{-1} + \frac{dj_{\mathrm{i}}(t)}{dt}\right] + I_{\mathrm{o}}j_{\mathrm{i}}(t-1)N_{\mathrm{SUB}} \\ & + I_1\left[N_{\mathrm{SUB}}^2 + 2j_{\mathrm{i}}(t)N_{\mathrm{SUB}}\Lambda\right] + I_2N_{\mathrm{SUB}}^2\Lambda + I_3N_{\mathrm{SUB}}V_{\mathrm{gj}} + I_4 2j_{\mathrm{i}}(t)\Lambda V_{\mathrm{gj}} \\ & + 2I_5N_{\mathrm{SUB}}\Lambda V_{\mathrm{gj}} + (I_6\Lambda + I_7)V_{\mathrm{gj}}^2,\end{aligned} \tag{6.55}$$

where

$$I_o = \int_0^{\tau(t)} \left(j_i(t-t'-1) + V_{gj}\right)dt',$$

$$I_1 = \int_0^{\tau(t)} \left\{ \int_0^{t'} e^{N_{SUB}t''}\left[j\left(t-1-t''\right) + V_{gj}\right]dt'' \right\} \left[j_i(t-t'-1) + V_{gj}\right]dt',$$

$$I_2 = \int_0^{\tau(t)} \left\{ \int_0^{t'} e^{N_{SUB}t''}\left[j\left(t-1-t''\right) + V_{gj}\right]dt'' \right\}^2 \left[j_i(t-t'-1) + V_{gj}\right]dt',$$

$$I_3 = \int_0^{\tau(t)} \left(1 - 2e^{N_{SUB}t'}\right)\left[j_i(t-t'-1) + V_{gj}\right]dt',$$

$$I_4 = \int_0^{\tau(t)} \left(1 - e^{N_{SUB}t'}\right)\left[j_i(t-t'-1) + V_{gj}\right]dt',$$

$$I_5 = \int_0^{\tau(t)} \left(1 - e^{N_{SUB}t'}\right)\left\{ \int_0^{t'} e^{N_{SUB}t''}\left[j\left(t-1-t''\right) + V_{gj}\right]dt'' \right\} \left[j_i(t-t'-1) + V_{gj}\right]dt',$$

$$I_6 = \int_0^{\tau(t)} \left(1 - e^{N_{SUB}t'}\right)^2 \left[j_i(t-t'-1) + V_{gj}\right]dt',$$

$$I_7 = \left(1 - e^{-N_{SUB}\tau}\right)e^{-N_{SUB}\tau} r_\rho.$$

The upper limit of the previous integrals is the two-phase residence time, which is given by Eq. (6.49). However, when later the equations are linearized this residence time will be considered constant and equal to its steady-state value, which satisfies:

$$e^{N_{SUB}\tau} = 1 + N_{SUB}\frac{1-\bar{j}}{\bar{j} + V_{gj}} \tag{6.56}$$

so that

$$\tau = \frac{1}{N_{SUB}} \ln\left[1 + \frac{N_{SUB}\left(1-\bar{j}\right)}{\bar{j} + V_{gj}}\right]. \tag{6.57}$$

where $\bar{j}$ is the steady-state dimensionless total flux.

6.5 Delay Drift-Flux Model

The analysis of Eq. (6.55) is not a trivial task. The usual method is to linearize around the steady state and then classify the roots of the Laplace-transformed equations. This task will be performed in Sects. 6.6 and 6.7. However, now we can make a pause to collect the findings so far and linger over the physics hidden behind the intricacy of the equations before proceeding with the comprehensive analysis.

What follows is the drift-flux extension of the delay theory of a boiling channel developed for the homogeneous model (Clausse et al. 1996; Delmastro et al. 2001). Accordingly, Eqs. (6.44), (6.51), and (6.53) will be used to grasp the mechanics behind the boiling channel instability without the complexity of the full Eq. (6.55). Let us assume that j_{i} oscillates approximately around the steady-state value $\bar{j} = N_{\mathrm{SUB}}/N_{\mathrm{PCH}}$ with a certain angular frequency ω:

$$j_{\mathrm{i}}(t) = \bar{j} + \delta j \sin \omega t. \tag{6.58}$$

Then, from Eq. (6.44)

$$\lambda(t) = \int_{t-1}^{t} j_{\mathrm{i}}(t')dt' = \bar{j} + \frac{\delta j}{\omega}[\cos \omega(t-1) - \cos \omega t]. \tag{6.59}$$

Regarding that

$$\cos \omega(t-1) = \cos \omega\left(t - \frac{1}{2} - \frac{1}{2}\right) = \cos \omega\left(t - \frac{1}{2}\right)\cos\frac{\omega}{2} + \sin \omega\left(t - \frac{1}{2}\right)\sin\frac{\omega}{2}, \tag{6.60}$$

$$\cos \omega t = \cos \omega\left(t - \frac{1}{2} + \frac{1}{2}\right) = \cos \omega\left(t - \frac{1}{2}\right)\cos\frac{\omega}{2} - \sin \omega\left(t - \frac{1}{2}\right)\sin\frac{\omega}{2}, \tag{6.61}$$

the oscillation of the subcooling length can be written as

$$\lambda(t) = \bar{j}_{\mathrm{i}} + \frac{2\delta j}{\omega} \sin\left(\frac{\omega}{2}\right) \sin \omega\left(t - \frac{1}{2}\right). \tag{6.62}$$

For low frequencies, $\left(\sin\frac{\omega}{2}\right)/\left(\frac{\omega}{2}\right) \cong 1$, and so:

$$\lambda(t) \cong j_{\mathrm{i}}\left(t - \frac{1}{2}\right). \tag{6.63}$$

On the other hand, the steady-state value of the two-phase residence time τ is given by Eq. (6.57). Note that for large N_{SUB}, τ is small. Assuming that this is the same in general for $\tau(t)$, Eq. (6.49) can be then approximated by:

$$1 \cong \lambda(t) + [j_{\mathrm{i}}(t-1) + V_{\mathrm{gj}}] \int_0^{\tau(t)} e^{N_{\mathrm{SUB}} t'} dt',$$

that is,

$$1-\lambda(t)\cong\frac{\left[j_i(t-1)+V_{gj}\right]\left[e^{N_{SUB}\tau(t)}-1\right]}{N_{SUB}}.$$

The exit density and velocity can then be estimated evaluating Eqs. (6.52) and (6.53) for $t-t_1=\tau(t)$, resulting:

$$\rho_e(t)=e^{-N_{SUB}\tau(t)}\cong\frac{j_i(t-1)+V_{gj}}{j_i(t-1)+V_{gj}+N_{SUB}[1-\lambda(t)]}, \tag{6.64}$$

$$\begin{aligned} v_e(t)&=j_i+N_{SUB}[1-\lambda(t)]+V_{gj}\left(1-e^{N_{SUB}\tau(t)}\right)\\ &\cong j_i(t)+N_{SUB}[1-\lambda(t)]\left[\frac{j_i(t-1)}{j_i(t-1)+V_{gj}}\right]. \end{aligned} \tag{6.65}$$

Let us assume now a quasi-static approximation of the momentum Eq. (6.43), i.e., we drop the $\frac{\partial v_m}{\partial t}$ term, which is valid for low frequencies. For simplicity let us assume that large friction pressure losses are concentrated at the inlet and the exit of the channel, i.e., the distributed friction is replaced by a concentrated coefficient K_e at the outlet and the acceleration terms are neglected. Also, if the friction is sufficiently large compared to the inverse of the Froude number, the gravity term can be neglected too. In such case, Eq. (6.43) integrates to

$$\Delta p=K_ij_i(t)^2+K_e\rho_e(t)v_{me}^2(t). \tag{6.66}$$

Combining Eqs. (6.63)–(6.66) gives:

$$\Delta p=K_ij_n^2+K_e\frac{\left(j_{n-2}+V_{gj}\right)}{j_{n-2}+V_{gj}+N_{SUB}\left(1-j_{n-1}\right)}\left\{j_n+N_{SUB}\left(1-j_{n-1}\right)\left(\frac{j_{n-2}}{j_{n-2}+V_{gj}}\right)\right\}^2, \tag{6.67}$$

where

$$\begin{aligned} j_n&=j_i(t)\\ j_{n-1}&=j_i\left(t-\frac{1}{2}\right)\\ j_{n-2}&=j_i(t-1) \end{aligned}$$

The pressure drop can be calculated from the steady-state value (i.e., $j_n=j_{n-1}=j_{n-2}=\bar{j}$) which gives

$$\Delta p=K_i\bar{j}^2+K_e\bar{j}^2\left(\frac{\bar{j}_e+V_{gj}}{\bar{j}+V_{gj}}\right) \tag{6.68}$$

$$\bar{j}_{\mathrm{e}} = \bar{j} + N_{\mathrm{SUB}}\left(1 - \bar{j}\right) \tag{6.69}$$

The delay equation (6.67) is of course a strong simplification of the complex dynamics of a boiling channel. Nonetheless, it is useful to capture the nature of boiling channel instabilities and, as we shall see, it leads to the same stability criterion as the complete drift-flux differential equations for large K_{i}, K_{e}, and Fr.

According to Eq. (6.67), the time series of the inlet velocity can be readily estimated with the iterative map:

$$j_n = \frac{-A + \sqrt{B^2 - 4AC}}{2A},$$
$$A = K_{\mathrm{i}} + K_{\mathrm{e}}\rho_{\mathrm{e}}, \quad B = 2K_{\mathrm{e}}\rho_{\mathrm{e}}j^*, \quad C = K_{\mathrm{e}}\rho_{\mathrm{e}}j^{*2} - \Delta p,$$
$$j^* = N_{\mathrm{SUB}}(1 - j_{n-1})\left(\frac{j_{n-2}}{j_{n-2} + V_{\mathrm{gj}}}\right),$$
$$\rho_{\mathrm{e}} = \frac{j_{n-2} + V_{\mathrm{gj}}}{j_{n-2} + N_{\mathrm{SUB}}(1 - j_{n-1}) + V_{\mathrm{gj}}}.$$

The stability of the map can be determined by perturbing the inlet velocity around the steady state,

$$j_n = \bar{j} + \delta j_n.$$

Linearizing in δj leads to

$$a_{\mathrm{o}}\delta j_{n-2} + a_1\delta j_{n-1} + a_2\delta j_n = 0 \tag{6.70}$$

where

$$a_{\mathrm{o}} = 2(K_{\mathrm{i}} + K_{\mathrm{e}})\bar{j},$$
$$a_1 = -K_{\mathrm{e}}N_{\mathrm{SUB}}\left(\frac{\bar{j}^2}{\bar{j} + V_{\mathrm{gj}}}\right),$$
$$a_2 = K_{\mathrm{e}}N_{\mathrm{SUB}}\left(1 - \bar{j}\right)\left[1 - \left(\frac{V_{\mathrm{gj}}}{\bar{j} + V_{\mathrm{gj}}}\right)^2\right].$$

The general solution of the linear difference Eq. (6.70) is of the form q^n, with eigenvalues q given by the characteristic equation:

$$a_2q^2 + a_1q + a_{\mathrm{o}} = 0$$

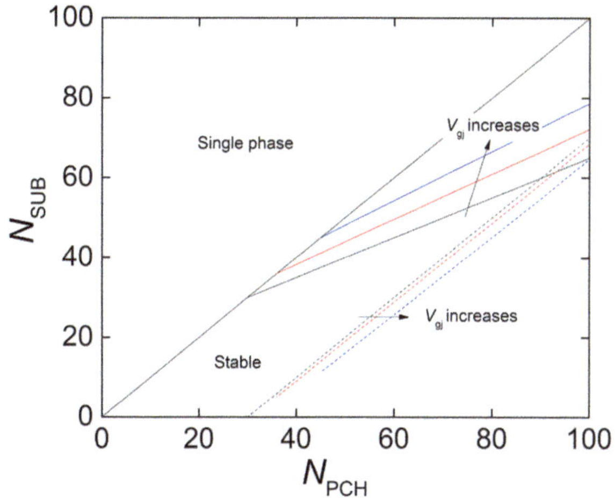

Fig. 6.5 Linear stability map of the delay model of a boiling channel ($K_{\mathrm{i}} = 70, K_{\mathrm{e}} = 5$) for $V_{\mathrm{gj}} = 0,\ 0.02,\ 0.05$. Excursive Ledinegg instability (*solid*), density-wave instabilities (*dashed*)

whose solution is

$$q = \frac{-a_1 \pm \sqrt{a_1^2 - 4a_2 a_{\mathrm{o}}}}{2a_{\mathrm{o}}}. \tag{6.71}$$

If the radicand of Eq. (6.71) is negative, the solution is oscillatory, and it is stable if

$$|q|^2 = \frac{a_1^2 - a_1^2 + 4a_2 a_{\mathrm{o}}}{4a_{\mathrm{o}}^2} = \frac{a_2}{a_{\mathrm{o}}} < 1$$

which yields to the following condition for density waves instabilities,

$$N_{\mathrm{SUB}} - N_{\mathrm{PCH}} < \frac{2\left(\frac{K_{\mathrm{i}}+K_{\mathrm{e}}}{K_{\mathrm{e}}}\right)}{1 - \left(\frac{V_{\mathrm{gj}}}{\bar{j}+V_{\mathrm{gj}}}\right)^2}. \tag{6.72}$$

The dashed lines in Fig. 6.5 are the stability margin given by Eq. (6.72) in the parameter plane (N_{PCH}, N_{SUB}) for various values of V_{gj}. For $V_{\mathrm{gj}} = 0$, Eq. (6.72) leads to the well-known Ishii's criterion (for $K_{\mathrm{e}} \gg 1$), which entails that increasing the inlet friction and/or decreasing the exit friction stabilizes the boiling channel (Ishii 1971). The extended criterion given by Eq. (6.72) predicts that increasing V_{gj} also has a stabilizing effect, which is in agreement with Ishii's findings too.

On the other hand, if the radicand of Eq. (6.71) is positive, q is real. In such case, the temporal evolution is not oscillatory, and it will be excursive if $|q| > 1$. Since $a_1 < 0$, the stability margin of this excursive behavior is given by

$$\sqrt{a_1^2 - 4a_2 a_{\mathrm{o}}} = 2a_{\mathrm{o}} + a_1,$$

which leads to

$$a_o + a_1 + a_2 = 0. \tag{6.73}$$

Inserting the expressions of a_o, a_1, and a_2 in Eq. (6.73) gives the following condition for excursive instability,

$$N_{SUB} < \frac{K_i + K_e}{K_e} + \frac{N_{PCH}}{2}\left[1 + V_{gj}\left(1 - \frac{V_{gj}(1 + V_{gj})}{\left(\frac{N_{SUB}}{N_{PCH}} + V_{gj}\right)^2}\right)\right]. \tag{6.74}$$

Equation (6.74) can be also obtained by making zero the derivative of Δp with respect to $\bar{j}$ in Eq. (6.68), which is called the Ledinegg stability limit (Ledinegg 1938). The solid lines in Fig. 6.5 show the stability threshold given by Eq. (6.74) in the parameter plane (N_{SUB},N_{PCH}) for various values of V_{gj}.

6.6 Flow Excursion

6.6.1 Homogeneous Equilibrium Model

We now return to the complete equations of the boiling channel derived in Sect. 6.4 and begin the analysis of the Ledinegg instability, which results in a sudden and severe decrease in the flow. To simplify the mathematics let us start with the homogeneous model (HEM), i.e., $V_{gj} = 0$. Then, assuming steady state, Eq. (6.55) may be easily integrated to obtain the flow excursion equation of Achard et al. (1985),

$$\Delta p_{HEM} = N_{SUB}\bar{j}(1 - \bar{j}) + Fr^{-1}\bar{j}(1 + \tau) + \Lambda\left[\bar{j}^2 + \frac{1}{2}N_{SUB}\bar{j}(\bar{j} - 1)^2\right] + K_i\bar{j}^2. \tag{6.75}$$

Figure 6.6 shows three curves corresponding to Eq. (6.75) for $N_{SUB} = 1, 5, 10$, $\Lambda = 11$, and $Fr = 1$. The flow is unstable in the region where the slope of the flow curve is negative and so, for $N_{SUB} = 10$, the flow will experience a discontinuous jump from a to b upon a reduction of the pressure drop. Looking at Fig. 6.6 it becomes clear that the stability condition is given by:

$$\frac{\partial \Delta p_{HEM}}{\partial j} = Fr^{-1} + 2\bar{j}K_i + 2\bar{j}\Lambda + \frac{Fr^{-1}\log\left[1 - \left(1 - \frac{1}{\bar{j}}\right)N_{SUB}\right]}{N_{SUB}} + \frac{1}{2}\left[2 + \Lambda + 3\bar{j}^2\Lambda - 4\bar{j}(1 + \Lambda)\right]N_{SUB} - \frac{Fr^{-1}}{\bar{j} + (1 - \bar{j})N_{SUB}} = 0. \tag{6.76}$$

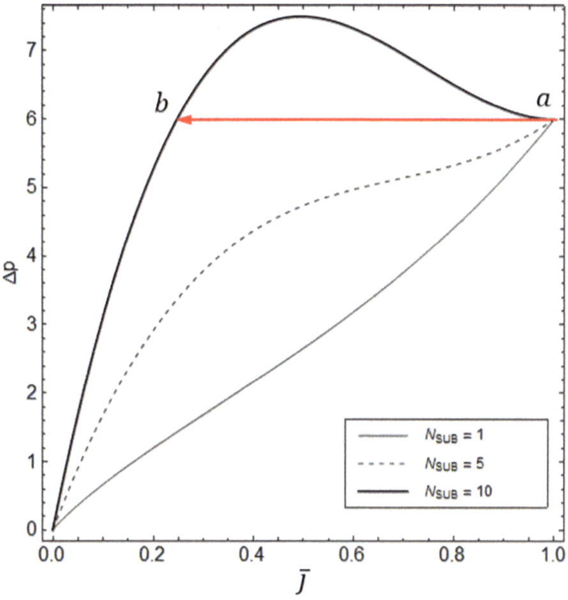

Fig. 6.6 Pressure drop vs. flow curves (Eq. 6.75)

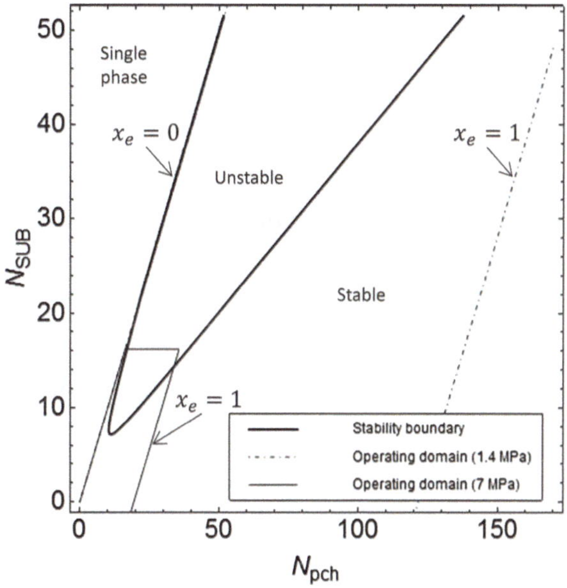

Fig. 6.7 Flow excursion stability map: effect of system pressure

The dimensionless stability curve does not change with the system pressure but the operating range of reachable values of N_{PCH} and N_{SUB} does. Figure 6.7 shows the stability map obtained from Eq. (6.76) with $\Lambda = 11$, $Fr = 1$, $K_i = 0$, in the plane (N_{SUB}, N_{PCH}) for low and high operating pressures. It can be seen that the

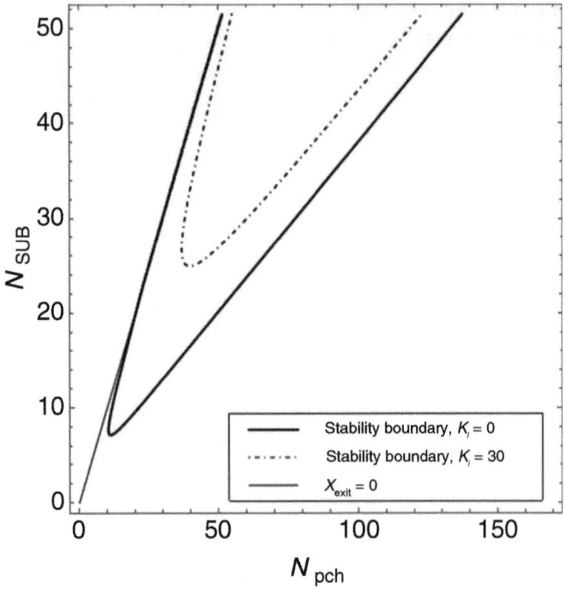

Fig. 6.8 Effect of inlet restriction on flow excursion ($\Lambda = 11,\ Fr = 1$)

operating domain of low pressure boilers is more vulnerable to Ledinegg excursions, a characteristic that led to frequent boiler explosions in the nineteenth century. The effect of the inlet restriction, $K_i = 30$, is illustrated in Fig. 6.8 and, as is well known in engineering practice, it makes the boiler significantly more stable.

6.6.2 *Drift-Flux Model*

We now incorporate the drift-flux (DF) terms, i.e., terms dependent on V_{gj}, for completeness,

$$\Delta p = \Delta p_{HEM} + \Delta p_{DF}, \tag{6.77}$$

where

$$\Delta p_{DF} = a_1 V_{gj} + a_2 V_{gj}^2 + a_3 V_{gj}^3 \tag{6.78}$$

and

$$a_1 = \frac{1}{2N_{SUB}} \left\{ 4j(1 - j - j\tau) + 2\left[-1 - 2j(-1 + \Lambda) + j^2(-1 + 2(1 + \tau)\Lambda) + \frac{\tau}{Fr} \right] N_{SUB} \right.$$
$$\left. + (1 - j)(-2 + (1 - j)\Lambda) N_{SUB}^2 \right\}, \tag{6.79}$$

Fig. 6.9 Effect of V_{gj} on the Ledinegg stability boundary ($\Lambda = 11,\ Fr = 1,\ K_i = 0$)

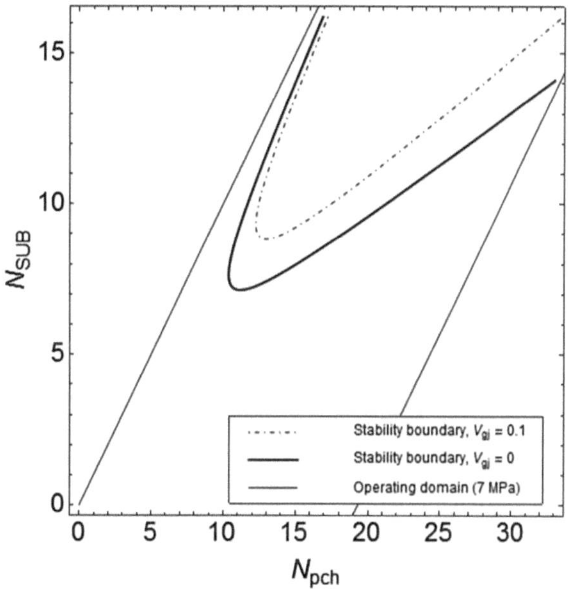

$$a_2 = \frac{1}{N_{SUB}}\{-2(-1+j+2j\tau) + [-1-\Lambda+j(1+\tau+\Lambda+3\tau\Lambda]N_{SUB}\} + \left\{\frac{(1-j)N_{SUB}r_\rho(j+V_{gj})}{[j+(1-j)N_{SUB}+V_{gj}]^2} + \frac{(1-j)^2N_{SUB}+2(j+V_{gj})(-1+j+j\tau+\tau V_{gj})}{2(j+V_{gj})}\right\}, \quad (6.80)$$

$$a_3 = \tau\left(1+\Lambda-\frac{2}{N_{SUB}}\right). \quad (6.81)$$

The results for $V_{gj} = 0.1$ are shown in Fig. 6.9. The stability curve is qualitatively similar to that of the homogeneous model, showing a stabilizing effect of V_{gj}. The effect is in agreement with the delay drift-flux model as shown in Fig. 6.5.

6.7 Density Wave Instability

6.7.1 Homogeneous Equilibrium Model

As was shown in Sect. 6.5 with the delayed drift-flux model, the density wave instability is caused by the relation of the flow at the inlet of a heated channel and the time it takes for a fluid element to reach the exit of the channel, thus modifying the density, velocity, and the associated two-phase pressure loss there. In this section we will analyze the behavior of Eq. (6.55) around the steady state in order to determine the boiling channel stability in the plane (N_{SUB} , N_{PCH}).

The corresponding perturbation of the integral-differential Eq. (6.55) with $V_{gj} = 0$ was derived by Achard et al. (1981), resulting:

$$\begin{aligned}&\left[\bar{j}(\bar{j}\Lambda + N_{SUB}) + Fr^{-1}\right]\delta I_0 + N_{SUB}^2(\delta I_1 + \Lambda\delta I_2) + \bar{j}N_{SUB}[2\Lambda\,\delta I_1 + \tau\,\delta j_i(-1+t)]\\&+\left[\bar{j}^2(K_i+\Lambda) + Fr^{-1}\right]\delta\lambda + 2\,(K_i+\Lambda)\delta j_i(t) + (1+\tau)\frac{\partial}{\partial t}\delta j_i(t) = 0.\end{aligned} \tag{6.82}$$

where the perturbation of the corresponding integral terms is

$$\delta I_0 = \bar{j}\,\delta\tau + \int_0^\tau \delta j_i(-1+t-t')dt'$$

$$\begin{aligned}\delta I_1 = \frac{\bar{j}}{N_{SUB}}&\left[-\int_0^\tau \delta j_i(-1+t-t')dt' + \int_0^\tau e^{N_{SUB}t'}\delta j_i(t-1-t')dt' + N_{SUB}(1-j_i)\delta\tau\right.\\&\left.+N_{SUB}\int_0^\tau\int_0^{t'} e^{N_{SUB}t''}\delta j_i\left(t-1-t''\right)dt''dt'\right]\end{aligned}$$

$$\begin{aligned}\delta I_2 = \frac{\bar{j}^2}{N_{SUB}^2}&\left[\left[\int_0^\tau \delta j_i(t-1-t')dt' - 2\int_0^\tau e^{N_{SUB}t'}\delta j_i(t-1-t')dt'\right.\right.\\&+\int_0^\tau e^{2N_{SUB}t'}\delta j_i(t-1-t')dt'\\&+2N_{SUB}\int_0^\tau e^{N_{SUB}t'}\int_0^{t'} e^{N_{SUB}t''}\delta j_i\left(t-1-t''\right)dt''dt'\\&\left.-2N_{SUB}\int_0^\tau\int_0^{t'} e^{N_{SUB}t''}\delta j_i\left(t-1-t''\right)dt''dt' = N_{SUB}^2\frac{\left(1-\bar{j}\right)^2}{\bar{j}}\delta\tau\right]\end{aligned}$$

where, from Eq. (6.44) it follows that

$$\delta\lambda(t) = \int_0^1 \delta j_i(t-t')dt', \tag{6.83}$$

and Eq. (6.57) leads to

$$\delta\tau(t) = -\frac{e^{-\tau N_{SUB}}}{\bar{j}}\left(\delta\lambda + \int_0^\tau e^{N_{SUB}t'}\delta j_i(-1+t-t')dt'\right). \tag{6.84}$$

6.7.2 Transfer Function

Inserting Eqs. (6.83) and (6.84) into (6.82), replacing $\delta j(t) = e^{i\omega t}$ into it and taking the Laplace transform yields the transfer function

$$\phi(i\omega) = \frac{Q(i\omega)}{i\omega(i\omega + N_{\mathrm{SUB}})^2(i\omega + 2N_{\mathrm{SUB}})} = 0, \tag{6.85}$$

where $Q(i\omega) = a + be^{-i\omega} + ce^{-i\omega(1+\tau)}$ is the numerator that contains the time delays for the subcooled and two-phase regions. The coefficients of the numerator are as follows:

$$\begin{aligned}
a = {} & 2jN_{\mathrm{SUB}}^3(-jK_i + (1 + \Lambda - j\Lambda)N_{\mathrm{SUB}}) - \frac{2(1-j)jN_{\mathrm{SUB}}^4}{j + (1-j)N_{\mathrm{SUB}}}Fr^{-1} \\
& - \left(jN_{\mathrm{SUB}}^2(5jK_i - (5 + 4K + 9\Lambda - 5j\Lambda)N_{\mathrm{SUB}}) + \frac{5(1-j)jN_{\mathrm{SUB}}^3}{j + (1-j)N_{\mathrm{SUB}}}Fr^{-1}\right) i\omega \\
& - \left(2jN_{\mathrm{SUB}}(2jK_i - (2 + 5K + 7\Lambda - 2j\Lambda)N_{\mathrm{SUB}} + (1+\tau)N_{\mathrm{SUB}}^2\right) \\
& + \left.\frac{4(1-j)jN_{\mathrm{SUB}}^2}{j + (1-j)N_{\mathrm{SUB}}}Fr^{-1}\right)(i\omega)^2 \\
& - \left(j(jK_i + (-1 - 8K_i + (-9 + j)\Lambda)N_{\mathrm{SUB}} + 5(1+\tau)N_{\mathrm{SUB}}^2\right) \\
& + \left.\frac{(1-j)jN_{\mathrm{SUB}}}{j + (1-j)N_{\mathrm{SUB}}}Fr^{-1}\right)(i\omega)^3 \\
& + 2j(K_i + \Lambda - 2(1+\tau)N_{\mathrm{SUB}})(i\omega)^4 - j(1+\tau)(i\omega)^5, \\
b = {} & 2jN_{\mathrm{SUB}}^3(jK_i + (-1 + (-1 + j)\Lambda)N_{\mathrm{SUB}}) + \frac{2(1-j)jN_{\mathrm{SUB}}^4}{j + (1-j)N_{\mathrm{SUB}}}Fr^{-1} \\
& + \left(jN_{\mathrm{SUB}}^2(j(5K_i - \Lambda) + 7(-1 + (-1 + j)\Lambda)N_{\mathrm{SUB}})\right. \\
& + \left.\frac{(2j + 5(1-j)N_{\mathrm{SUB}})jN_{\mathrm{SUB}}^2}{j + (1-j)N_{\mathrm{SUB}}}Fr^{-1}\right) i\omega \\
& - \left(jN_{\mathrm{SUB}}(j(-4K_i + \Lambda) + (5 - 7(-1 + j)\Lambda)N_{\mathrm{SUB}} - 2\tau N_{\mathrm{SUB}}^2\right) \\
& + \left.\frac{(-3j - 4(1-j)N_{\mathrm{SUB}})jN_{\mathrm{SUB}}}{j + (1-j)N_{\mathrm{SUB}}}Fr^{-1}\right)(i\omega)^2 \\
& + \left(j(jK_i + (-1 + 2(-1 + j)\Lambda)N_{\mathrm{SUB}} + 3\tau N_{\mathrm{SUB}}^2) + jFr^{-1}\right)(i\omega)^3 + j\tau N_{\mathrm{SUB}}(i\omega)^4, \\
c = {} & -\left(N_{\mathrm{SUB}}^2[j + (1-j)N_{\mathrm{SUB}}][j\Lambda + (2 + \Lambda - j\Lambda)N_{\mathrm{SUB}}] + 2N_{\mathrm{SUB}}^2 Fr^{-1}\right) i\omega \\
& + \left(N_{\mathrm{SUB}}[j + (1-j)N_{\mathrm{SUB}}][j\Lambda + (1 + \Lambda - j\Lambda)N_{\mathrm{SUB}}] - 3N_{\mathrm{SUB}}Fr^{-1}\right)(i\omega)^2 \\
& - Fr^{-1}(i\omega)^3.
\end{aligned}$$

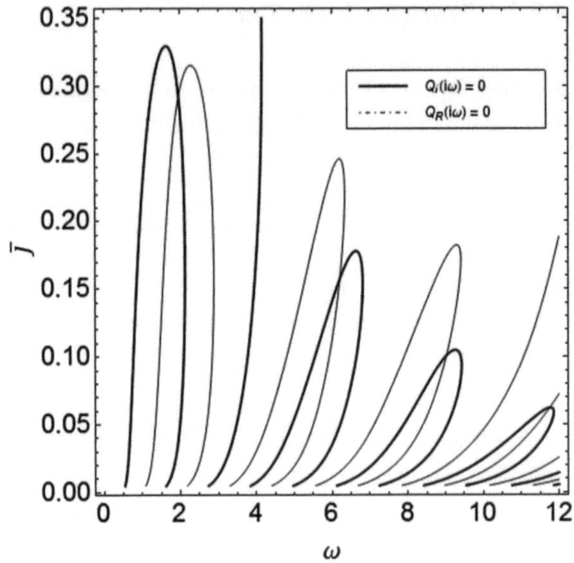

Fig. 6.10 Zeroes of the transfer function

Ishii (1971) proposed to perform the stability analysis based on the conditions

$$Q_{\rm i}(i\omega) = Q_{\rm R}(i\omega) = 0. \tag{6.86}$$

Due to the exponentials in $Q(i\omega)$ Eq. (6.86) has an infinite number of roots. The D-partition condition may be visualized in Fig. 6.10 where the functions $Q_{\rm i}(i\omega) = 0$ and $Q_{\rm R}(i\omega) = 0$ are plotted for $N_{\rm SUB} = 4$, $\Lambda = 11$, $Fr = 1$. The intersection points in the Figure are the first four roots corresponding to the four stability obtained by Ishii (1971) shown in Figs. 6.11 and 6.12.

The paths of the first two roots of the density wave instability, obtained by solving the D-partition condition with $\Lambda = 11$ and $Fr = 1$ in the plane ($N_{\rm PCH}$, $N_{\rm SUB}$) are shown in red in Fig. 6.11, superposed over Ishii's original results. The flow excursion curve of Fig. 6.8 is also included for completeness. The agreement of the results of Achard et al. (1981) with the earlier results of Ishii (1971) is a verification of the general linear stability approach using two different but equivalent analytical techniques.

The plane of dimensionless parameters ($N_{\rm PCH}$, $N_{\rm SUB}$) used in Fig. 6.11 was first identified by Ishii as the most significant to characterize the density wave instability. As was shown in the previous section they are also appropriate for the flow excursion as they scale the effects of pressure, power, geometry, etc., effectively. The results of Ishii (1971) for the effect of the inlet restriction on the density wave oscillation are also shown in Fig. 6.11. It is well known that the inlet restriction stabilizes the oscillations, as it does the flow excursion, cf. Fig. 6.8.

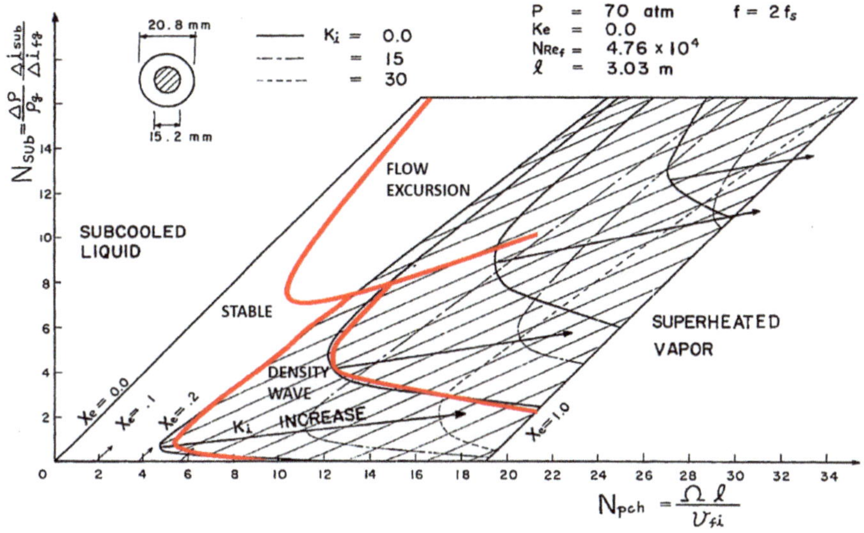

Fig. 6.11 Density wave stability map, effect of inlet restriction: (*Red*) present model vs. (*Black*) Ishii's model [Reprinted from Ishii (1971) with permission of the author]

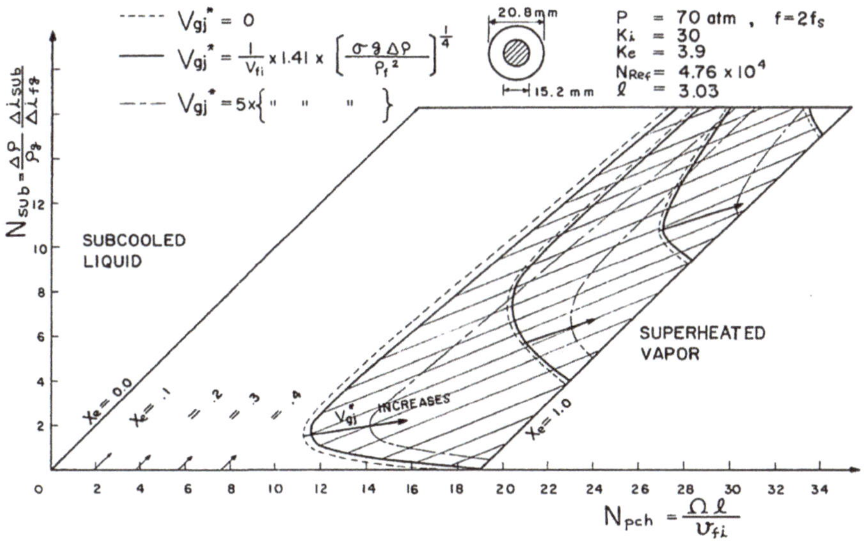

Fig. 6.12 Density wave stability map, effect of V_{gj} [Reprinted from Ishii (1971) with permission of the author]

6.7.3 *Drift-Flux Model*

The complete linear perturbation analysis of Eq. (6.55) including V_{gj} has been performed by Rizwan-Uddin and Dorning (1986) and will not be repeated here. Nonetheless, equivalent results obtained by Ishii for the effect of V_{gj} are shown in Fig. 6.12. The effect of drift flux is to make the flow more stable when compared to the homogeneous flow model, the same as with the Ledinegg excursion shown Fig. 6.9. Also, the same trend is found with the delayed drift-flux model as was shown in Sect. 6.5 and Fig. 6.5. It is worth noting that while the effect of drift flux is small for high pressure, it is significant at atmospheric conditions.

6.8 Summary and Discussion

The stable DFM was derived rigorously from the FFM. It is one of the most practical two-phase flow models because it retains the capability of very long material wave propagation and it is stable, i.e., the local instabilities are removed by the kinematic condition and the long ones by the fixed flux condition. Several stable but transient problems that require accurate wave propagation prediction have been illustrated.

Next it was shown that if the fixed flux condition is removed and the mixture momentum equation is added, the dynamic DFM is obtained and it becomes possible to predict global instabilities. The integral momentum equation of Achard et al. (1985) with drift flux was applied to the flow excursion and the density wave oscillation. The results were presented in terms of the dimensionless numbers N_{SUB} and N_{PCH} proposed by Ishii (1971). The density wave stability results of Ishii were confirmed with the equations of Achard, showing the validity of the analysis irrespective of the details of the mathematical approach.

The roles of the fixed flux and drift-flux approximations with respect to the stability of the TFM are now apparent, namely, are counterparts allowing separate local and global material wave instability analyses. Furthermore, applying both approximations together results in the unconditionally stable DFM of Wallis (1969) that was described at the beginning of the chapter. Thus, we have gone full circle around the various TFM simplifications shown in Fig. 1.1 that enable different two-phase flow instability analyses.

References

Achard, J.-L., Drew, D. A., & Lahey, R. T., Jr. (1981). The effect of gravity and friction on the stability of boiling flow in a channel. *Chemical Engineering Communications, 11*, 59–79.

Achard, J.-L., Drew, D. A., & Lahey, R. T., Jr. (1985). The analysis of nonlinear density-wave oscillations in boiling channels. *Journal of Fluid Mechanics, 166*, 213–232.

Bernier, R. J. N. (1982). Unsteady two-phase flow instrumentation and measurement. Ph.D. 533 Thesis, California Institute of Technology, Pasadena, CA.

Clausse, A., Delmastro, D., & Juanico, L. (1996). A simple delay model for density-wave oscillations. *Latin American Journal of Applied Research, 26*, 185–191.

Delmastro, D., Juanico, L., & Clausse, A. (2001). A delay theory for boiling flow stability analysis. *International Journal of Multiphase Flow, 27*, 657–671.

Hewitt, G. F. (1982). Void fraction. In G. Hetsroni (Ed.), *Handbook of multiphase systems*. New York: McGraw-Hill.

Ishii, M. (1971). *Thermally induced flow instabilities in two-phase thermal equilibrium*. Ph.D. Thesis, School of Mechanical Engineering, Georgia Institute of Technology.

Ishii, M. (1975). *Thermo-fluid dynamic theory of two-phase flow* (Collection de la Direction des Etudes et Researches d'Electricite de France). Paris, France: Eyrolles.

Ishii, M. (1977). *One-dimensional drift flux model and constitutive equations for relative motion between phases in various two-phase flow regimes* (Argonne National Lab. Report, ANL-77-47).

Ishii, M., & Hibiki, T. (2006). *Thermo-fluid dynamics of two-phase flow*. New York: Springer.

Lahey, R. T., Jr., & Moody, F. J. (1977). *The thermal-hydraulics of a boiling water nuclear reactor*. La Grange Park: American Nuclear Society.

Ledinegg, M. (1938). Instability of flow during natural and forced circulation. *Die Wärme, 61*(8), 891–898.

Miles, G. D., Shledovsky, L., & Ross, J. (1943). *Journal of Physical Chemistry, 49*, 93.

Rizwan-Uddin, & Dorning, J. J. (1986). Some nonlinear dynamics of a heated channel. *Nuclear Engineering and Design, 93*, 1–14.

Ruspini, L. C., Marcel, C. P., & Clausse, A. (2014). Two-phase flow instabilities: A review. *International Journal of Heat and Mass Transfer, 71*, 521–548.

Wallis, G. B. (1969). *One-dimensional two-phase flow*. New York: McGraw-Hill.

Wallis, G. (2013). Novak Zuber and the drift flux model. *Multiphase Science and Technology, 25* (2–4), 107–112.

Wallis, G. B., & Heasley, J. H. (1961, August). Oscillations in two phase flow systems. *Journal of Heat Transfer, 83*, 363–369.

Wulff, W. (1985, June). Kinematics of two-phase mixture level motion in BWR pressure vessels. In *Proceedings of Specialists Meeting on Small-Break LOCA Analysis in LWRs, Pisa, Italy*.

Zuber, N., & Findlay, J. A. (1965). Average volumetric concentration in two-phase flow systems. *Journal of Heat Transfer, 9*, 453–468.

Zuber, N., & Staub, F. W. (1967). An analytical investigation of the transient response of the volumetric concentration in a boiling forced-flow system. *Nuclear Science and Engineering, 30*, 268–278.

Chapter 7
Drift-Flux Model Nonlinear Dynamics and Chaos

Abstract After analyzing the linear stability of the Drift-Flux Model for boiling channels, in this chapter we address the question of the nonlinear evolution of density waves after their initial growth. In particular, we are interested in the occurrence of sustained oscillations under operating conditions close to the linear stability margins.

First, a nonlinear discrete mapping derived with the Drift-Flux Model with transport delays is used to get a preliminary understanding of the development of stable and unstable limit cycles at high and low N_{SUB} numbers, respectively. With this simple model it is possible to visualize the effect of the drift velocity, V_{gj}, on the limit cycles.

Then, the Moving Nodes Model (MNM), which was originally developed for homogeneous two-phase flow (DiMarco et al., International Journal of Heat and Technology 8: 125–141, 1990; Clausse and Lahey, Proceedings of the 9th International Heat Transfer Conference, Jerusalem, 1990; Chaos, Solitons & Fractals 1:167–178, 1991) and was applied extensively, particularly in the nuclear industry, is extended to incorporate drift flux. The MNM is applied to simulate and analyze the nonlinear dynamics of a system comprised of a boiling channel coupled with an adiabatic riser, which is relevant to recent advanced water-cooled nuclear reactor designs, among other applications. A complex nonlinear dynamic behavior is encountered at operating conditions of incipient boiling at high N_{SUB}. The flow reaches sustained chaotic oscillations alternating with limit cycles of different periods and quasi-periodic oscillations.

7.1 Introduction

The nonlinear dynamics of the drift-flux model have been mainly studied in boiling channels, where different types of transient behaviors are encountered: flow excursions, sustained limit cycles, and quasi-periodic and strange attractors. The most studied kind of boiling flow instabilities are the so-called density waves. As was seen in the previous chapter, the mechanisms of generation of this phenomenon are the delays in the propagation of disturbances and the feedback processes imposed by the boundary conditions.

M.L. de Bertodano et al., *Two-Fluid Model Stability, Simulation and Chaos*,
DOI 10.1007/978-3-319-44968-5_7

Fukuda and Kobori (1979) presented the classical taxonomy of different types of density-wave (DW) oscillations. These authors classified the different kinds of phenomena according to which is the dominant mechanism, namely:

- Type 1: dominated by gravity, which appears mostly in natural circulating systems under low quality conditions.
- Type 2: dominated by friction, which appears mostly in forced circulation systems under high quality conditions.
- Type 3: dominated by inertia, which appears mostly in long channels.

The second type is the most common instability described in the literature. During these oscillations, changes in flow or void fraction in the two-phase region result in out-of-phase single and two-phase pressure-drop variations. The first type is relevant in nuclear reactor safety, either during loss of flow accidents or the operation of advanced passive cooled designs (Marcel et al. 2013). A good recent review of the analytical treatment of DW oscillations can be found in Ruspini et al. (2014).

To analyze the nonlinear behavior of density waves two approaches have been proposed: perturbation analysis and numerical time-domain simulations. The perturbation analysis was presented by Achard et al. (1985) for the homogeneous model, based on Hopf bifurcation techniques, valid near the marginal-stability boundaries. They found that the nonlinear terms can change significantly the stability limits for both normal and high order modes. Also they discovered cases where a finite amplitude forced perturbation may render unstable a linearly stable condition. The latter was an unpleasant surprise considering that all safety analyses were performed up to that date using the linear analysis. This shows the importance of the nonlinear effects in the stability of two-phase flow. Rizwan-Uddin and Dorning (1986) extended Achard's analysis to include the drift-flux model.

The direct numerical modeling of the drift-flux equations is the most popular tool for predicting the nonlinear behavior of boiling channels. Rizwan-Uddin and Dorning (1988, 1990) numerically integrated the model developed by Achard et al. (1985), which was presented in Chap. 6, and found that a boiling channel may develop chaotic oscillations if it is subjected to forced periodic oscillations. The drift-flux model can also be integrated using classical finite-difference or finite-volumes schemes, discretizing the space into a fixed spatial grid (Ambrosini et al. 2000).

A simpler numerical scheme was presented by Clausse and Lahey (1990) where the spatial coordinate is discretized into moving nodes whose position become the state variable of the equations, thus called the Moving Nodes Model (MNM). This model was originally developed for homogeneous two-phase flow and was applied extensively, particularly in the nuclear industry, incorporating neutronic kinetics and drift flux. Milestones of the model can be found in Clausse and Lahey (1990, 1991), DiMarco et al. (1990), Chang and Lahey (1997), Garea et al. (1999), and Theler et al. (2012). Garea et al. (1999) presented the first extension of the MNM including drift flux. Recently, Paruya et al. (2012) in a comparative study suggest that the MNM is more efficient and has a better convergence for the simulation of DW oscillations than the classical fixed-grid discretization.

The simplified delay drift-flux model presented in Sect. 6.5 is also a useful tool to understand the nonlinear behavior of boiling channels (Clausse et al. 1995). This model was analyzed using Hoft bifurcation methods, confirming the occurrence of subcritical bifurcations at low inlet subcooling (Juanico et al. 1998; Delmastro et al. 2001). A risk parameter was proposed as a design criterion to assess the magnitude of subcritical instabilities and the amplitude of limit cycles.

7.2 Nonlinear Mapping of the Boiling Channel Dynamics

In Sect. 6.5 we presented a delay drift-flux model of a boiling channel and showed that it can be used to understand the mechanisms that make the flow unstable, namely, the interplay between the void-fraction propagation along the channel and the integral momentum balance of the system. The simplicity of the mapping given by Eq. (6.67) can be exploited to get a preliminary understanding of what happens after the channel flow becomes linearly unstable.

Let us recall that the system consists of a single channel subjected to constant pressure drop and constant inlet temperature, with frictions concentrated at the inlet and the outlet. We are assuming that those frictions are sufficiently large that it is possible to neglect all the other terms in the momentum equation, namely, inertia, acceleration, gravity, and distributed friction. Under those conditions the external pressure head is balanced by the inlet and outlet friction pressure drops, resulting:

$$\Delta p = K_{\mathrm{i}} j_n^2 + K_{\mathrm{e}} \frac{(j_{n-2} + V_{\mathrm{gj}})}{j_{n-2} + V_{\mathrm{gj}} + N_{\mathrm{SUB}}(1 - j_{n-1})} \left\{ j_n + N_{\mathrm{SUB}}(1 - j_{n-1}) \left(\frac{j_{n-2}}{j_{n-2} + V_{\mathrm{gj}}} \right) \right\}^2, \tag{7.1}$$

where j is the inlet velocity evaluated a different times, that is:

$$j_n = j_{\mathrm{i}}(t)$$
$$j_{n-1} = j_{\mathrm{i}}\left(t - \frac{1}{2}\right)$$
$$j_{n-2} = j_{\mathrm{i}}(t-1)$$

The external pressure head Δp can be related to the steady-state value (i.e., $j_n = j_{n-1} = j_{n-2} = \bar{j}$) which gives

$$\Delta p = K_{\mathrm{i}} \bar{j}^2 + K_{\mathrm{e}} \bar{j}^2 \left(\frac{\bar{j}_{\mathrm{e}} + V_{\mathrm{gj}}}{\bar{j} + V_{\mathrm{gj}}} \right) \tag{7.2}$$

$$\bar{j}_{\mathrm{e}} = \bar{j} + N_{\mathrm{SUB}}\left(1 - \bar{j}\right), \tag{7.3}$$

and as was defined in Chap. 6:

$$\bar{j} = \frac{N_{\mathrm{SUB}}}{N_{\mathrm{PCH}}}. \tag{7.4}$$

Therefore, given the operation parameters N_{SUB} and N_{PCH}, and the friction coefficients K_{i} and K_{e}, the discrete time series of the inlet velocity can be readily estimated with the nonlinear second-order iterative map:

$$\begin{aligned} j_n &= \frac{-A + \sqrt{B^2 - 4AC}}{2A}, \\ A &= K_{\mathrm{i}} + K_{\mathrm{e}}\rho_{\mathrm{e}}, \quad B = 2K_{\mathrm{e}}\rho_{\mathrm{e}} j^*, \quad C = K_{\mathrm{e}}\rho_{\mathrm{e}} j^{*2} - \Delta p, \\ j^* &= N_{\mathrm{SUB}}(1 - j_{n-1})\left(\frac{j_{n-2}}{j_{n-2} + V_{\mathrm{gj}}}\right), \\ \rho_{\mathrm{e}} &= \frac{j_{n-2} + V_{\mathrm{gj}}}{j_{n-2} + N_{\mathrm{SUB}}(1 - j_{n-1}) + V_{\mathrm{gj}}}. \end{aligned} \tag{7.5}$$

Starting with two initial values of the inlet velocity, j_1 and j_2, the subsequent values j_n, for $n > 2$, can be calculated by computing the map given by Eq. (7.5). Usually j_1 and j_2 are choosen close to the steady state, whose stability conditions are:

$$N_{\mathrm{SUB}} - N_{\mathrm{PCH}} < \frac{2\left(\frac{K_{\mathrm{i}} + K_{\mathrm{e}}}{K_{\mathrm{e}}}\right)}{1 - \left(\frac{V_{\mathrm{gj}}}{\bar{j} + V_{\mathrm{gj}}}\right)^2}$$

for density waves, and:

$$N_{\mathrm{SUB}} < \frac{K_{\mathrm{i}} + K_{\mathrm{e}}}{K_{\mathrm{e}}} + \frac{N_{\mathrm{PCH}}}{2}\left\{1 + V_{\mathrm{gj}}\left[1 - \frac{V_{\mathrm{gj}}(1 + V_{\mathrm{gj}})}{\left(\frac{N_{\mathrm{SUB}}}{N_{\mathrm{PCH}}} + V_{\mathrm{gj}}\right)^2}\right]\right\}.$$

for the Ledinegg excursion.

A useful way to visualize the nonlinear dynamics of discrete mappings is the phase plane where j_n is plotted against j_{n-1}. In particular, regarding Eq. (6.63), this plot is significant for because, within this approximation, j_{n-1} is exactly the subcooled length λ evaluated at the time step n. Figure 7.1 (top) shows a stable evolution around the steady state, $\bar{j} = N_{\mathrm{SUB}}/N_{\mathrm{PCH}}$, with $K_{\mathrm{i}} = 70$, $K_{\mathrm{e}} = 5$, $N_{\mathrm{SUB}} = 40$, $V_{\mathrm{gj}} = 0$, and $N_{\mathrm{PCH}} = 65$ (i.e., $\bar{j} = 0.615$). It can be seen that the state of the system spirals in toward the steady state, which indicates that there are complex eigenvalues with modules lower than unity (see Sect. 6.5). On the other hand, for $N_{\mathrm{PCH}} = 74$, i.e., $\bar{j} = 0.54$ (Fig. 7.1 bottom), the steady state is unstable and the trajectory follows an expanding

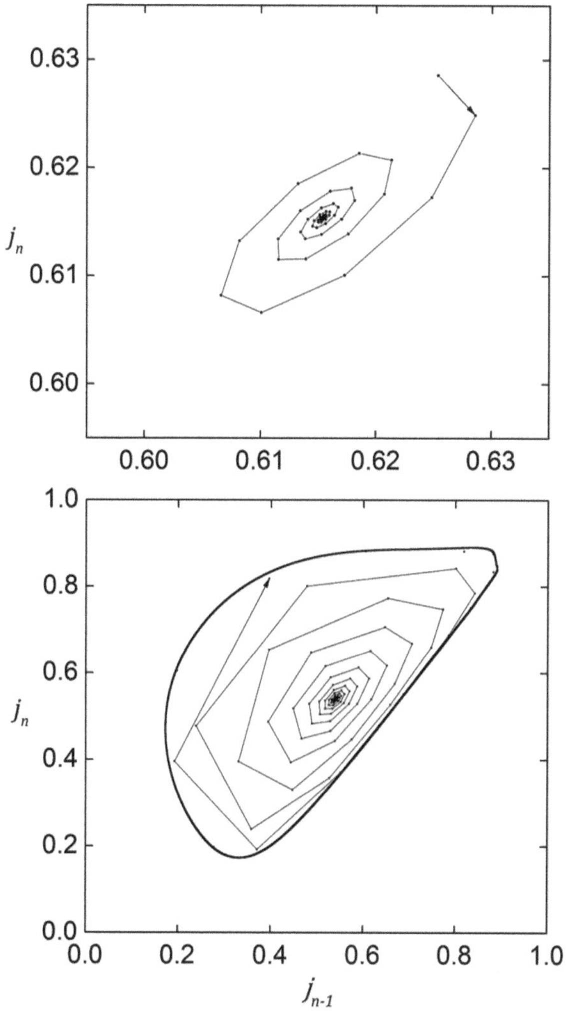

Fig. 7.1 Phase-space plot of the temporal evolution of the inlet velocity given by the delay model of a boiling channel $\left(K_i = 70,\ K_e = 5,\ N_{SUB} = 40,\ V_{gj} = 0\right)$. *Top*: $N_{PCH} = 65$ (stable), *Bottom*: $N_{PCH} = 74$ (unstable)

spiral which eventually is contained by nonlinear mechanisms, and the system finally reaches a limit cycle.

The amplitude of the limit cycle depends on how far is the corresponding steady state from the stability threshold. The linear stability map for the same set of parameters as Fig. 7.1 was depicted in Fig. 6.5. For $N_{SUB} = 40$ the stability threshold is at $N_{PCH} = 70$. Figure 7.2 shows the limit cycles obtained following a path of increasing N_{PCH} while keeping N_{SUB} constant (i.e., a horizontal path in Fig. 6.5). In practice this can be done, for example, by increasing the input power.

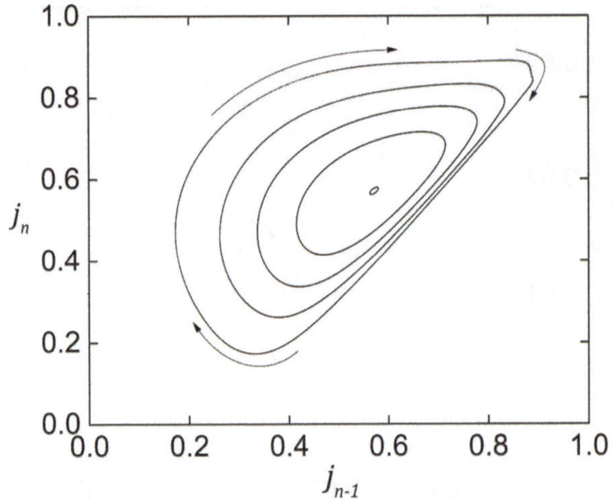

Fig. 7.2 Limit cycles described by the inlet velocity of a boiling channel $(K_\mathrm{i} = 70,\ K_\mathrm{e} = 5,$ $N_\mathrm{SUB} = 40,\ V_\mathrm{gj} = 0)$. Following increasing amplitude, $N_\mathrm{PCH} = 70.1,\ 71,\ 72,\ 73,\ 74$

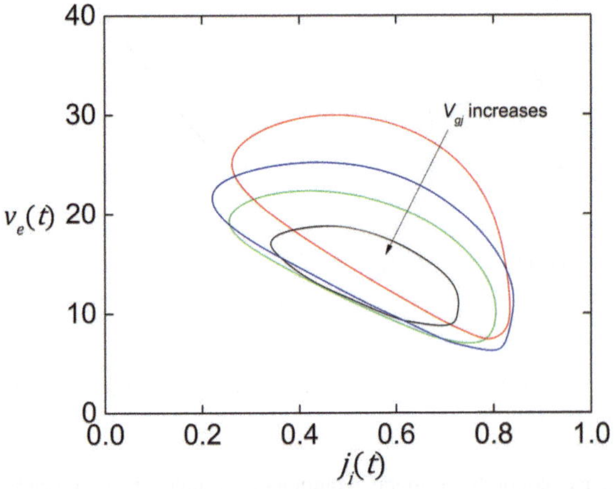

Fig. 7.3 Effect of the drift flux on the nonlinear dynamics of a boiling channel $(K_\mathrm{i} = 70,\ K_\mathrm{e} = 5,$ $N_\mathrm{SUB} = 40, N_\mathrm{PCH} = 73,\ V_\mathrm{gj} = 0, 0.1, 0.015, 0.2)$

It can be seen that the amplitude of the limit cycles increases as N_PCH increases beyond the linear stability margin at $N_\mathrm{PCH} = 70$. The "nose" that appears at the upper right of the larger cycles is caused by the presence of a second steady state. In this case, this point, close to $\bar{j} = 1$, is Ledinegg unstable. Here we can appreciate the nonlinear manifestation of the Ledinegg instability. From the point of view of system dynamics, the Ledinegg excursion is a saddle point, i.e., two real eingenvalues, one negative and one positive, respectively, attract and repel the trajectory along the directions of the corresponding eigenvectors.

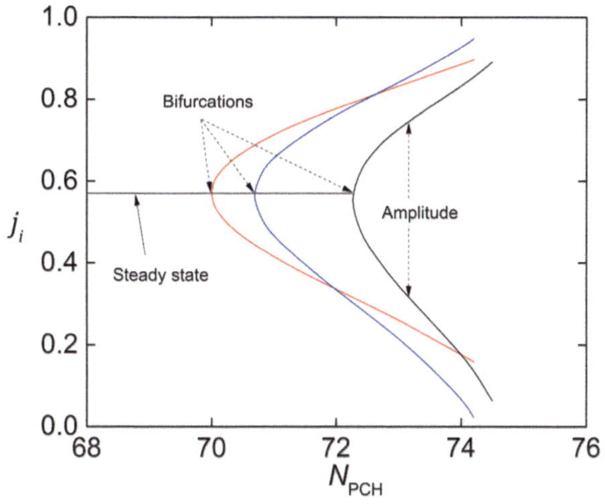

Fig. 7.4 Bifurcation diagram of the limit cycles obtained with the delay drift-flux model ($K_i = 70$, $K_e = 5$, $N_{SUB} = 40$, $N_{PCH} = 73$) for different $V_{gj} = 0$ (*red*), 0.1(*blue*), 0.2(*black*)

The effect of V_{gj} on the nonlinear dynamics is shown in Fig. 7.3, where the exit velocity $v_e(t)$ is plotted against the simultaneous value of the inlet velocity $j_i(t)$. Increasing V_{gj} shifts the center of the cycles toward lower velocities. However, the amplitude of the oscillations increases with V_{gj} up to $V_{gj} = 0.1$. This effect can be better appreciated in the bifurcation plots shown in Fig. 7.4, where the variation of the maximum and minimum inlet velocity of the cycle is depicted for each N_{PCH}. The lower value of N_{PCH} in each curve indicates the linear stability threshold. It can be seen that, although V_{gj} makes the channel linearly stable, the amplitude of the limit cycles increases faster for higher V_{gj}.

The delay model also reveals another nonlinear feature of the boiling channel dynamics, namely, the fact that a linearly stable steady state might become unstable if the perturbation is sufficiently large. In effect, stable limit cycles like the ones shown in Fig. 7.2 occurs only for high inlet subcoolings (i.e., high N_{SUB}). For $N_{SUB} < 2(K_i + K_e)/K_e$, there are no limit cycles bounding the unstable steady states. Instead, unstable limit cycles appear around stable steady states. These unstable cycles are called subcritical bifurcations because they occur before the critical stability margin is reached, as opposed to the limit cycles which are called supercritical bifurcations. This feature can be found also for the complete set of two-phase flow equations (Achard et al. 1985; DiMarco et al. 1990). Figure 7.5 shows an example of a subcritical bifurcation exhibited by the delay model in the stable region with $N_{SUB} = 20$. For initial states inside the basin of attraction delimited by the unstable limit cycle the trajectory spirals in toward the stable steady state, whereas for initial states outside the cycle the system spirals out. In such case, eventually, flow reversals occur, which provide other nonlinear boundaries not considered in the present model.

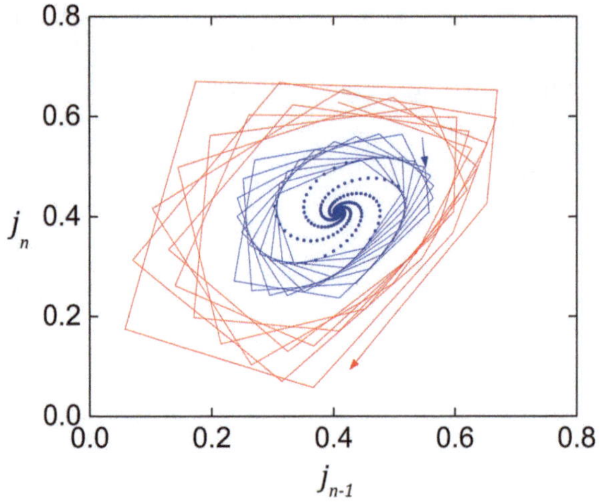

Fig. 7.5 Unstable limit cycle in the stable region ($K_i = 70$, $K_e = 5$, $N_{SUB} = 20$, $N_{PCH} = 49$, $V_{gj} = 0$). Initial states inside the limit cycle spiral into the steady state (*blue*), whereas initial states outside the limit cycle spiral out (*red*)

7.3 Model of a Boiling Channel with Moving Nodes

A useful model of the nonlinear dynamics of a boiling channel is the so-called Moving Nodes Model (MNM). This model was originally developed for homogeneous two-phase flow and was further applied extensively, particularly in the nuclear industry, incorporating neutronic kinetics and drift flux. Milestones of the model can be found in DiMarco et al. (1990), Clausse and Lahey (1990, 1991), Chang and Lahey (1997), Garea et al. (1999), and Theler et al. (2012). In what follows, a drift-flux extension of the MNM will be introduced.

We start with the dimensionless drift-flux model given by Eqs. (6.41)–(6.43). Consider the channel in Fig. 6.4, and divide the single-phase region into N_s segments of variable length whose extremes are defined by enthalpy nodes, i.e., moving locations inside the channel where the enthalpy is constant. Garea et al. (1999) recommended that N_s should be even in order to represent the correct dynamics. Let us define the enthalpy ground state such that the dimensionless enthalpy vanishes at liquid saturation, h_l, i.e., $h_i = \frac{h_{in} - h_l}{h_{12}}$. The boiling boundary, $z = \lambda(t)$, is then defined by the condition $h = 0$. Taking the enthalpy unit as the enthalpy increase along the channel in steady state, given by the power input divided by the flow rate, the dimensionless enthalpy at the inlet is given by:

$$h_i = -\frac{N_{SUB}}{N_{PCH}} \tag{7.6}$$

The position of the n-th enthalpy node, $\lambda_n(t)$, defining the boundary between segments n and $n+1$ is then defined as the spatial point where the fluid enthalpy is:

$$h_n = \frac{N_{\mathrm{SUB}}}{N_{\mathrm{PCH}}}\left(\frac{n}{N_{\mathrm{s}}} - 1\right) \tag{7.7}$$

where N_{s} is the number of subcooled nodes. Note that these values of enthalpy are constant provided that the power and the inlet enthalpy are constant, and therefore the enthalpy jump between any two sequential nodes is always $\Delta h = -h_{\mathrm{i}}/N_{\mathrm{s}}$.

The temporal evolution of $\lambda_n(t)$ is governed by the energy equation in the single-phase region, which is given by:

$$\frac{\partial h}{\partial t} + j_{\mathrm{i}}\frac{\partial h}{\partial x} - \frac{N_{\mathrm{SUB}}}{N_{\mathrm{PCH}}} = 0 \tag{7.8}$$

A set of ordinary differential equations can be derived by applying the method of residuals to Eq. (7.8). The basic procedure is to propose a family of solutions with time-dependent parameters. Then, ordinary differential equations for the parameters are obtained by asking that the spatial averages of Eq. (7.8) between nodes (which are called residuals) are identically zero (Finlayson 1972). Integrating Eq. (7.8) between nodes $n-1$ and n using piecewise linear spatial profiles of enthalpy gives:

$$\int_{\lambda_{n-1}}^{\lambda_n} \frac{\partial h}{\partial t}\, dx + (h_n - h_{n-1})j_{\mathrm{i}} = \frac{N_{\mathrm{SUB}}}{N_{\mathrm{PCH}}}(\lambda_n - \lambda_{n-1}) \tag{7.9}$$

Using Leibnitz's rule, the first term can be written as follows:

$$\int_{\lambda_{n-1}}^{\lambda_n} \frac{\partial h}{\partial t}\, dx = \frac{d}{dt}\left(\int_{\lambda_{n-1}}^{\lambda_n} h\, dx\right) - h_n\frac{d\lambda_n}{dt} + h_{n-1}\frac{d\lambda_{n-1}}{dt} \tag{7.10}$$

Assuming a linear enthalpy profile between nodes:

$$\int_{\lambda_{n-1}}^{\lambda_n} h\, dx = \frac{(\lambda_n - \lambda_{n-1})(h_n - h_{n-1})}{2} \tag{7.11}$$

Combining Eqs. (7.9)–(7.11) yields:

$$\frac{d\lambda_n}{dt} = 2j_{\mathrm{i}} - 2N_{\mathrm{s}}(\lambda_n - \lambda_{n-1}) - \frac{d\lambda_{n-1}}{dt} \tag{7.12}$$

On the other hand, integrating the mass conservation equation along the whole channel gives:

$$\frac{dm}{dt} = j_{\mathrm{i}} - \rho_{\mathrm{e}}v_{\mathrm{e}} \tag{7.13}$$

where m is the mass of the channel, ρ_e is the exit mixture density, and v_e is the exit mixture velocity. In order to relate m with ρ_e and λ_{N_s}, the method of residuals is applied again assuming a density profile with the same spatial dependence as the steady-state and time-dependent parameters. The density profile in the two-phase region at steady state is given by Eq. (6.52):

$$\rho_m(x) = e^{-N_{SUB}\tau(x)}. \tag{7.14}$$

where $\tau(x)$ is defined by:

$$x = \overline{\lambda}_{N_s} + \left(\bar{j}_i + V_{gj}\right) \int_0^{\tau(x)} e^{N_{SUB}t'}\, dt' = \overline{\lambda}_{N_s} + \frac{\left(\bar{j}_i + V_{gj}\right)\left[e^{N_{SUB}\tau(x)} - 1\right]}{N_{SUB}} \tag{7.15}$$

Canceling $\tau(x)$ from Eqs. (7.14) and (7.15) gives the steady-state profile of the mixture density:

$$\rho_m(x) = \left[1 + \frac{N_{SUB}\left(x - \overline{\lambda}_{N_s}\right)}{\bar{j}_i + V_{gj}}\right]^{-1} \tag{7.16}$$

Equation (7.16) can be used as model of the spatial dependence of $\rho_m(x,t)$. Regarding that $\rho_m = 1$ at $x = \lambda_{N_s}(t)$, and $\rho_m = \rho_e(t)$ at $x = 1$, the following density profile can be proposed to perform the spatial averaging:

$$\rho_m(x,t) = \left[1 + \left(\frac{1}{\rho_e} - 1\right)\left(\frac{x - \lambda_{N_s}}{1 - \lambda_{N_s}}\right)\right]^{-1} \tag{7.17}$$

Equation (7.17) entails that we are representing the boiling region with a single variable-length segment. With this approximation, the instantaneous mass inside the channel is then given by:

$$m = \lambda_{N_s} + \int_{\lambda_{N_s}}^{1} \frac{dx}{1 + \left(\frac{1}{\rho_e} - 1\right)\left(\frac{x - \lambda_{N_s}}{1 - \lambda_{N_s}}\right)} = \lambda_{N_s} + \frac{(1 - \lambda_{N_s})\ln(1/\rho_e)}{1/\rho_e - 1} \tag{7.18}$$

Equation (7.17) for $\rho_m(x,t)$, together with Eq. (6.35) for $v_m(x,t)$ and Eq. (6.47) for $j(x,t)$, can be used to derive the differential equation for $j_i(t)$ by integrating the momentum conservation Eq. (6.43) along the channel. The complete set of terms was integrated by several authors using different approaches. Here we will present the dynamics of a channel with large frictions losses concentrated at the inlet and the outlet, so that the acceleration and distributed friction terms may be neglected, thus simplifying the analysis of the results, so that:

$$\Delta p_I + \Delta p_F + \Delta p_G = \Delta p \tag{7.19}$$

where:

$$\Delta p_{\mathrm{I}} = \frac{d}{dt}\left(j_{\mathrm{i}}\lambda_{N_{\mathrm{s}}} + \int_{\lambda_{N_{\mathrm{s}}}}^{1} \rho_{\mathrm{m}} v_{\mathrm{m}} dx \right) \tag{7.20}$$

$$\Delta p_{\mathrm{F}} = K_{\mathrm{i}}\, j_{\mathrm{i}}^{2} + K_{\mathrm{e}}\rho_{\mathrm{e}} v_{\mathrm{e}}^{2} \tag{7.21}$$

$$\Delta p_{\mathrm{G}} = \frac{m}{Fr} \tag{7.22}$$

are the inertia, friction, and gravity terms, respectively.

Equation (7.14) can be combined with Eqs. (6.35) and (6.47) to obtain the spatial profile of the mixture velocity v_{m}, that is:

$$v_{\mathrm{m}} = j_{\mathrm{i}} + N_{\mathrm{SUB}}(x - \lambda) - V_{\mathrm{gj}}\left(\frac{1}{\rho_{\mathrm{m}}} - 1\right). \tag{7.23}$$

The integral in Eq. (7.20) then becomes:

$$\int_{\lambda_{N_{\mathrm{s}}}}^{1} \rho_{\mathrm{m}} v_{\mathrm{m}} dx = (m - \lambda_{N_{\mathrm{s}}}) j_{\mathrm{i}} + \frac{N_{\mathrm{SUB}}(1 - \lambda_{N_{\mathrm{s}}})(1 - m)}{1/\rho_{\mathrm{e}} - 1} - V_{\mathrm{gj}}(1 - \lambda_{N_{\mathrm{s}}})\left[1 - \frac{\ln(1/\rho_{\mathrm{e}})}{1/\rho_{\mathrm{e}} - 1}\right] \tag{7.24}$$

Combining Eqs. (7.18)–(7.24) leads to:

$$\frac{dI_{\mathrm{c}}}{dt} = \Delta p - K_{\mathrm{i}}\, j_{\mathrm{i}}^{2} - K_{\mathrm{e}}\rho_{\mathrm{e}} v_{\mathrm{e}}^{2} - \frac{m}{Fr} \tag{7.25}$$

where:

$$I_{\mathrm{c}} = m\, j_{\mathrm{i}} + (1 - m)\left[\frac{N_{\mathrm{SUB}}(1 - \lambda_{N_{\mathrm{s}}})}{1/\rho_{\mathrm{e}} - 1} - V_{\mathrm{gj}}\right] \tag{7.26}$$

is the total channel momentum.

The model of the boiling channel then consists of $N_{\mathrm{s}} + 2$ state variables, namely, λ_n with $n = 1, \ldots, N_s$, m and I_{c}, i.e., $N_{\mathrm{s}} + 2$ ordinary differential equations, namely, Eqs. (7.12), (7.13), and (7.25); and three auxiliary variables, j_{i}, defined by the algebraic Eq. (7.26), ρ_{e} defined by Eq. (7.18), and v_{e} from Eq. (7.23) evaluated at $x = 1$. This set of nonlinear algebraic-differential equations can be solved using any of the numerical solvers of ordinary differential equations available nowadays. It is important to remark that the system of equations is stiff, because in general the time constant of the momentum Eq. (7.25) is much shorter than the transport Eqs. (7.12) and (7.13).

Figure 7.6 shows the stability map in the parameter plane ($N_{\mathrm{PCH}}, N_{\mathrm{SUB}}$) resulting from the MNM using eight subcooled nodes, for a channel with $K_{\mathrm{i}} = 70$, $K_{\mathrm{e}} = 5$ and $Fr > 1$ (forced circulation). Increasing V_{gj} stabilizes the channel as expected.

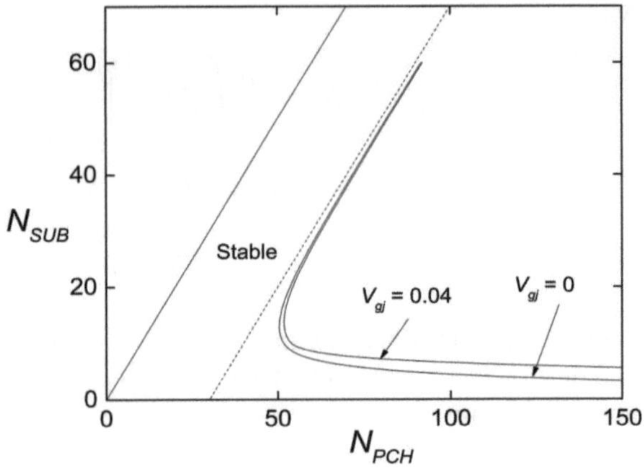

Fig. 7.6 Stability map of density-wave oscillations of a boiling channel with $N_s = 8$, $K_i = 70$, $K_e = 5$ and $Fr > 1$ (forced circulation). Compare with Fig. 6.5. Above the *solid line* given by $N_{sub} = N_{PCH}$ there is no boiling. The *dashed line* corresponds to the Ishii criterion

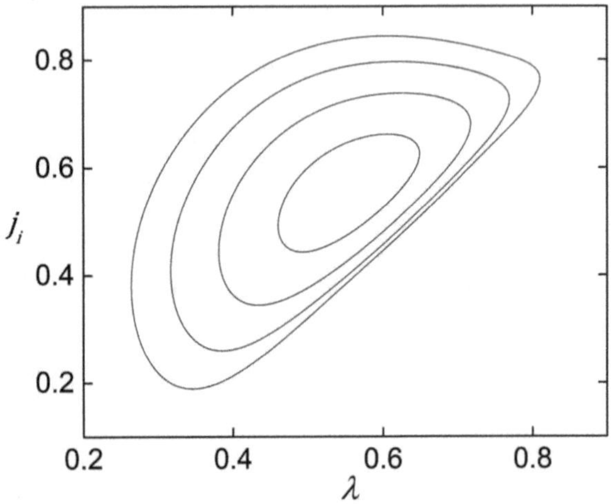

Fig. 7.7 Limit cycles obtained with the MNS model with $N_s = 8$, $K_i = 70$, $K_e = 5$, $N_{SUB} = 40$, $V_{gj} = 0$ and $Fr > 1$. Following increasing amplitude, the cycles correspond to N_{PCH} = 71.94, 72.07, 72.33, 72.46. Compare with Fig. 7.2

Comparing with the stability map predicted by the linear analysis shown in Fig. 6.12 it can be seen that the MNM captures the shape of the threshold not only for high N_{SUB}, but also for low N_{SUB}. It is worth mentioning that the number of subcooled nodes, N_s, should be sufficient to capture the most unstable mode of the enthalpy wave. Nonetheless, the average length of the subcooled segments should not differ too much from the length of the two-phase region; otherwise numerical instabilities give unrealistic results. Garea et al. (1999) extended the method to multiple boiling nodes dividing the region in segments of equal size. The discretization more consistent with the MNM would be to define the position of the inner boiling nodes based on jumps of equal enthalpy increase. However, this procedure is not straightforward as in the single-phase region because the exit enthalpy is variable.

Limit cycles are encountered in a narrow band adjacent to the right of the stability margin at large N_{SUB}. Figure 7.7 shows the limit cycles obtained following

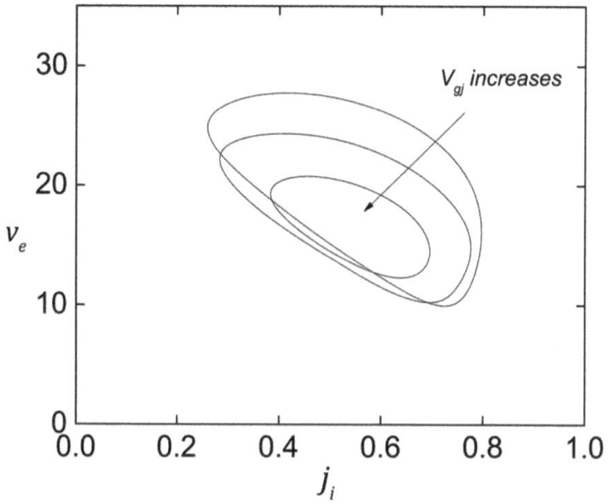

Fig. 7.8 Effect of the drift flux on the nonlinear dynamics of a boiling channel using the MNS, with $N_s = 8$, $K_i = 70$, $K_e = 5$, $N_{SUB} = 40$, $N_{PCH} = 7$, $Fr > 1$, and $V_{gj} = 0$, 0.05 and 0.075. Compare with Fig. 7.3

a path of increasing N_{PCH} while keeping N_{SUB} constant, analogous to the delay model in Sect. 7.2. It can be seen that the shape of the cycles is very similar to those shown in Fig. 7.2 for similar N_{PCH}.

The effect of increasing V_{gj} on the limit cycles is shown in Fig. 7.8, and comparing with Fig. 7.3 it can be seen that the MNS results are also similar. The effect of the gravity term is also very important in boiling channel instabilities. This term becomes relevant for low Fr numbers. A special case of practical interest is natural circulation, where the external pressure drop is given by the weight of a downcomer column of constant height, e.g., the secondary side of a steam generator of a nuclear power plant (Fig. 7.9). In such case, the dimensionless pressure drop is given by:

$$\Delta p = \frac{\lambda_D}{Fr} \tag{7.27}$$

where λ_D is the ratio between downcomer and the boiling channel length.

Figure 7.10 shows the stability map of a boiling channel under natural circulation conditions, with $N_s = 8$, $K_i = 70$, $K_e = 5$, $N_{SUB} = 40$, λ_D. In such case, the Froude number is determined by the steady-state condition, $dI_c/dt = 0$, of the momentum equation, Eq. (7.25), together with and Eq. (7.27), which yields:

$$Fr = \frac{\lambda_D - m}{K_i \bar{j}^2 + K_e \bar{\rho}_e \bar{v}_e^2}. \tag{7.28}$$

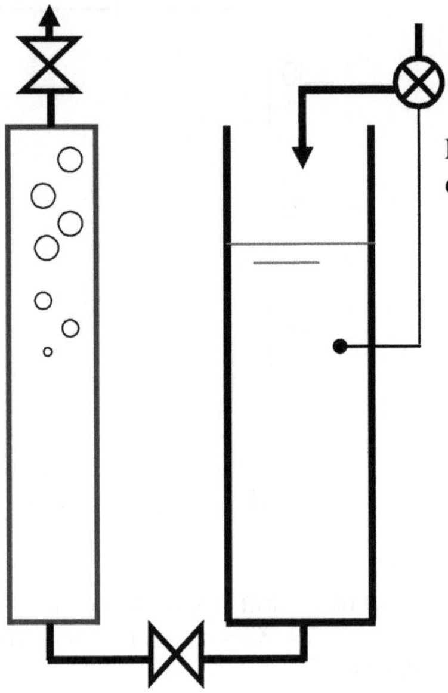

Fig. 7.9 Boiling channel driven by natural circulation

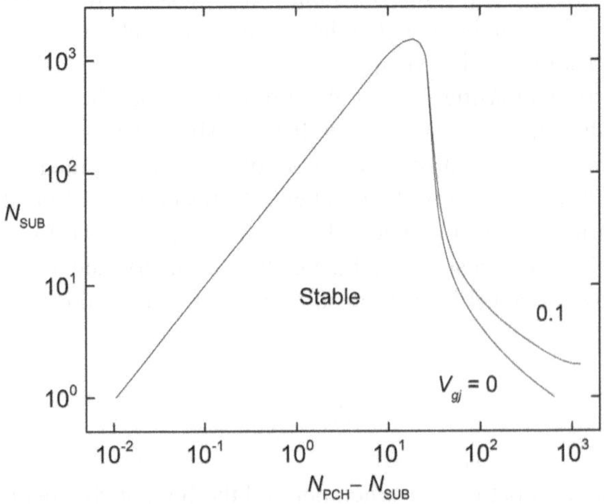

Fig. 7.10 Stability map of density-wave oscillations of a boiling channel driven by natural circulation ($N_s = 8$, $K_i = 70$, $K_e = 5$, $\lambda_D = 1$)

Equation (7.28) shows that λ_D should be larger than m in order to activate natural circulation in the channel. For low N_{SUB} and high N_{PCH} the effect of the gravity term is not significant, and the usual stability margin is obtained. At low Fr, the Ledinegg boundary moves to higher N_{SUB} and even vanishes. Furthermore, the most important feature is the appearance of an unstable region at low values of $(N_{PCH} - N_{SUB})$ which forms a very narrow strip adjacent to the right side of the line of boiling inception, given by $N_{SUB} = N_{PCH}$. In Fig. 7.10, the variable in the horizontal axis, $(N_{PCH} - N_{SUB})$, indicates the distance from the boiling-inception line. Therefore, at low power, as soon as incipient boiling starts, the channel becomes unstable. This instability is relevant during the startup of natural circulating systems, e.g., some advanced nuclear reactor designs. Also, the lower power stability margin is seldom affected by V_{gj}, as can be seen in Fig. 7.5.

7.4 Dynamics of a Boiling Channel with an Adiabatic Riser

For low *Fr* numbers the presence of a riser section above the heated channel can significantly affect the flow dynamics, which leads to more complex behaviors arising from the interaction of the buoyancy term with density waves propagating along the riser. The first autonomous chaotic oscillations reported in boiling channels were found in this kind of coupled system (Clausse and Lahey 1991). The riser is usually included in the design to increase the flow rate for a given heat input, which permits reaching higher nominal powers, but the price paid in return is a contraction of the range of stable steady states. However, with the state-of-the-art instrumentation and control technologies it is possible to safely operate boiling systems with risers, e.g., several advanced water nuclear reactor designs. In the present section we will describe an extension of the original channel-riser MNM (Clausse and Lahey 1991) to account for drift-flux effects. The adiabatic riser of length λ_R is divided in N_R segments of equal fixed length (Fig. 7.11).

Integrating the continuity equation over a generic riser segment r leads to:

$$\frac{dm_r}{dt} = \rho_{r-1} v_{r-1} - \rho_r v_r \tag{7.29}$$

where ρ_r and v_r are the mixture density and velocity at the end of segment r, and m_r is the segment mass per unit of cross-section area.

Since the riser is adiabatic, the mixture velocity v_r is uniform along the whole run, and it is given by:

$$v_r = j_i + N_{SUB}(1 - \lambda_{N_s}) - V_{gj}\left(\frac{1}{\rho_r} - 1\right) \tag{7.30}$$

For simplicity, it is assumed that the riser and the boiling channel have the same cross-section area in Eq. (7.29). Otherwise, each member of the equation should be multiplied by the corresponding flow area. Combining Eqs. (7.29) and (7.30) leads to:

$$\frac{dm_r}{dt} = \left[j_i + N_{SUB}(1 - \lambda_{N_s}) + V_{gj}\right](\rho_{r-1} - \rho_r) \tag{7.31}$$

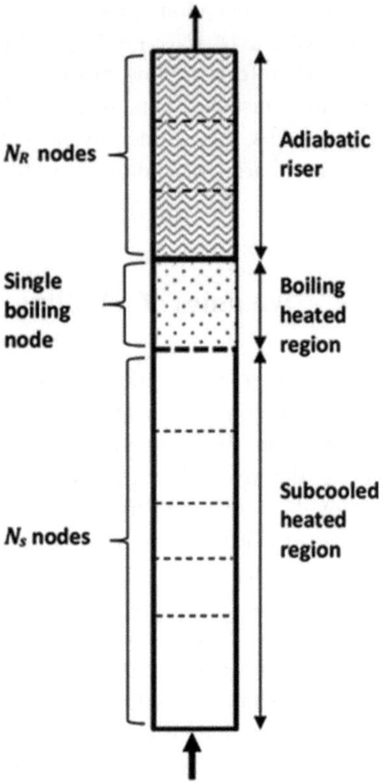

Fig. 7.11 Coupled system of a boiling channel with an adiabatic riser

The mass m_r can be related to the densities ρ_{r-1} and ρ_r using a procedure analogous to that used for the channel mass m in the derivation of Eq. (7.18), which gives:

$$m_r = \int_{x_{r-1}}^{x_r} \rho_{\mathrm{m}} dx = \frac{\lambda_{\mathrm{R}}}{N_{\mathrm{R}}} \left[\frac{\ln(\rho_r/\rho_{r-1})}{1/\rho_r - 1/\rho_{r-1}} \right] \tag{7.32}$$

where x_r is the position of the riser r-node located at the end of the r-segment.

To simplify the algebra for the present case, the frictional losses are assumed to be concentrated at the riser outlet. For more complicated cases, the friction term should be integrated along the riser. The pressure drop introduced by the riser is then given by:

$$\Delta p_{\mathrm{R}} = \frac{d}{dt}\left(\int_1^{1+\lambda_{\mathrm{R}}} \rho_{\mathrm{m}} v_{\mathrm{m}} dx \right) + \frac{1}{Fr} \int_1^{1+\lambda_{\mathrm{R}}} \rho_{\mathrm{m}} dx + K_{\mathrm{R}} \rho_{N_{\mathrm{R}}} v_{N_{\mathrm{R}}}^2 \tag{7.33}$$

The integral in the temporal derivative in Eq. (7.33) is calculated using Eq. (7.30), that is:

$$\begin{aligned}\int_1^{1+\lambda_R} \rho_m v_m dx &= \left[j_i + N_{SUB}(1-\lambda_{N_s}) + V_{gj}\right]\int_1^{1+\lambda_R} \rho_m dx - V_{gj}\lambda_R \\ &= m_R\left[j_i + N_{SUB}(1-\lambda_{N_s}) + V_{gj}\right] - V_{gj}\lambda_R\end{aligned} \tag{7.34}$$

Combining Eqs. (7.32)–(7.34) gives:

$$\Delta p_R = \frac{d}{dt}\left\{\left[j_i + N_{SUB}(1-\lambda_{N_s}) + V_{gj}\right]m_R - V_{gj}\lambda_R\right\} + \frac{m_R}{Fr} + K_R\rho_{N_R}v_{N_R}^2 \tag{7.35}$$

where

$$m_R = \sum_{r=1}^{N_R} m_r. \tag{7.36}$$

7.4.1 Summary of MNM Equations for the Channel-Riser System

The complete model of the boiling channel coupled with an adiabatic riser consists of $N_s + N_R + 2$ state variables, namely, the position of the subcooled enthalpy nodes, λ_n with $n = 1, \ldots, N_s$, the mass of the riser nodes, m_r with $r = 1, \ldots, N_R$, the heated-channel mass m, and the total momentum of the system, I. The corresponding ordinary differential equations are as follows:

$$\frac{d\lambda_n}{dt} = 2j_i - 2N_s(\lambda_n - \lambda_{n-1}) - \frac{d\lambda_{n-1}}{dt} \tag{7.37}$$

with $n = 1, \ldots, N_s$ and $\lambda_o = 0$.

$$\frac{dm_r}{dt} = \left[j_i + N_{SUB}(1-\lambda_{N_s}) + V_{gj}\right](\rho_{r-1} - \rho_r) \tag{7.38}$$

with $r = 1, \ldots, N_R$.

$$\frac{dm}{dt} = j_i - \rho_e v_e \tag{7.39}$$

$$\frac{dI}{dt} = \Delta p - K_i j_i^2 - K_e\rho_e v_e^2 - K_R\rho_{N_R}v_{N_R}^2 - \frac{m}{Fr} - \frac{m_R}{Fr} \tag{7.40}$$

and the following auxiliary algebraic equations complete the set:

$$j_{\rm i} = \frac{I - (1-m)\left[\frac{N_{\rm SUB}(1-\lambda_{N_{\rm s}})}{1/\rho_{\rm e}-1} - V_{\rm gj}\right] - m_{\rm R}\left[N_{\rm SUB}(1-\lambda_{N_{\rm s}}) + V_{\rm gj}\right] - V_{\rm gj}\lambda_{\rm R}}{m + m_{\rm R}}, \quad (7.41)$$

$$v_{\rm e} = j_{\rm i} + N_{\rm SUB}(1-\lambda_{N_{\rm s}}) - V_{\rm gj}\left(\frac{1}{\rho_{\rm e}} - 1\right), \quad (7.42)$$

$$m_{\rm R} = \sum_{r=1}^{N_{\rm R}} m_r \quad (7.43)$$

The mixture density at the channel exit, $\rho_{\rm e}$, and at the riser nodes, ρ_r, are related to the state variables by Eqs. (7.18) and (7.32). However, these equations are implicit in $\rho_{\rm e}$ and ρ_r. There are several ways to retrieve the latter. Equations (7.18) and (7.32) can be inserted in Eqs. (7.38) and (7.39), transforming the densities into state variables. Alternatively, Eqs. (7.18) and (7.32) can be solved iteratively in each step. A simpler way, regarding that Eqs. (7.18) and (7.32) are already approximations based on the method of residuals, is to rewrite both equations with relations explicit in $\rho_{\rm e}$ and ρ_r. A good approximation can be obtained by noting that $\xi \ln \xi/(\xi - 1) \cong \sqrt{\xi}$ for $0 \le \xi \le 1$ (see Fig. 7.12). Accordingly, setting $\xi = \rho_{\rm in}/\rho_{\rm out}$, where $\rho_{\rm in}$ and $\rho_{\rm out}$ are densities at the inlet and outlet of a given two-phase interval, Eqs. (7.18) and (7.32) become:

$$\rho_{\rm e} = \left(\frac{m - \lambda_{N_{\rm s}}}{1 - \lambda_{N_{\rm s}}}\right)^2 \quad (7.44)$$

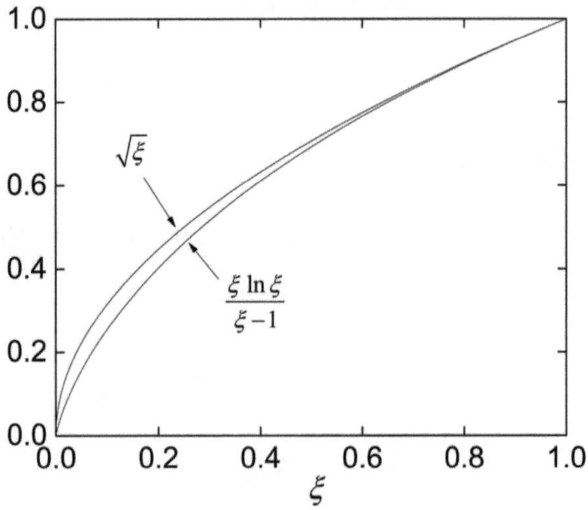

Fig. 7.12 Approximation used to derive Eqs. (7.42) and (7.43)

$$\rho_r = \frac{1}{\rho_{r-1}} \left(\frac{m_r N_\mathrm{R}}{\lambda_\mathrm{R}} \right)^2; \quad r = 1, \ldots, N_\mathrm{R} \quad \text{and} \quad \rho_1 = \rho_\mathrm{e} \tag{7.45}$$

The set of nonlinear ordinary differential Eqs. (7.37)–(7.40), together with the algebraic Eqs. (7.41)–(7.45), can then be solved with any numerical package for stiff problems.

Given that the MNM is an initial value problem, a set of initial conditions for each state variable should be provided. It is convenient to start with a small perturbation from the steady state. The latter is determined by the parameters N_SUB, N_PCH, Fr, V_gj, K_i, K_e, K_R, λ_R, N_s, and N_R. Accordingly, the steady state is calculated as:

$$\bar{j} = \frac{N_\mathrm{SUB}}{N_\mathrm{PCH}} \tag{7.46}$$

$$\bar{\lambda}_n = \frac{n}{N_\mathrm{s}} \bar{j} \tag{7.47}$$

$$\bar{\rho}_\mathrm{e} = \frac{\bar{j} + V_\mathrm{gj}}{\bar{j} + V_\mathrm{gj} + N_\mathrm{SUB}\left(1 - \bar{j}\right)} \tag{7.48}$$

$$\bar{v}_\mathrm{e} = \bar{j} + N_\mathrm{SUB}\left(1 - \bar{j}\right) - V_\mathrm{gj}\left(\frac{1}{\bar{\rho}_\mathrm{e}} - 1\right) \tag{7.49}$$

$$\bar{m} = \bar{j} + \left(1 - \bar{j}\right)\sqrt{\bar{\rho}_\mathrm{e}} \tag{7.50}$$

$$\bar{m}_r = \frac{\lambda_\mathrm{R} \bar{\rho}_\mathrm{e}}{N_\mathrm{R}} \tag{7.51}$$

$$\bar{m}_\mathrm{R} = \lambda_\mathrm{R} \rho_\mathrm{e} \tag{7.52}$$

$$\Delta p = K_\mathrm{i} \bar{j}^2 + K_\mathrm{e} \bar{\rho}_\mathrm{e} \bar{v}_\mathrm{e}^2 + \frac{\bar{m} + \bar{m}_\mathrm{R}}{Fr} \tag{7.53}$$

7.4.2 Low Power Oscillations at Low Fr *Numbers in a Heated Channel with Adiabatic Riser*

In order to picture how the presence of an adiabatic riser influences the nonlinear dynamics of a heated channel, Eqs. (7.35)–(7.43) were solved using the FORTRAN Livermore Solver for Ordinary Differential Equations, LSODE (Radhakrishnan and Hindmarsh 1993). A heated channel with a riser 30 % of the channel length was analyzed. The set of parameters are listed in Table 7.1. To prevent numerical instabilities, it is recommended to choose the values of N_s and N_R as to keep similar steady-state lengths for every segment of the system (i.e., riser, subcooled, and boiling segments). The stability map of the system is shown in Fig. 7.13. Note that the Froude number is low, so the operating conditions can be imposed by natural circulation. The external pressure drop is equivalent to having a downcomer with

Table 7.1 Characteristic parameters of the channel-riser system

Parameter	Fr	K_i	K_e	K_R	λ_R	N_s	N_R
Value	0.003	60	10	0	0.3	10	2

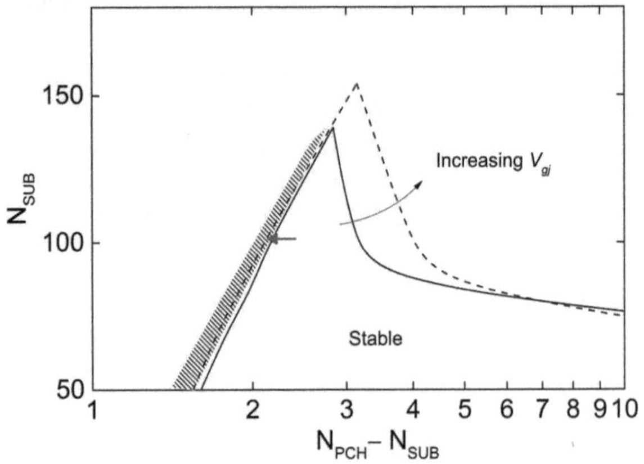

Fig. 7.13 Stability map of the boiling channel with an adiabatic riser for $V_{gj} = 0$ (*solid*) and 0.05 (*dashed*). The control parameters given in Table 7.1. The shadowed zone corresponds to complex nonlinear dynamics of limit cycles, quasi-periodic tori and chaotic attractors. The *red arrow* shows the path studied in Figs. 7.14, 7.15, 7.16, and 7.17

length $\lambda_D \cong 1.35$, with variations in the second decimal place depending on the input power.

There is a region of instability at low N_{PCH} (i.e., low power) like in the case of the channel without riser shown in Fig. 7.10, which merges with the density-wave instability margin coming from the high-N_{PCH} side. However, the maximum stable N_{SUB} is significantly lower when a riser is added to the system. Also, it should be mentioned that when the length of the riser is large, the pressure can vary significantly along the system. Under these conditions, the pressure dependence of the fluid properties cannot be neglected, specially the specific enthalpies. In such cases, flashing can occur, which requires a different treatment. Readers interested in this phenomenon might refer to Riznic and Ishii (1989), Ruspini et al. (2014) and references therein.

The nonlinear dynamics of the low-power low-Fr instability in a boiling channel with a riser is quite complex. Similar behaviors are present without a riser, but in this case the flow oscillations lead to either periods of flow reversal or transient excursions of the boiling boundary out of the heated length, which invalidate the moving-nodes model assumptions. Actually, the presence of a riser does not preclude these excursions per se, but only some special combinations of the control parameters yield sustained nonlinear oscillations bounded within the range of validity of the model. Nevertheless, experimental confirmations of the nonlinear features of this kind of instabilities, either with or without flow reversals and boiling-boundary excursions, have been reported (Delmastro and Clausse 1994).

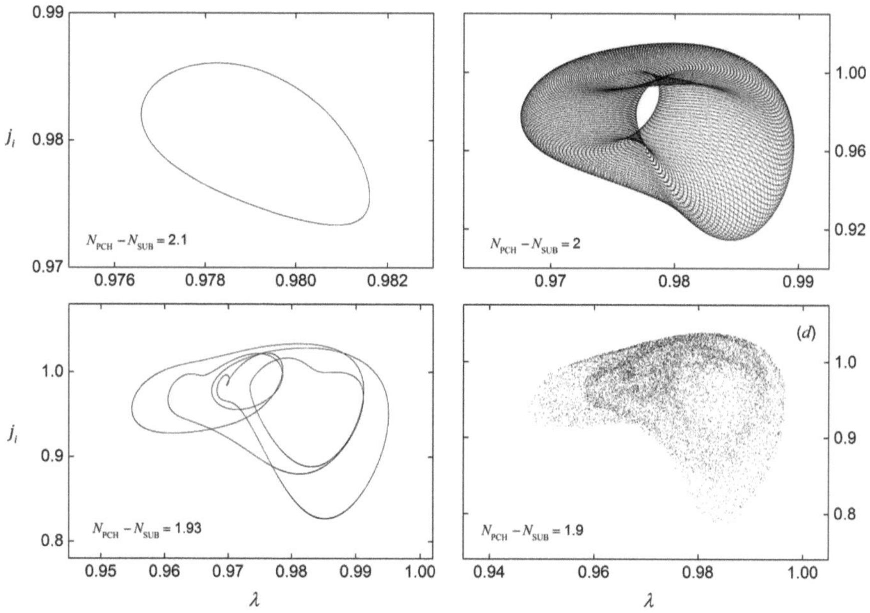

Fig. 7.14 Sequence of phase-plane invariants for the parameters listed in Table 7.1 and $N_{SUB} = 100$, $V_{gj} = 0$

Figure 7.14 shows the sequence of phase-plane plots produced by the MNM for the parameters listed in Table 7.1, for $V_{gj} = 0$. The control parameter of the sequence is $(N_{PCH} - N_{SUB})$, which corresponds to a right-to-left horizontal trajectory in Fig. 7.13 along $N_{PCH} = 100$. The trajectory starts at $(N_{PCH} - N_{SUB}) = 2.1$, just to the left of the stability margin, and it proceeds by decreasing the control parameter into the unstable region (i.e., the dashed region in Fig. 7.14). Each graphic displays the asymptotic invariants reached after 5000 time steps, plotting the stroboscopic values of the inlet velocity and subcooled length for integer intervals of time, i.e., intervals separated by the characteristic time lag t_o defined in Eq. (6.36), namely:

$$t_o = \frac{L_H A_{xs}}{Q} \rho_1 (h_1 - h_i) \tag{7.54}$$

where L_H and A_{xs} are the length and the cross-section area of the heated channel, Q is the input power, and the subscripts 1 and i refer to the saturated liquid and the inlet conditions.

The oscillation starts with a limit cycle whose particular feature is that the velocity tends to drop as the boiling boundary approaches the end of the channel (i.e., $\lambda = 1$). This behavior should be contrasted with the opposite trend encountered in Fig. 7.7 for higher N_{PCH}, and it is due to the reduction of the buoyancy force, which is the main driving mechanism at low Froude numbers. Decreasing N_{PCH} the oscillation becomes quasi-periodic ($N_{PCH} - N_{SUB} = 2$), followed by a window of limit cycles of higher period ($N_{PCH} = 1.93$), and finally chaotic aperiodic attractors ($N_{PCH} - N_{SUB} = 1.9$).

When drift flux is included in the model richer nonlinear features arise. Fig. 7.15 shows the sequence of phase-plane plots produced for the conditions listed in Table 7.1 and $V_{gj} = 0.05$. Decreasing N_{PCH} the oscillations reach first a limit

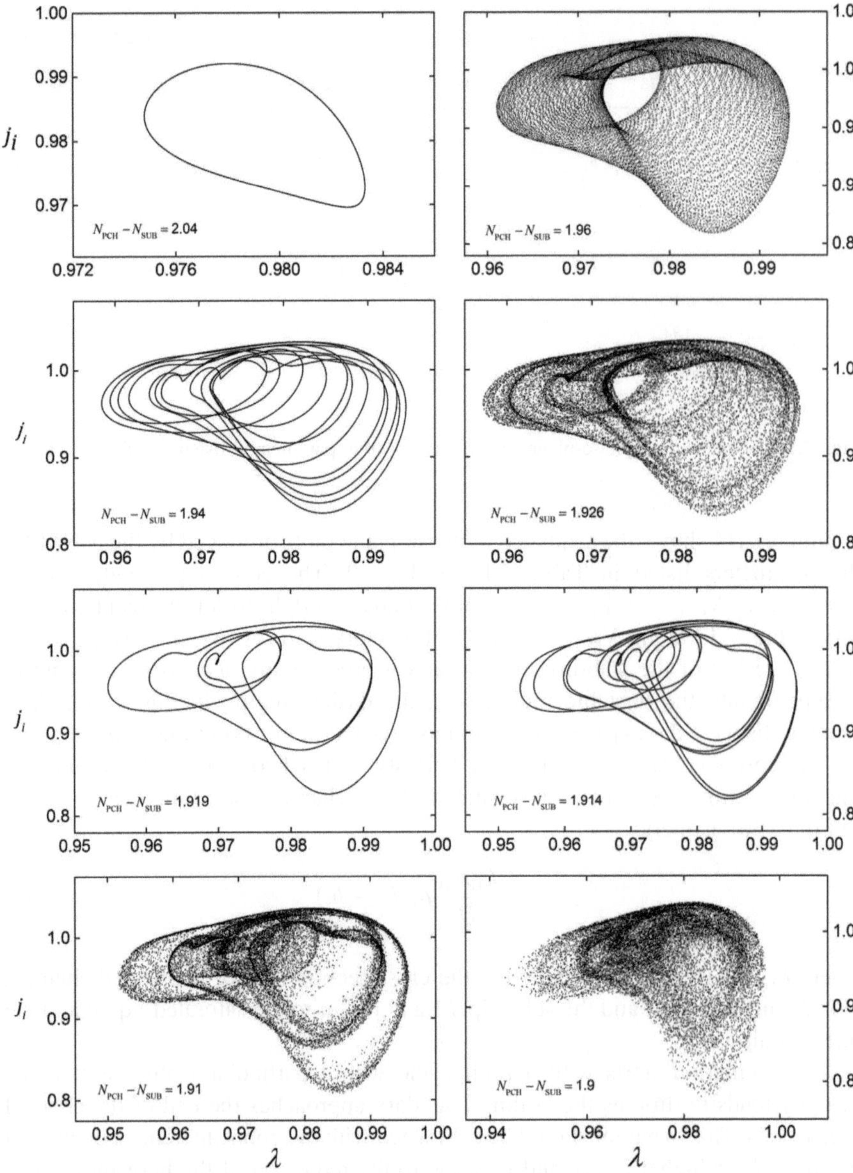

Fig. 7.15 Sequence of phase-plane invariants for the parameters listed in Table 7.1 and $N_{SUB} = 100$, $V_{gj} = 0.05$

cycle ($N_{PCH} - N_{SUB} = 2.04$) and then a quasi-periodic torus ($N_{PCH} - N_{SUB} = 1.96$), similar to the homogeneous model. For lower N_{PCH}, limit cycles with different periods (12, 6, 10 loops) are encountered, alternating with bands of chaotic oscillations.

Further insight of the road map of the sequence may be achieved by means of Poincaré maps. These are stroboscopic views produced by plotting the intersections of the trajectories in the phase space with a lower dimensional subspace, generally defined by one of the variables taking a singular constant value. In the present case, the stroboscopic visualization is defined by the boiling boundary passing through its steady-state value. Accordingly, every Poincaré time, t_p, at which the condition $\lambda(t_p) = N_{SUB}/N_{PCH}$ is met, the instantaneous value of the other state variables is recorded. In order to capture the invariant structure, this procedure should start after the system reaches a stationary state, disregarding the initial transient. Figure 7.16 shows the bifurcation diagram obtained by plotting the values of the inlet velocity, j_i, at the Poincaré section as function of the control parameter. This bifurcation diagram resembles that of the logistic map (Fig. B.36). Following a path from right to left, i.e., decreasing N_{PCH}, the system starts with a single level representing the stable steady state $\bar{j}$. At the stability boundary there is a Hopf bifurcation, generating two Poincaré branches corresponding to the intersection of a single-loop limit cycle with the Poincaré section. Reducing N_{PCH} further the oscillations become quasi-periodic, revealed as bands in the diagram. Finally, for $(N_{PCH} - N_{SUB}) \lesssim 1.94$ regions of limit cycles with several loops alternate with bands of quasi-periodic and chaotic oscillations.

The geometric structure of the nonlinear oscillations can be revealed by projecting the Poincaré map on a plane of state variables. Figure 7.17 shows the Poincaré maps recorded for the Poincaré sections indicated by the arrows in Fig. 7.16, depicting the simultaneous values of inlet velocity and channel mass. Limit cycles display finite sets of points depending on the number of loops that the

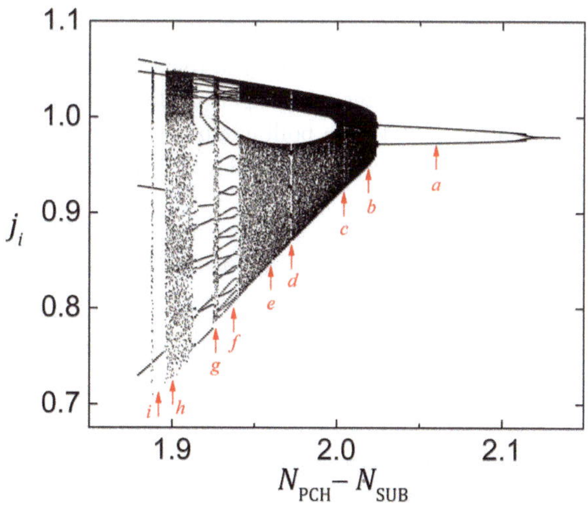

Fig. 7.16 Bifurcation diagram for the parameters listed in Table 7.1 and $N_{SUB} = 100$, $V_{gj} = 0.05$. The points correspond to the values that the variable j_i takes each time that the boiling boundary passes through its steady-state position $\bar{j} = N_{SUB}/N_{PCH}$. The *arrows* indicate the sections corresponding to the Poincaré maps depicted in Fig. 7.17

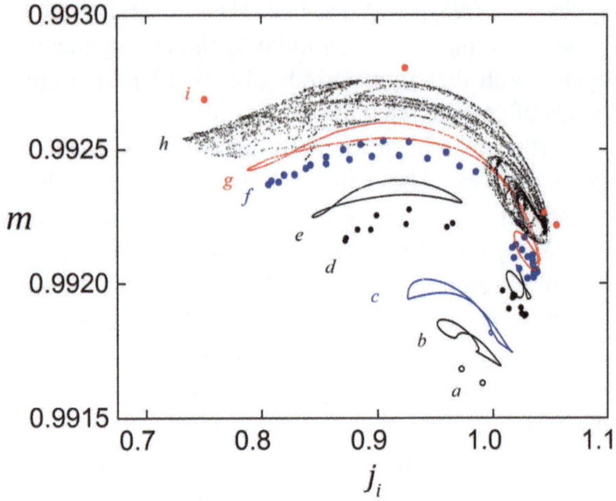

Fig. 7.17 Poincaré map for the parameters listed in Table 7.1 and $V_{gj} = 0.05$. The points correspond to the simultaneous values of the inlet velocity j_i and the channel mass m every time that the boiling boundary passes through its steady-state position $\bar{j} = N_{SUB}/N_{PCH}$. Each attractor is labeled with a letter corresponding to the arrows shown in Fig. 7.16

trajectory takes to complete its orbit (*a*, *d*, *f* and *i*). Note that the limit cycle *d* appears in a very narrow window interrupting a band of quasi-periodic attractors. Quasi-periodic oscillations produce closed curves in the map (*b*, *c*, *e* and *g*), which correspond to the intersection of the torus formed by the bundle of trajectories with the Poincaré section. Note that the section of the torus *e* is divided in two, corresponding to the two bands of the bifurcation diagram with $(N_{PCH} - N_{SUB})$ varying between 1.94 and 1.99. In turn, the torus *g* is found in a narrow band interrupting a region of limit cycles. Finally, chaotic attractors are revealed in a Poincaré map by the appearance of fractal dust sets, as the one labeled *h* in Fig. 7.16.

7.4.3 Experimental Validation of Quasi-periodic Oscillations

It is now shown that the nonlinear dynamic features predicted by the MNM are in qualitative agreement with experimental measurements performed by Delmastro and Clausse (1994) in a boiling channel coupled with a riser of similar characteristics as the one specified in Table 7.1. The diagram of the experimental setup is shown in Fig. 7.18.

The inlet velocity $j_i(t)$ was measured by means of the pressure drop in an orifice plate, and the movement of the boiling boundary was calculated as:

$$\lambda(t) = \int_{t-1}^{t} j_i(t')\, dt' \tag{7.55}$$

The experiments were performed at atmospheric pressure maintaining the water inlet temperature constant at 82.6 °C, corresponding to $N_{SUB} = 48$, decreasing the input power in small steps starting with an initial power $Q = 2500\,\text{W}$. In each step

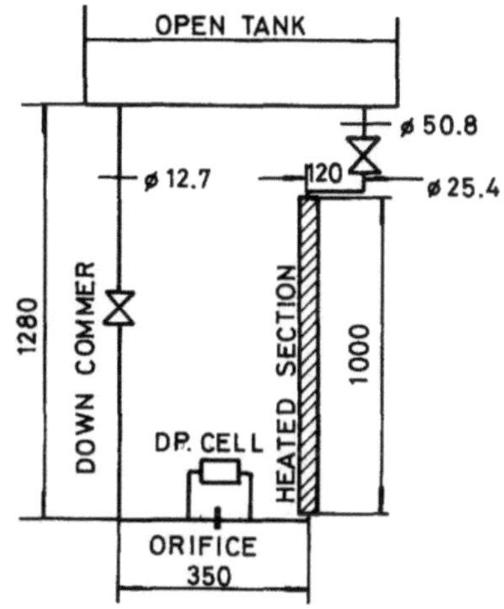

Fig. 7.18 Diagram of the experimental setup (Delmastro and Clausse 1994)

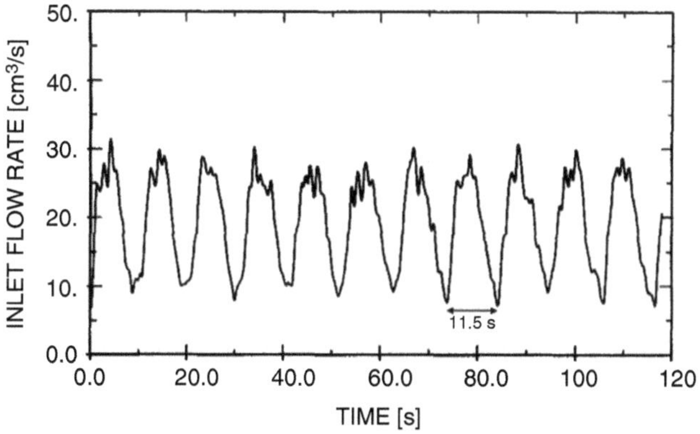

Fig. 7.19 Measurement of the sustained oscillation of the inlet flow rate in a boiling channel coupled with a riser under natural circulation. Input power = 1360 W, $N_{SUB} = 48$

the system was allowed to reach the corresponding stationary state (either stable steady state or oscillatory attractor). Decreasing the power below a certain critical value the flow became unstable and began to oscillate. Figures 7.19 and 7.20 show the inlet flow oscillation recorded at $Q = 1360\,\mathrm{W}$ and the corresponding attractor in the phase plane.

With the present simplified MNM it is not possible to represent the exact geometric and friction characteristics of the experimental test section. However, simulations with similar values of friction and riser length show oscillations comparable with the experimental observations which, while not accurate in amplitude, exhibit similar stability behavior and practically the same period. Figures 7.21 and 7.22 show the sustained oscillations and the corresponding phase-plane attractor resulting from a numerical simulation with the present MNM for a heated

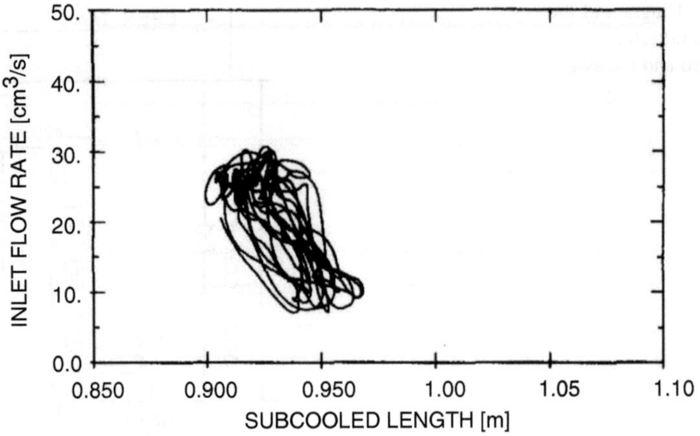

Fig. 7.20 Phase-plane trajectory of the oscillation shown in Fig. 7.19

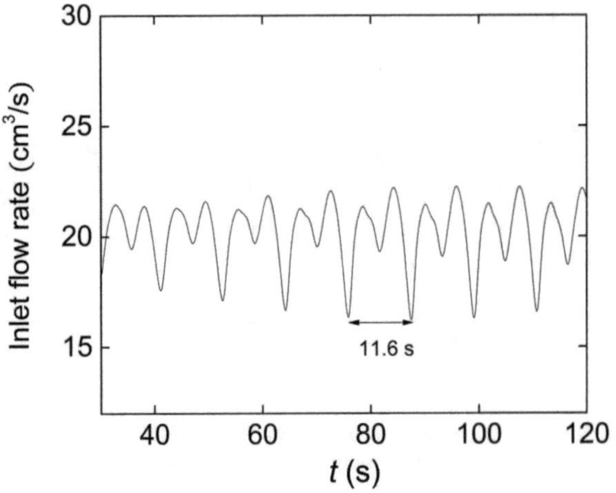

Fig. 7.21 Sustained oscillations of the inlet velocity calculated with the MNM for a heated channel with a riser under conditions consistent with the natural circulation experiment (Delmastro and Clausse 1994). $N_{\mathrm{SUB}} = 50,\ N_{\mathrm{PCH}} = 51.282,\ K_{\mathrm{i}} = 60,\ K_{\mathrm{e}} = 10,\ \lambda_{\mathrm{R}} = 0.3,\ N_{\mathrm{s}} = 10,\ N_{\mathrm{R}} = 2,$ $V_{\mathrm{gj}} = 0.05$

channel and a riser with $N_{\mathrm{SUB}} = 50$ and $V_{\mathrm{gj}} = 0.05$. The theoretical oscillations under those conditions are quasi-periodic.

The time reference t_{o} corresponding to the experiment is:

$$t_{\mathrm{o}} = \frac{L_{\mathrm{H}} A_{xs}}{Q} \rho_1 (h_1 - h_{\mathrm{i}}) = \frac{L_{\mathrm{H}} A_{xs}}{Q} \frac{h_{12}}{v_{12}} N_{\mathrm{SUB}} \tag{7.56}$$

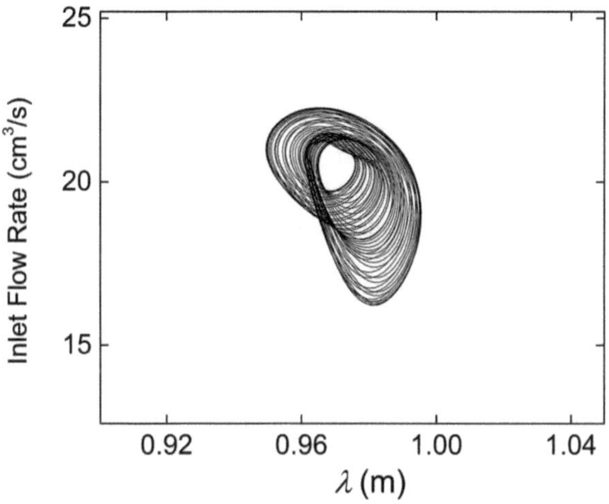

Fig. 7.22 Phase-plane trajectory of the sustained oscillations calculated for a heated channel with a riser under analogous conditions with the natural circulation experiment (Delmastro and Clausse 1994). $N_{SUB} = 50,\ N_{PCH} = 51.282,\ K_i = 60,\ K_e = 10,\ \lambda_R = 0.3,\ N_s = 10,\ N_R = 2,\ V_{gj} = 0.05$

For water at atmospheric pressure, $v_{12} = 1.67\,\mathrm{m}^3/\mathrm{kg}$ and $h_{12} = 2256\,\mathrm{kJ/kg}$, and the channel dimensions are $L_H = 1\,\mathrm{m}$ and $A_{xs} = 2.1\,\mathrm{cm}^2$, giving $t_o = 10\,\mathrm{s}$. The corresponding reference flow rate is 21 cm^3/s. The plots in Figs. 7.21 and 7.22 are presented in metric units using these reference values, for better comparison with the experiment. It can be seen that the period of oscillations has an excellent agreement, whereas the experimental amplitude (58 %) is larger than the theoretical one (16 %). The shape of the phase-plane attractors is similar.

7.5 Summary and Discussion

The Drift-Flux Model was applied to analyze the nonlinear dynamics of boiling channel flows. The homogeneous delay approximation of Clausse et al. (1996) was extended to account for drift-flux effects. A case of density-wave oscillations in a boiling channel subjected to forced circulation is illustrated, showing the establishment of stable limit cycles whose amplitude grows as the operating conditions become more unstable. Furthermore, the model predicts the appearance of subcritical bifurcations in the linearly stable region at low N_{SUB}, where a stable steady state might become unstable if the perturbation is large.

Next the Moving Nodes Model (MNM), used for modeling boiling channel dynamics of nuclear reactors, was extended to incorporate drift flux. The model was applied to a single boiling channel, either under forced circulation (type 2 instabilities) or natural circulation (type 1 instability). Finally, the scope was broadened to derive a complete model of a boiling channel coupled with an adiabatic riser. This system is relevant to recent advanced water-cooled nuclear reactor designs, among other applications. A complex nonlinear dynamic behavior is encountered at operating conditions of incipient boiling at high N_{SUB} under natural circulation. The flow reaches sustained chaotic oscillations interspersed with limit cycles of different periods and quasi-periodic oscillations. Pointcaré maps, displaying stroboscopic snaps of the state variables at the instants when the boiling boundary passes through its steady-state value, were used to characterize the nonlinear invariants.

References

Achard, J.-L., Drew, D. A., & Lahey, R. T., Jr. (1985). The analysis of nonlinear density-wave oscillations in boiling channels. *Journal of Fluid Mechanics, 166*, 213–232.

Ambrosini, W., Di Marco, P., & Ferreri, J. C. (2000). Linear and nonlinear analysis of density wave instability phenomena. *International Journal of Heat and Technology, 18*, 27–36.

Chang, C., & Lahey, R. T., Jr. (1997). Analysis of chaotic instabilities in boiling systems. *Nuclear Engineering and Design, 167*, 307–334.

Clausse, A., Delmastro, D., & Juanicó, L. (1995). A simple delay model for two-phase flow dynamics. In *International Conference on Nuclear Reactor Thermal Hydraulics, NURETH-7, Saratoga, USA* (Vol. 4, pp. 3232–3240).

Clausse, A., Delmastro, D., & Juanico, L. (1996). A simple delay model for density-wave oscillations. *Latin American Journal of Applied Research, 26*, 185–191.

Clausse, A., & Lahey, R. T., Jr. (1990). An investigation of periodic and strange attractors in boiling flows using chaos theory. In *Proceedings of the 9th International Heat Transfer Conference, Jerusalem* (Vol. 2, pp. 3–8).

Clausse, A., & Lahey, R. T., Jr. (1991). The analysis of periodic and strange attractors during density wave oscillations in boiling flows. *Chaos, Solitons & Fractals, 1*, 167–178.

Delmastro, D., & Clausse, A. (1994). Experimental phase trajectories in boiling flow oscillations. *Experimental Thermal and Fluid Science, 9*, 47–52.

Delmastro, D., Juanico, L., & Clausse, A. (2001). A delay theory for boiling flow stability analysis. *International Journal of Multiphase Flow, 27*, 657–671.

DiMarco, P., Clausse, A., Lahey, R. T., Jr., & Drew, D. (1990). A nodal analysis of instabilities in boiling channels. *International Journal of Heat and Technology, 8*, 125–141.

Finlayson, B. A. (1972). *The method of weighted residuals and variational principles with application in fluid mechanics, heat and mass transfer*. New York: Academic Press.

Fukuda, K., & Kobori, T. (1979). Classification of two-phase flow instability by density-wave oscillation model. *Journal of Nuclear Science and Technology, 16*, 95–108.

Garea, V., Drew, D., & Lahey, R. T., Jr. (1999). A moving-boundary nodal model for the analysis of the stability of boiling channels. *International Journal of Heat and Mass Transfer, 42*, 3575–3584.

Juanico, L., Delmastro, D., & Clausse, A. (1998). A fully analytical treatment of Hopf bifurcations in a model of boiling channel. *Latin American Applied Research, 28*, 165–173.

Marcel, C. P., Acuña, F. M., Zanocco, P. G., & Delmastro, D. F. (2013). Stability of self-pressurized, natural circulation, low thermo-dynamic quality, nuclear reactors: The stability performance of the CAREM-25 reactor. *Nuclear Engineering and Design, 265*, 232–243.

Paruya, S., Maiti, S., Karmakar, A., Gupta, P., & Sarkar, J. (2012). Lumped parameterization of boiling channel—Bifurcations during density wave oscillations. *Chemical Engineering Science, 74*, 310–326.

Radhakrishnan, K., & Hindmarsh, A. C. (1993). *Description and use of LSODE, the Livermore solver for ordinary differential equations* (Lawrence Livermore National Laboratory Report UCRL-ID-113855).

Riznic, J. R., & Ishii, M. (1989). Bubble number density and vapor generation in flashing flow. *International Journal of Heat and Mass Transfer, 32*, 1821–1833.

Rizwan-Uddin, & Dorning, J. J. (1986). Some nonlinear dynamics of a heated channel. *Nuclear Engineering and Design, 93*, 1–14.

Rizwan-Uddin, & Dorning, J. (1988). A chaotic attractor in a periodically forced two-phase flow system. *Nuclear Science and Engineering, 100*, 393–404.

Rizwan-Uddin, & Dorning, J. (1990). Chaotic dynamics of a triply forced two phase flow system. *Nuclear Science and Engineering, 105*, 123–135.

Ruspini, L. C., Marcel, C. P., & Clausse, A. (2014). Two-phase flow instabilities: A review. *International Journal of Heat and Mass Transfer, 71*, 521–548.

Theler, G., Clausse, A., & Bonetto, F. (2012). A moving boiling-boundary model of an arbitrary-powered two-phase flow loops. *Mecanica Computacional, 31*, 695–720.

Chapter 8
RELAP5 Two-Fluid Model

Abstract In this chapter, we consider the full 1D TFM of RELAP5 for bubbly vertical flows and assess its linear stability behavior and material wave propagation capabilities in light of the linear stability analyses of Chap. 5, i.e., the characteristics and the dispersion relation. The incomplete virtual mass implementation is the key to the model's void propagation velocity fidelity and regularization, i.e., hyperbolization. We also analyze the numerical convergence.

RELAP5/MOD3.3 (Information Systems Laboratories, RELAP5/MOD3.3 code manual, Vol. 1: Code structure, system models, and solution methods, 2003) is a well-known TFM nuclear reactor safety code used for the analysis of Loss of Coolant Accidents (LOCA) and is representative of other codes used by industry. A linear stability assessment of the RELAP5 code for vertical bubbly flow demonstrates that the RELAP5 TFM is almost unconditionally hyperbolic, i.e., locally stable, because of artificial regularization by a simplified virtual mass force. In spite of this artificial device, a comparison with experimental data shows that the TFM preserves the capability to model the kinematic wave speed correctly. This is a necessary condition for the prediction of the global instabilities addressed in Chaps. 6 and 7.

In industrial practice the KH instability is removed by artificial correlations and numerical viscosity, but a filter may be used instead. A low pass filter, which has a precise cutoff wavelength, is proposed to replace numerical FOU regularization. It offers two advantages with respect to FOU; it is not mesh dependent and it allows finer nodalizations so that numerical convergence may be tested under all circumstances. In addition, higher order numerical schemes may be easier to implement.

8.1 Introduction

When the TFM codes were first developed, almost forty years ago, TFM modeling and nonlinear stability analysis were not as advanced as they are today. Therefore, while the fundamental derivation of the TFM itself is rigorous, the constitutive relations used in industrial codes are often incomplete. For example, TFM codes do not have constitutive models for short wave momentum transfer, e.g., some of the interfacial forces that were analyzed in Chap. 5 are missing and so are the Reynolds stresses. This lack of subscale models makes the TFM codes prone to be KH

M.L. de Bertodano et al., *Two-Fluid Model Stability, Simulation and Chaos*,
DOI 10.1007/978-3-319-44968-5_8

unstable when they shouldn't and ill-posed when that happens, which is the reason for the methods of regularization used to this day, e.g., the present generation of TFM codes employ artificial differential terms and FOU schemes with coarse meshes.

In this chapter characteristic analysis shows how the short wave growth beyond KH is suppressed by artificial means. Nevertheless, in spite of these devices, it is shown that RELAP5 is still capable of predicting the material wave speed accurately and the kinematic wave propagation predictions are validated with the bubbly flow data of Bernier (1982). Therefore, while the local instabilities are arbitrarily removed, the global instabilities, discussed in Chaps. 6 and 7, may be predicted in principle. The problem that remains is that the artificial devices used to regularize the TFM may interfere with these predictions.

One of these artificial devices is the FOU numerical scheme, which is numerically robust but either suppress the KH instability severely or does not converge if the flow becomes KH unstable. Furthermore when used in conjunction with coarse meshes it produces significant numerical viscosity. A low pass filter which achieves both regularization and convergence with a reduced amount of numerical viscosity is analyzed with linear theory.

8.2 Material Waves

8.2.1 RELAP5 Adiabatic Two-Fluid Model

The RELAP5/MOD3.3TFM equations (Information Systems Laboratories 2003) are similar to those of Chap. 5, and we will preserve that notation. However, there are some significant wave propagation differences that need to be discussed. First the RELAP5 TFM is compressible and makes use of the energy equations together with Steam Tables to calculate the densities. This allows RELAP5 to make realistic acoustic two-phase wave predictions (Lafferty et al. 2010) that are not assessed in this book. Second, RELAP5 does not consider the interfacial pressure or the collision force but does consider a simplified virtual mass force. This has a significant impact on the material wave predictions and is the object of the present chapter. There are, of course, countless other differences, but these are the most significant ones from the standpoint of fluid dynamic stability. There is one caveat: the effects of evaporation and condensation on TFM stability, e.g., Kocamustafaogullari (1985), have been neglected. The RELAP5 one-dimensional mass and momentum conservation equations, equivalent to Eqs. (4.1)–(4.4), are

$$\boxed{\frac{\partial}{\partial t}\rho_1\alpha_1 + \frac{\partial}{\partial x}\rho_1\alpha_1 u_1 = -\Gamma,} \tag{8.1}$$

$$\boxed{\frac{\partial}{\partial t}\rho_2\alpha_2 + \frac{\partial}{\partial x}\rho_2\alpha_2 u_2 = \Gamma,} \tag{8.2}$$

$$\boxed{\frac{D_1}{Dt}\rho_1\alpha_1 u_1 = -\alpha_1\frac{\partial p}{\partial x} + \rho_1\alpha_1 g - \frac{2f_1}{D_\mathrm{h}}\rho_1|u_1|u_1 + M^\mathrm{D} + M^\mathrm{VM} - \Gamma(u_\mathrm{i} - u_1),} \tag{8.3}$$

$$\boxed{\frac{D_2}{Dt}\rho_2\alpha_2 u_2 = -\alpha_2\frac{\partial p}{\partial x} + \rho_2\alpha_2 g - M^\mathrm{D} - M^\mathrm{VM} + \Gamma(u_\mathrm{i} - u_2).} \tag{8.4}$$

where Γ is the mass transfer rate, p is the averaged pressure of the two phases, and f_1 is the Fanning friction factor. The mass transfer rate is zero for the adiabatic flow which is analyzed here. RELAP5 uses a simplified formulation of the virtual mass that neglects the spatial derivatives of the objective form of the acceleration. According to the manual (Information Systems Laboratories 2003), "[t]he reason for this change is that inaccuracies in approximating the spatial derivative portion of the term for the relatively coarse nodalizations used in system representations can lead to nonphysical characteristics in the numerical solution." On the other hand, the virtual mass force is completely neglected in TRACE V5 (USNRC 2008) "because of no evidence of significance in reactor safety problems, combined with significant increase in complexity of the solution procedure" (of the complete form). The developers go on to state that inclusion of the virtual mass force will be "reconsidered if higher order numerical methods are introduced into a future version of TRACE and a well-posed set of PDEs becomes important." The difference between the two codes is not trivial since the virtual mass model allows the prediction of the proper sound speed in bubbly flows, Sect. 5.2.2, whereas the lack of it results in the proper sound speed in stratified flows, Sect. 2.2.2. The difference is also important for the prediction of the material wave speed as we shall demonstrate in the next section. We now compare the complete virtual mass model in Eq. (5.5) with that implemented in RELAP5:

$$M^\mathrm{VM} = -\alpha_1\alpha_2\rho_\mathrm{m}C_\mathrm{VM}\left(\frac{\partial u_1}{\partial t} - \frac{\partial u_2}{\partial t}\right) \tag{8.5}$$

Aside from neglecting the spatial derivatives, the other differences between Eqs. (8.5) and (5.5) are the mixture density that replaces the liquid density, i.e., $\rho_\mathrm{m} = \alpha_1\rho_1 + \alpha_2\rho_2$, and the additional factor, α_1. This last term ensures the proper limiting forms at both extremes of the void fraction spectrum, i.e., liquid and vapor flows, since RELAP5 applies virtual mass to all flow regimes. The virtual mass coefficient is given by:

$$C_\mathrm{VM} = \frac{1}{2}\left(\frac{1 + 2\alpha_\mathrm{min}}{1 - \alpha_\mathrm{min}}\right). \tag{8.6}$$

Equation (8.6) was originally derived by Zuber (1964) using the dispersed phase void fraction, α_2, in place of $\alpha_{\min}$. Note that Eq. (8.6) approaches $C_{\mathrm{VM}} = 1/2$ in the limit $\alpha_{\min} \rightarrow 0$, which is the value used throughout Chap. 5.

The question is: how does the modified virtual mass model affect the material waves analyzed in Chap. 5? The material characteristics dictate the motion of void waves that do not necessarily move at either of the phasic velocities. We have seen in Chap. 6 how several global two-phase phenomena such as level swell, density wave oscillations, etc., are governed by void wave propagation, so it is important to get them right. Hence the characteristics of the RELAP5 TFM are analyzed next.

8.2.2 Characteristics

Expanding the derivatives in Eqs. (8.3) and (8.4), the original system of partial differential equations can be cast into non-conservative vector equation form

$$\boldsymbol{A}\frac{\partial}{\partial t}\underline{\phi} + \boldsymbol{B}\frac{\partial}{\partial x}\underline{\phi} = \underline{F}, \tag{8.7}$$

where $\underline{\phi} = [\alpha_2, u_1, u_2, p]^T$ is the vector of the independent variables. The coefficient matrices, obtained from Eqs. (8.1) through (8.5), are

$$\boldsymbol{A} = \begin{bmatrix} 1 & 0 & 0 & 0 \\ -1 & 0 & 0 & 0 \\ 0 & \alpha_2(c_{\mathrm{VM}}\alpha_1\rho_{\mathrm{m}} + \rho_2) & -\alpha_2 c_{\mathrm{VM}}\rho_1 & 0 \\ 0 & -\alpha_2 c_{\mathrm{VM}}\rho_1 & \alpha_1(\rho_1 + \alpha_2 c_{\mathrm{VM}}\rho_{\mathrm{m}}) & 0 \end{bmatrix}, \tag{8.8}$$

$$\boldsymbol{B} = \begin{bmatrix} u_2 & \alpha & 0 & 0 \\ -u_1 & 0 & \alpha_1 & 0 \\ 0 & \alpha_2\rho_2 u_2 & 0 & \alpha_2 \\ 0 & 0 & \alpha_1\rho_1 u_1 & \alpha_1 \end{bmatrix}. \tag{8.9}$$

The vector $\underline{F}$ contains the algebraic source terms, e.g., interfacial drag, that do not factor directly into the characteristic analysis. The characteristic equation to the system defined by Eq. (8.7) is:

$$\mathrm{Det}[\boldsymbol{A} - \lambda\boldsymbol{B}] = 0. \tag{8.10}$$

Equation (8.10) was previously used to determine the characteristics of a TFM using the closures of Pauchon and Banerjee (PB) (1986) provided in Sects. 5.2.2–5.3.1 and 8.2.1, respectively. Therefore, we can compare the results of the RELAP5/MOD3.3 characteristics studied here directly with the more complete

Fig. 8.1 Comparison of the nondimensional characteristics for PB and RELAP5/MOD3.3, reprinted from Fullmer and Lopez de Bertodano (2015) with permission from ANS

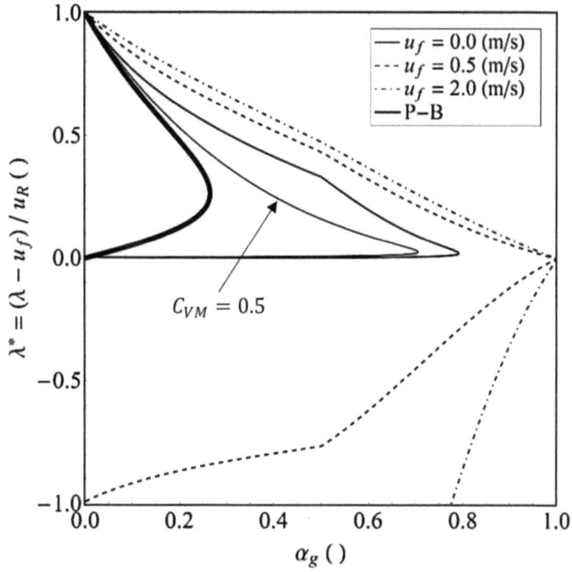

results of Chap. 5. Without additional constitutive modeling, the neglect of both the interfacial pressure and virtual mass—such as in the TRACE V5 TFM—leads to complex characteristics for all but single phase or homogeneous conditions. For the present systems, only the two material characteristics due to the incompressible assumption. When compressibility is included the acoustic roots are always real, cf. Sect. 2.2.2. Therefore, the classification of the system, i.e., well-posed or ill-posed, is solely determined by the material roots. In the nondimensional form $\lambda^* = (\lambda - u_{\mathrm{f}})/u_{\mathrm{R}}$, the characteristics of Pauchon and Banerjee (1986) reduce to an algebraic expression of the void fraction only, cf. Sect. 5.4.2. This is not the case for the nondimensional characteristics of the RELAP5/MOD3.3 model. The two models are compared in Fig. 8.1 for $u_{\mathrm{R}} = 0.5\,\mathrm{m/s}$ and $u_{\mathrm{f}} = 0 - 2\,\mathrm{m/s}$.

Several important observations become immediately apparent:

1. The primary (i.e., faster) roots of the two models are quite similar for low liquid velocity, dilute bubbly flows. A significant source of discrepancy in this region is simply the difference in the virtual mass coefficients. This can be seen when Eq. (8.6) is replaced by $C_{\mathrm{VM}} = 0.5$ in the RELAP5/MOD3.3 model.
2. The difference between the two models increases as the liquid velocity increases. The primary root remains relatively well behaved. The secondary root, on the other hand, diverges from the PB model in a rather dramatic fashion. Fortunately, this discrepancy is of relatively minor consequence, however, as the secondary root has a much larger attenuation rate and is very difficult to observe experimentally (Park et al. 1990, 1998). As the relative velocity increases for a given liquid velocity, the agreement of the secondary root improves.

3. The RELAP5/MOD3.3 model is well-posed, i.e., the characteristics are real, for higher void fractions than the PB model. While it is not unconditionally well-posed, the ill-posed region is isolated to a narrow range of low liquid velocity and high gas void fraction conditions.

While there are some clear differences between the two models, the agreement of the primary roots is quite a remarkable and fortuitous result—especially considering the nontrivial simplifications made in the RELAP5/MOD3.3 model. It is worth reiterating that further simplification by neglecting the virtual mass altogether eliminates the real characteristics and such a comparison is not possible.

8.2.3 Bernier's Experiment

While the RELAP5/MOD3.3 characteristics have been sorted out, it is worthwhile to validate them with the experimental kinematic wave speed data of Bernier (1982). In this experiment, liquid water is pumped in to the bottom of a 10.16 cm diameter PVC pipe. Slightly downstream of the pump is a sparger, constructed of an array of 3.18 cm diameter brass tubing drilled with 0.40 mm diameter holes, which injects 5 mm bubbles into the flow. It was visually verified that the bubble size was relatively uniform. The two-phase mixture runs approximately 2.6 m vertically up where it exits at atmospheric pressure into a discharge tank. In the test section, two impedance meters are separated by a distance of approximately 0.84 m. The cross-correlation of the two signals is used to determine the time lag which gives the material wave speed.

In the previous section, the relative velocity was fixed at $u_R = 0.5$ m/s for comparison to the PB characteristics. However, when comparing to data a better estimate is required. Here, the relative velocity will be determined as a function of the void fraction using the Drift-Flux Model (Zuber and Findlay 1965; Ishii 1977). When the gas drift velocity is expressed in void-weighted mean form (Ishii and Hibiki 2006; Ishii 1977), it can be manipulated to give an expression for the gas velocity,

$$u_2 = \frac{C_0 j_1 + V_{gj}}{1 - C_0 \alpha_2} \tag{8.11}$$

The liquid flux is taken from the experimental data so that the liquid velocity is simply given by $u_1 = j_1/(1 - \alpha_2)$. Then, after the Drift-Flux Model closure relations V_{gj} and C_0 are prescribed, the characteristics become independent of all variables except the void fraction.

Although RELAP5/MOD3.3 is based on a TFM, which was previously outlined, the Drift-Flux Model is used for closure of the interfacial drag force in vertical bubbly and slug flows. For large diameter pipes ($D > 8$ cm), the drift velocity of Zuber and Findlay (1965),

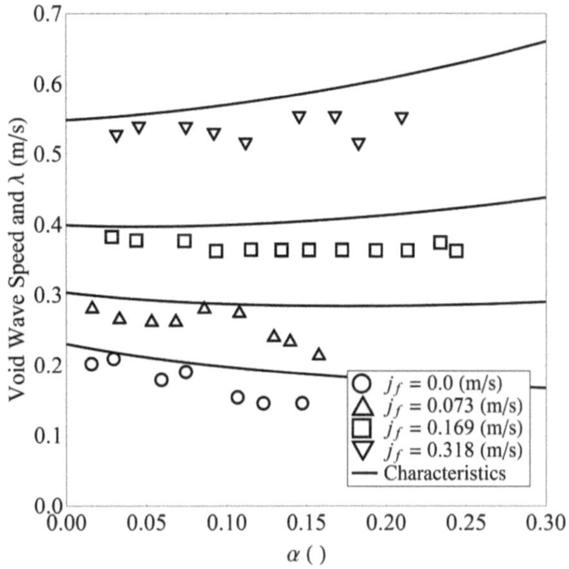

Fig. 8.2 RELAP5/MOD3.3 characteristics using Eq. (8.12) compared with Bernier data, reprinted from Fullmer and Lopez de Bertodano (2015) with permission from ANS

$$V_{gj}^{ZF} = 1.41\left[\frac{\sigma g(\rho_1 - \rho_2)}{\rho_1^2}\right]^{1/4}, \tag{8.12}$$

is used for low liquid flux conditions (ISL 2003). The low liquid flux condition is defined by $j_g^+ \leq 0.5$, where $j_g^+ = \sqrt{2}j_g/V_{gj}^{ZF}$. For higher liquid fluxes, $j_g^+ \leq 1.768$, the drift velocity is given by the correlation of Kataoka and Ishii (1987) and linear interpolation is used in between. The distribution parameter is set to unity, $C_0 = 1$, consistent with the observations of Bernier (1982) that the cross-sectional bubble distribution remained relatively uniform for the reported conditions.

The resulting closed-form characteristics of the RELAP5/MOD3.3 model are compared to the Bernier (1982) data in Fig. 8.2. In general, the comparison is very good, considering the simplifications made to the RELAP5/MOD3.3 TFM, though there is a slight overprediction by the model. As it turns out, the largest source of this error is not the simplified model itself, but the choice of drift velocity.

The correlation of Zuber and Findlay (1965) corresponds to churn-turbulent bubbly flow in which bubble–bubble interactions produce large cap bubbles (Ishii and Hibiki 2006). A more accurate description of the flow described in the Bernier (1982) experiment—where the liquid flux is relatively low and the bubbles are carefully introduced with a sparger—is distorted bubbly flow. Ishii (1977) gives a correlation,

$$V_{gj} = V_{gj}^{ZF}\alpha_1^{1.75}. \tag{8.13}$$

For the drift velocity in distorted bubbly flow which is very similar to the churn-turbulent drift velocity, Eq. (8.12) used previously except that a void fraction

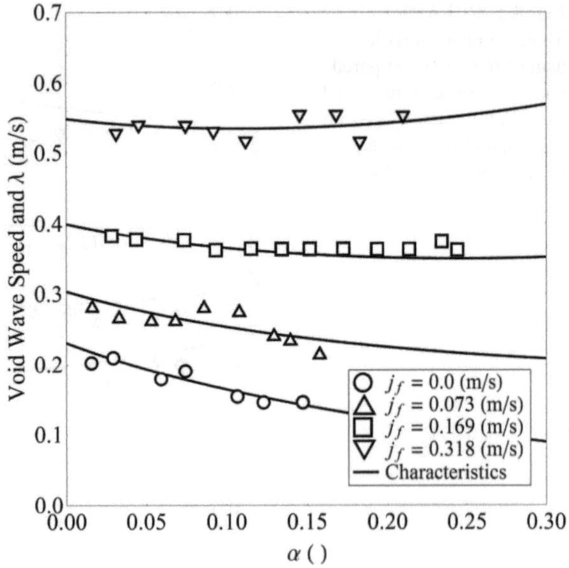

Fig. 8.3 RELAP5/MOD3.3 characteristics using Eq. (8.13) compared with Bernier data, reprinted from Fullmer and Lopez de Bertodano (2015) with permission from ANS

dependence has been introduced. In fact, Ishii (1977) shows that even in the churn-turbulent regime, the drift velocity is void fraction dependent. However, its dependence is much weaker, $V_{gj} \propto (1-\alpha_2)^{1/4}$, and is typically neglected. Figure 8.3 shows the same comparison as Fig. 8.2, here using Eq. (8.13) to determine the gas velocity. The agreement is outstanding again considering the simplified form of the TFM.

Fullmer and Lopez de Bertodano (2015) performed simulations of Bernier's experiment using RELA5/MOD3.3. The kinematic waves were created with a 0.5 s small pulse in the gas velocity. In accordance with the previous findings, the void fraction dependence of the dispersed bubbly drift velocity, Eq. (8.13), is added to the code. The results of the slightly modified code identified as RELAP5/MODdb are compared to the Bernier data (1981) in Fig. 8.4. The agreement between the simulation results and the data is again very good. However, it is apparent that it deteriorates with increasing liquid flux. This behavior was not observed in the characteristic analysis. The difference stems from the distribution parameter. In the characteristic analysis C_0 was set to unity for all liquid fluxes based on experimental observations. In the RELAP5/MOD3.3 code, the distribution parameter depends on the total mass flux. Of course, this could also be modified to produce a better fit for this particular data set. However, the uniform distribution up to $j_1 = 0.32\,\mathrm{m/s}$ is most likely due to the careful bubble injection through the sparger, which is not typical for most reactor applications. Finally, for the two higher liquid flux cases, an abrupt transition occurs where the wave speed increases drastically. This transition in the code occurs at $j_2^+ = 0.5$, beyond which interpolation with the drift velocity model of Kataoka and Ishii (1987) begins.

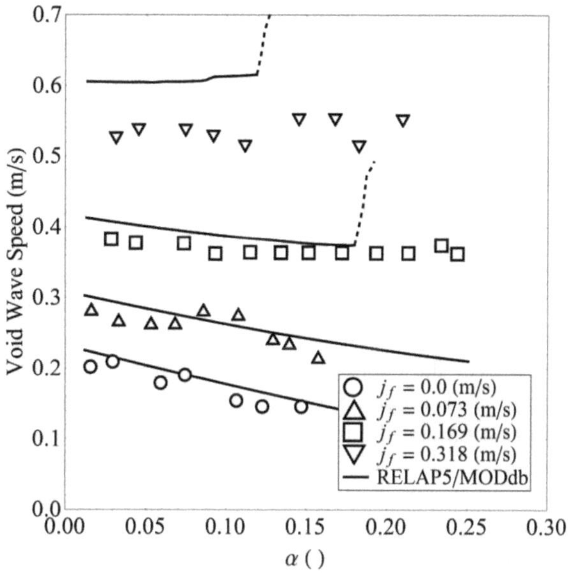

Fig. 8.4 RELAP5/MOD3.3 simulations using Eq. (8.13) compared with Bernier data, reprinted from Fullmer and Lopez de Bertodano (2015) with permission from ANS

8.3 Low Pass Filter Regularization of the TFM

In theory, stabilization of the TFM via physical models is preferable to artificial regularization, but physical stabilization for all flow regimes appears impractical for several reasons. Many of the local nonlinear stabilization mechanisms for the various flow regimes are not well understood and those that are understood may be difficult to constitute due to the type of local-instant data needed. Additionally, the prevalent static flow regime maps in industrial codes already account for the KH instability implicitly. The simplest way to remove local TFM instabilities from a dynamics point of view would be to use the Drift-Flux Model. However, when the TFM has been adopted and artificial regularizations are necessary, as in present-day nuclear reactor TFM safety codes, the practice is to use a variety of artificial and numerical regularizations that involve significant guess work. It would be desirable then to adopt a regularization scheme that reduces the guesswork and preserves the global instabilities capability. This means that in addition to modeling the material wave speed accurately it is also necessary to minimize the artificial or numerical damping of the long waves.

While artificial regularizations vary between codes, first-order upwinding (FOU) with a coarse mesh is by far the most widespread. A prevalent view of the ill-posedness of the 1D TFM is that it is not a substantial problem because, in practice, the discrete equations determine whether or not the model is ill-posed and regularization is achieved through numerical viscosity, i.e., the leading order term of the truncation error in a FOU scheme. However, one shortcoming of this combination of FOU with an ill-posed TFM is that the solution does not converge beyond KH because the numerical viscosity is proportional to the grid size, cf. Sect. 2.5.2.2.

In this section, we develop a low pass filter that allows convergence while eliminating the need for numerical viscosity. Furthermore, it separates the artificial regularization from the physics of the TFM more distinctly so it may be possible to preserve global instability accuracy.

Numerical viscosity produces a combination of linear growth and decay via a cutoff wavelength. This approach has the advantage of being flow regime and geometry independent. However, the cutoff wavelength is dependent on the grid size, which will be demonstrated herein. Each time the grid is refined a new cutoff and critical (i.e., most dangerous) wavelength appears, which results in different solutions for each different grid. Such a scenario renders a convergence study for code verification impossible. It has been argued by some that averaging the 3D TFM over the channel cross-section imposes a characteristic filter length on the resulting 1D model. Therefore, the validity of phenomena below this length scale is in question and the model should not be extrapolated into $\Delta x < D_{\mathrm{h}}$. This type of argument has produced several 'user guidelines' for the numerical TFM practitioner (ISL 2003). A guideline that limits the node size is problematic for three reasons: (1) even if the information for $\Delta x < D_{\mathrm{h}}$ should not be utilized for an engineering application, verification exercises are important to demonstrate confidence in the numerical solution scheme via convergence and consistency; (2) simple guidelines do not prevent different code users from violating them, inadvertently or intentionally; and (3) the resolution of the governing equations not only depends on the mesh size, but it also depends on the time step, flow conditions, material properties, and the order of accuracy of the solution scheme. Therefore, it is proposed to add higher order nonphysical terms to the governing equations instead (Fullmer et al. 2014). This method can be equivalently viewed as filtering the governing equations. The preferable filter method appears to be adding second-order artificial viscosity terms to each transport equation[1] with a clearly defined cutoff wavelength. A uniform artificial viscosity for the TFM was used by Holmås et al. (2008) and later refined to be wavenumber dependent using a spectral solution method. The artificial viscosity concept discussed later was extended independently by Vreman (2011) and Fullmer et al. (2014) to create a filter that prescribes the cutoff length scale precisely.

8.3.1 Dispersion Analysis

The incompressible isothermal one-dimensional TFM given by Eqs. (4.1)–(4.4) is simplified to the bare essential by setting $g_y = 0$ and $\sigma = 0$, which corresponds with adding diffusion terms to Eqs. (8.1)–(8.4) and removing mass transfer and other algebraic terms,

[1]One could also devise an artificial regularization method utilizing third-order derivatives, i.e., artificial surface tension, or higher derivatives.

$$\frac{D_1\alpha_1}{Dt} + \alpha_1\frac{\partial u_1}{\partial x} = \varepsilon_1\frac{\partial^2\alpha_1}{\partial x^2}, \tag{8.14}$$

$$\frac{D_2\alpha_2}{Dt} + \alpha_2\frac{\partial u_2}{\partial x} = \varepsilon_2\frac{\partial^2\alpha_2}{\partial x^2}, \tag{8.15}$$

$$\rho_1\frac{D_1 u_1}{Dt} = -\frac{\partial p_{2i}}{\partial x} + \rho_1\nu_1\frac{\partial^2 u_1}{\partial x^2}, \tag{8.16}$$

$$\rho_2\frac{D_2 u_2}{Dt} = -\frac{\partial p_{2i}}{\partial x} + \rho_2\nu_2\frac{\partial^2 u_2}{\partial x^2}. \tag{8.17}$$

Applying the simplifying assumptions, $\varepsilon_1 = \varepsilon_2 = \varepsilon$ and $\nu_1 = \nu_2 = \nu$, the dispersion relation is very similar to Eq. (3.13),

$$\frac{\omega_i}{k} = \frac{\widetilde{\rho u}}{\widetilde{\rho}} - \frac{i}{2}(\varepsilon+\nu)k \pm \frac{1}{\widetilde{\rho}}\sqrt{(1-\alpha)\alpha\left[-(u_2-u_1)^2\rho_1\rho_2 - \frac{\widetilde{\rho}^2(\varepsilon-\nu)^2}{4(1-\alpha)\alpha}k^2\right]}, \tag{8.18}$$

where $\alpha = \alpha_2$ and $\widetilde{\rho} = (1-\alpha)\rho_2 + \alpha\rho_1$. The results are shown in Fig. 8.5 using the conditions of Table 8.1 and several different combinations of ε and ν. Except where specifically noted, the linear stability results use the values given in Table 8.1 representing the condition of idealized bubbly flow at 7.6 MPa.

First, when $\varepsilon_k = \nu_k = 0$ the ill-posed result is obtained. The logarithmic coordinates were chosen because for the very simple dispersion relation given by

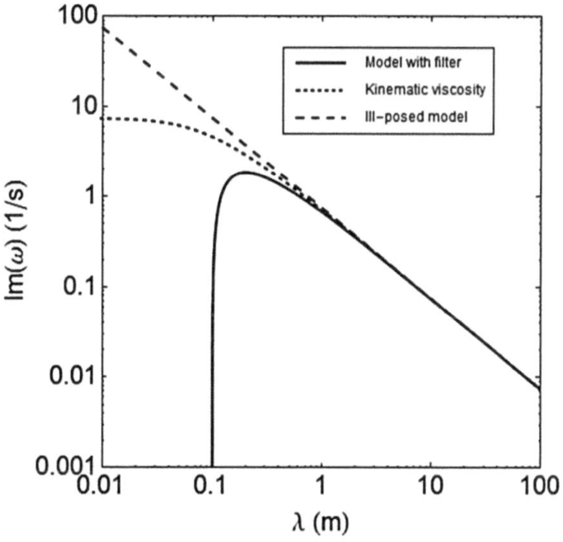

Fig. 8.5 Growth rates of the dispersion analysis, reprinted from Fullmer and Lopez de Bertodano (2015) with permission from ANS

Table 8.1 Properties used for the linear stability analysis

ρ_2 (kg/m^3)	ρ_1 (kg/m^3)	u_2 (m/s)	u_1 (m/s)	α_2	D_H (m)
40	780	2.25	2.0	0.10	0.05

Eq. (8.18) the slope is exactly -1. When either the continuity or momentum viscosities are nonzero, the zero wavelength growth rate becomes finite in the limit $\lambda \to 0$. Figure 8.5 shows the result for either $\nu_1 = \nu_2 = 0.005\,\mathrm{m^2\,s^{-1}}$ or $\varepsilon_1 = \varepsilon_2 = 0.005\,\mathrm{m^2\,s^{-1}}$. While either would create a model that is regularized in a rigorous mathematical sense (the growth rate does not increase unboundedly at $\lambda \to 0$), this does little to help the original nonconvergence issue. With such a numerical discretization, the maximum growth rate will approach an asymptote; but, unfortunately the magnitude and wavelength at which it occurs will continue to be mesh dependent until a very fine mesh is used. The maximum growth rate can be increased or decreased by a corresponding adjustment of viscosities. Even for large values though, the growth will remain finite in the zero wavelength limit.

However, when both ν_k and ε_k are nonzero the linear growth reaches a maximum at a critical wavelength then decreases until the growth turns to decay at a cutoff wavelength. The location cutoff wavelength can be adjusted arbitrarily by modifying the viscosities. With this method, the ill-posed Euler TFM is regularized in a similar fashion to numerical viscosity—for obvious reasons. The important difference is that artificial viscosity coefficients can be set in the differential model to fix the cutoff wavelength, which becomes the asymptotic value of the discrete model as $\Delta x, \Delta t \to 0$.

8.3.2 *Numerical Viscosity*

To investigate the effect of numerical viscosity on the resolution of the 1D TFM, a first-order finite difference discretization of Eqs. (8.1) to (8.4) in non-conservative form and without source terms will be employed. The numerical method is assumed to have a uniform staggered grid with FOU for donored variables, cell centering where applicable and forward Euler time stepping. The hypothetical staggered grid is shown in Fig. 3.3. The resulting Finite Difference Equations (FDEs) are similar to the semi-implicit RELAP5 (ISL 2003) solution method and are given by:

$$\frac{\alpha_{k,i}^{n+1} - \alpha_{k,i}^{n}}{\Delta t} + \frac{\hat{\alpha}_{k,\mathrm{R}}^{n} u_{k,\mathrm{R}}^{*} - \hat{\alpha}_{k,\mathrm{L}}^{n} u_{k,\mathrm{L}}^{*}}{\Delta x} = 0, \tag{8.19}$$

and

$$\frac{u_{k,j}^{n+1} - u_{k,j}^{n}}{\Delta t} + u_{k,j}^{n} \frac{\hat{u}_{k,\mathrm{R}}^{n} - \hat{u}_{k,\mathrm{L}}^{n}}{\Delta x} = -\frac{1}{\overline{\rho}_k} \frac{P_{\mathrm{R}}^{*} - P_{\mathrm{L}}^{*}}{\Delta x}. \tag{8.20}$$

The superscripts denote the time level, n for the old time values and $n+1$ for the new time values, and the subscripts denote the spatial location, as shown in Fig. 3.3. The variables Δx and Δt indicate the spatial discretization and the time step. Since a uniform grid is used, Δx is constant and equivalent for i to $i+1$ and j to $j+1$ spacing. Donored values are indicated by a hat and the overbar is used for simple averaging of adjacent values. Finally the superscript indicates the pressure–velocity coupling. In the TFIT code used for the numerical simulations in Sect. 8.3.4, the coupling is semi-implicit or pressure implicit as discussed in Sect. 3.4.4. To simplify the analysis, the old time values will be used here because it makes no difference to the present argument, i.e., $* = n$. This simplification does not cause any loss of generality since the coupling typically utilizes higher order differences, as discussed in the following section.

In space, the location of the 'right' and 'left' indices, R and L, respectively, depends on the equation. The continuity equation, Eq. (8.19), is cell centered and the right and left of the continuity control volume are located at cell faces or junctions. The opposite is true of the momentum equation: Eq. (8.20) is junction centered and the right and left of the momentum control volume are located at cell centers. Values that are stored at control volume faces are used when available, e.g., velocity in continuity and pressure in momentum. Values that are not must be donored, i.e., extrapolated, using the upwind method. Assuming that $u_k > 0$ in the vicinity of i and j, then the FOU method gives $\alpha_{k,\mathrm{R}} = \alpha_{k,i}$, $\alpha_{k,\mathrm{L}} = \alpha_{k,i-1}$, $u_{k,\mathrm{R}} = u_{k,j+1}$, and $u_{k,\mathrm{L}} = u_{k,j}$. In order to solve for the numerical viscosity, the convective term of the continuity equation is linearized as

$$\frac{\alpha_{k,i}^n u_{k,j+1}^n - \alpha_{k,i-1}^n u_{k,j}^n}{\Delta x} \rightarrow \overline{u}_k \frac{\alpha_{k,i}^n - \alpha_{k,i-1}^n}{\Delta x} + \alpha_{k,i} \frac{u_{k,j+1}^n - u_{k,j}^n}{\Delta x}. \tag{8.21}$$

Then by applying the FOU donoring scheme, i.e., the linearization of Eq. (8.21), and writing every variable in Eqs. (8.19) and (8.20) as they relate to i and j, respectively, the discrete equations take the form

$$\frac{\alpha_{k,i}^{n+1} - \alpha_{k,i}^n}{\Delta t} + \overline{u}_k \frac{\alpha_{k,i}^n - \alpha_{k,i-1}^n}{\Delta x} + \alpha_{k,i}^n \frac{u_{k,i+1/2}^n - u_{k,i-1/2}^n}{\Delta x} = 0, \tag{8.22}$$

and

$$\frac{u_{k,j}^{n+1} - u_{k,j}^n}{\Delta t} + u_{k,j}^n \frac{u_{k,j}^n - u_{k,j-1}^n}{\Delta x} = -\frac{1}{\overline{\rho}_k} \frac{P_{j+1/2}^n - P_{j-1/2}^n}{\Delta x}. \tag{8.23}$$

To calculate the numerical viscosity the discrete equations need to be transformed into modified differential equations by Taylor series expansion. Applying the Taylor series expansion for each term away from the (i,n) or (j,n) spatiotemporal locations, Eqs. (8.22) and (8.23) can be rearranged into the form

$$\left[\frac{\partial \alpha_k}{\partial t}+u_k\frac{\partial \alpha_k}{\partial x}+\alpha_k\frac{\partial u_k}{\partial x}\right]_i^n=0+\left\{\overline{u}_k\frac{\Delta x}{2}\left(\frac{\partial^2 \alpha_k}{\partial x^2}\right)_i^n-\frac{\Delta t}{2}\left(\frac{\partial^2 \alpha_k}{\partial t^2}\right)_i^n\right\}+O(2), \tag{8.24}$$

and

$$\left[\frac{\partial u_k}{\partial t}+u_k\frac{\partial u_k}{\partial x}+\frac{1}{\rho_k}\frac{\partial P}{\partial x}\right]_j^n=0+\left\{u_{k,j}^n\frac{\Delta x}{2}\left(\frac{\partial^2 u_k}{\partial x^2}\right)_i^n-\frac{\Delta t}{2}\left(\frac{\partial^2 u_k}{\partial t^2}\right)_j^n\right\}+O(2). \tag{8.25}$$

Equations (8.24) and (8.25) are the Modified Differential Equations (MDEs) which are actually being solved by the FDEs. The left-hand sides inside the square brackets are the original PDEs of the governing equations that have been recovered about the local i and j locations. However, there are now additional terms on the right-hand side. This is the truncation error which, for the FOU scheme has a leading error of order one, $O(1)$, which is given by the terms in the curly brackets. All other truncation error terms are at least of $O(2)$, i.e., $\Delta x^a \Delta t^b$ such at $a+b \geq 2$.

It is important to note that with the FOU finite difference representation, second derivatives of void fraction and velocity appear implicitly in the MDEs. Additionally, the gradients of the staggered terms do not contribute to the $O(1)$ error, i.e., velocity derivatives in Eq. (8.22) and pressure derivatives in Eq. (8.23). The $O(1)$ truncation error, specifically the coefficients of the derivatives, is commonly referred to as numerical viscosity owing to its similarity with physical viscosity. However, the numerical viscosity is grid and time step dependent and vanishes in the limit $\Delta x, \Delta t \rightarrow 0$. The artificial viscosity model presented in the next section will seek to fix this coefficient explicitly, thus acting like a nonvanishing and invariant numerical viscosity.

The von Neumann analysis of Eqs. (8.24) and (8.25) is very similar to that of Sect. 3.4.5 and reduces to solving the eigenvalues of the amplification matrix, $\boldsymbol{G}=\boldsymbol{M}^{-1}\boldsymbol{N}$,

$$\mathrm{Det}(\boldsymbol{N}-\xi\boldsymbol{M})=0, \tag{8.26}$$

where ξ are the eigenvalues of $\boldsymbol{G}$. The angular frequency is related to the eigenvalues by $\xi=\mathrm{e}^{-i\omega\Delta t}$. In general, the eigenvalues are also complex and can be expressed in polar form $\xi=|\xi|\mathrm{e}^{i\theta}$ where $\theta=\mathrm{Arg}[\xi]$. The modulus of the eigenvalues is given by $|\xi|=\sqrt{\xi\overline{\xi}}$ where $\overline{\xi}$ is the complex conjugate of ξ. The largest modulus of the eigenvalues gives the spectral radius of the growth matrix, G. Then, after some rearrangement, the growth rate is found by taking the natural logarithm of both sides of Eq. (8.26) with ξ in polar form,

$$\mathrm{Im}(\omega)=\frac{\ln(|\xi|)}{\Delta t}. \tag{8.27}$$

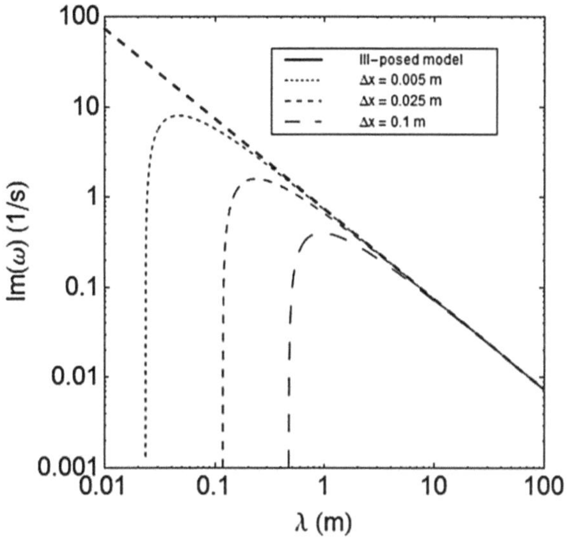

Fig. 8.6 Growth rates of the von Neumann analysis for several values of Δx and the differential limit, reprinted from Fullmer and Lopez de Bertodano (2015) with permission from ANS

Figure 8.6 shows the von Neumann discrete growth rates as a function of the wavelength of the perturbation, $\lambda = 2\pi/k$, beginning at $2\Delta x$ (the von Neumann analysis solution oscillates infinitely as $\lambda \rightarrow 0$, but any continuous function with $\lambda < 2\Delta x$ is not representable on the finite grid and, hence, is meaningless). For each Δx the time step is fixed by setting the gas Courant number equal to unity, i.e., $Co = u_2 \Delta t/\Delta x = 1$ or $\Delta t = \frac{\Delta x}{u_2} s$. The discrete growth rates are compared directly to the linear stability analysis of the differential model, outlined in the previous section, which can be viewed as the asymptote of the discrete analysis in the limit $\Delta x, \Delta t \rightarrow 0$. The classic ill-posedness of the Euler TFM is readily apparent; the growth rate increases exponentially as the wavelength of the perturbation is reduced. However, all of the discrete growth rates reach a maximum, the critical wavelength, and decline rapidly to a damping state. The point of neutral stability is referred to as the cutoff wavelength. Therefore, the otherwise ill-posed differential model has been regularized by the first-order numerical scheme. The robustness of the FOU scheme has been the unsung hero of the TFM in the nuclear reactor safety codes (Pokharna et al. 1997; Krishnamurthy and Ransom 1992).

Unfortunately, the degree of the regularization is not universal. Figure 8.6 also shows that as the grid is refined, i.e., Δx decreases, the critical and cutoff wavelengths decrease as well. This is simply because the regularization has been achieved by the numerical viscosity of the leading order second derivatives in the truncation error, Eqs. (8.24) and (8.25), which are grid dependent. Since the numerical scheme is consistent, the cutoff and critical wavelengths will continue to decrease for smaller Δx and Δt as the discrete equations approach the differential equations in the limit $\Delta x, \Delta t \rightarrow 0$. Practically speaking, this has severe implications on solution convergence, the foundation of code verification. It should be noted that Lax's equivalence theorem (Richtmeyer and Morton 1967) which states that numerical consistency and stability are necessary and sufficient conditions for convergence cannot be applied to this problem because the underlying governing equations are not stable. The following analysis does not provide an analog to Lax's theorem, i.e., it is not proved that the proposed conditions are *sufficient* for convergence. However, it is apparent that a fixed grid independent cutoff and critical wavelengths are *necessary* for convergence. While the heuristic approach of limiting the grid size to the hydraulic diameter, D_H, may suffice for engineering applications, convergence remains an essential objective, even if only for code verification.

8.3.3 Artificial Viscosity Model

Rather than setting the artificial viscosity coefficients arbitrarily, now a model will be sought which can be used to set the cutoff wavelength precisely at some prescribed dimension, ℓ. For this preliminary analysis, it is assumed that the artificial viscosity is the same for both phases and in all four equations, that is, $\varepsilon_k = \nu_k = \nu^A$. This assumption is not a necessary one and, in fact, it may be desirable to minimize ε_k with respect to ν_k or vice versa. However, since all of the viscosities are currently artificial ones, i.e., no viscosities—material or turbulent—are currently utilized for nuclear reactor safety codes, such an approach is not warranted at this point. Additionally, the assumption of uniform artificial viscosity allows for a simplified analysis and ensures that the total volumetric flux is uniform in space, i.e., $j = j(t)$, where $j = j_1 + j_2$ and $j_k = \alpha_k u_k$ (for the incompressible case). With the uniform artificial viscosity assumption, the growth rate resulting from Eq. (8.27) can be simplified to:

$$\mathrm{Im}(\omega) = \rho^* |u_R| k - \nu^A k^2, \tag{8.28}$$

where ρ^* is a dimensionless density parameter given by

$$\rho^* = \frac{\sqrt{\alpha_1 \alpha_2 \rho_1 \rho_2}}{\alpha_1 \rho_1 + \alpha_2 \rho_2}. \tag{8.29}$$

and u_R is the relative velocity. The first term of Eq. (8.28) should be recognizable as the imaginary component of the characteristic root of the basic equations which has been discussed extensively in the literature since the hypothesis of the ill-posed nature of the Euler 1D TFM (Gidaspow 1974). The second term is the viscous effect of the artificial damping that dominates in the short wavelength limit, $\lambda \to 0$ or $\lambda \to \infty$.

The intersection of the elliptic and parabolic roots determines the stability limit. Solving Eq. (8.28) for $\omega_I = 0$ gives

$$k_0 = \frac{2\pi}{\lambda_0} = \frac{\rho^* |u_R|}{\nu^A}. \tag{8.30}$$

Since the viscosity is not an independent parameter, but one that is being designed to give a prescribed cutoff, Eq. (8.30) is solved to give the artificial viscosity,

$$\nu^A = \frac{\rho^*}{2\pi} \ell |u_R|, \tag{8.31}$$

in terms of the filter length, ℓ. The filter defines the cutoff wavelength of the growth rate for all flow conditions, at least for the simplified TFM, and can be set to a characteristic dimension, e.g., a wavelength, mean bubble size, hydraulic diameter, etc. The user guideline which suggests that the 1D TFM should not be solved for length scales below the characteristic cross-sectional dimension given by D_H can be imposed by setting $\ell = 2D_H$. More elaborate filters could be derived to vary the length scale based on the grid size. One example would be $\ell = \min[0, 2(D_H - \Delta x)]$ which would reduce excessive damping caused by the combination of numerical and artificial viscosity, at least for a FOU scheme. Here, the simple definition $\ell = 2D_H$ will be used.

Returning to the von Neumann analysis, the discrete growth rates are again calculated with Eq. (8.27) now using the artificial viscosity of Eq. (8.31) and $\ell = 2D_H$. The conditions are given by Table 8.1 and the time step is again set by specifying the larger of the phasic Courant numbers to be one. The results of the von Neumann growth rates for several Δx and the dispersion analysis are shown in Fig. 8.7. Now, the growth rate of the differential model, i.e., obtained from the

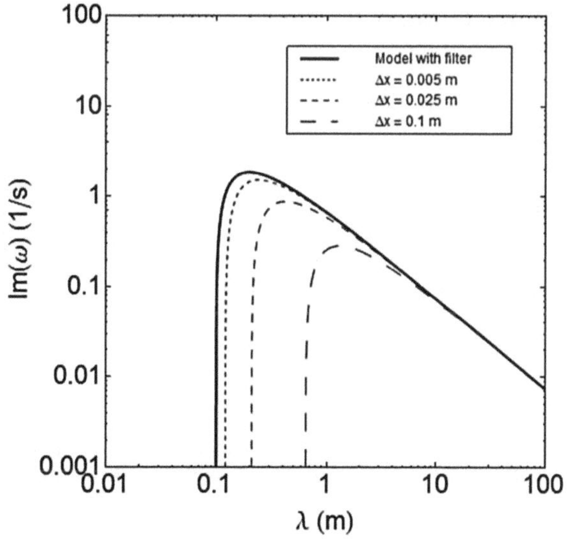

Fig. 8.7 Discrete and differential growth rates using the filter of Eq. (8.31) and $\ell = 2D_H$, reprinted from Fullmer and Lopez de Bertodano (2015) with permission from ANS

dispersion analysis, has a cutoff wavelength exactly at $2D_H$ which is approached by the discrete growth rates obtained from the von Neumann analysis as Δx is reduced. Comparing this with Fig. 8.6 the model with the filter now has a cutoff limit. The cutoff of a relatively large grid is still quite arbitrary and is now larger due to the added artificial viscosity. However, the cutoff and critical wavelengths do not decrease to zero as the grid is refined; there is a limit which has been set by the artificial viscosity incorporated in the underlying differential equations. It should be noted that this is still just a proof of consistency of the numerical scheme, not a proof of convergence. While a fixed cutoff wavelength may not be a sufficient condition to guarantee convergence of an unstable PDE, it is obviously a necessary one.

8.3.4 Water Faucet Problem

The water faucet problem was presented in Sect. 2.7 for the SWT equations and again in Sect. 3.5.2 for the TFM. In the latter case, the KH instability gave rise to a growing spike in the solution shown in Fig. 8.8 when using a second-order numerical method implemented in the TFIT code (Fullmer et al. 2013). The major difference between TFIT and RELAP5 is that TFIT does not have the artificial virtual mass term. In Sect. 3.5.3 the KH problem was overcome by modifying the water faucet problem so the relative velocity at the discontinuity was minimized. In the present case, the water faucet problem is not modified but the filter is used instead.

Figure 8.8 (left) shows the results of the TFIT code without the filter at $t = 0.3$ s. It is seen that as the mesh is refined the solution becomes steeper until it spikes due to the KH instability. Now consider the effect of the low pass filter proposed in the

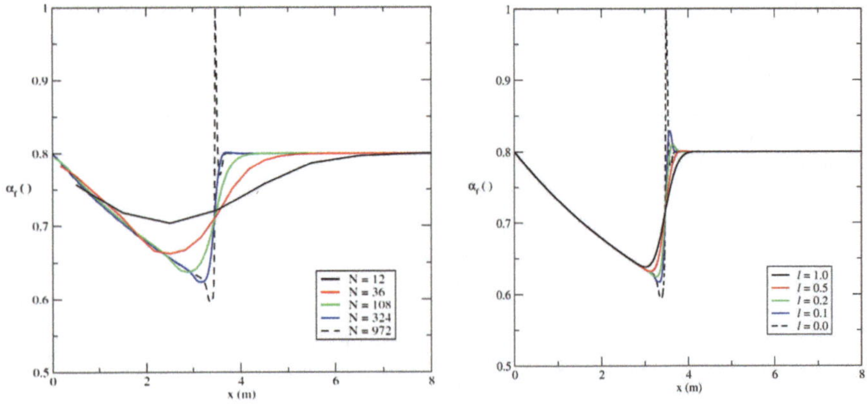

Fig. 8.8 Solution of the water faucet problem with the higher order TFIT code, no artificial viscosity (*left*), $N = 972$ using the artificial viscosity model of Eq. (8.31) with several filter lengths in m (*right*), reprinted from Fullmer and Lopez de Bertodano (2015) with permission from ANS

previous section, which makes the TFM parabolic–hyperbolic. Unlike the option of hyperbolizing the 1D TFM, i.e., making it unconditionally stable, the low-pass filter method of regularization retains the inherent KH instability of the model. However, the instability is retained only for longer wavelengths than the filter length scale, ℓ. Figure 8.8 (right) shows that the numerical and artificial viscosities are large enough at this point in the transient to suppress the growth of the KH wave, but the model is not unconditionally stable. This is demonstrated by arbitrarily adjusting the filter length for a grid number $N = 972$ which exhibits the KH instability at $t = 0.3\,\mathrm{s}$ for low values of the artificial viscosity, i.e., as the filter length is reduced the KH spike reappears. So the instability is suppressed but not precluded.

Depending on the problem being solved, the filter length may be set to some physically relevant parameter, e.g., a capillary length scale defining the extent to which surface tension effects are (in)significant or a wavelength associated to the physical waves. Otherwise, the choice $\ell = 2D_{\mathrm{H}}$ may be applied as an upper bound on the filter length scale rather than requiring the user to determine ℓ arbitrarily for every condition encountered in a simulation.

8.4 Summary and Discussion

The material wave characteristics of the RELAP5/MOD3.3 TFM have been analyzed and compared to the more complete model for bubbly flow of Chap. 5. Of the two material roots, the dominant root is predicted quite accurately by the RELAP5/MOD3.3 model, despite the simplified form of the virtual mass force and the neglect of other terms, for example, the interfacial pressure and the collision force which, while making a small difference in terms of fidelity, were shown in Chap. 5 to make a big contribution to the linear stability of the TFM. Also, the virtual mass force implementation significantly, albeit artificially, extends the stable (hyperbolic) region, although not exactly unconditionally. For low-flux vertical bubbly flow, the primary characteristic compares favorably to the experimental data of Bernier (1982). It was found that the comparison is further improved using the distorted bubbly drift velocity, rather than the churn-turbulent drift velocity, to determine the gas velocity for a given liquid flux. The assessment demonstrates that engineering ad hoc simplifications and artificialities which preserve wave propagation physics may be applied to the TFM after they have been verified with a more complete model and validated with experimental data. One major implication is that the RELAP5 TFM is capable to model global instabilities while suppressing local ones, in particular density wave oscillations addressed in Chaps. 6 and 7, at least in principle. However, the accuracy becomes questionable when coarse mesh FOU damping is employed.

Throughout much of this book, the ill-posedness of the 1D TFM has been demonstrated to be due to missing physics. The preferred method to correct such deficiencies is to include the missing or incomplete physics through additional constitutive modeling, e.g., surface tension (as in Chap. 3), bubble collision (as in Chap. 5), etc. In practice, however, including all relevant physic for every flow regime remains a difficult challenge and, oftentimes, an incomplete model which captures the most important physics (e.g., steady drag) may be deemed appropriate. The ill-posedness of an incomplete model must then be regularized in some fashion. Here we show that the TFM may be regularized with a filter instead of numerical diffusion from the FOU scheme. Such regularization is justified in practice because modeling the KH instability directly is incompatible, i.e., double counting, with the static flow regime maps used in RELAP5. Regularization is currently performed with the FOU scheme and coarse grids, which have an uncertainty that is difficult to quantify. A low pass filter, which has a precise cutoff wavelength, is proposed as an alternative to numerical FOU regularization. It offers two advantages with respect to FOU; it is not mesh dependent and it allows either higher order numerical schemes or finer nodalizations (i.e., reduced numerical diffusion) so that a complete numerical convergence test may be performed under all circumstances.

References

Bernier, R. J. N. (1982). *Unsteady two-phase flow instrumentation and measurement*. Ph.D. Thesis, California Institute of Technology, Pasadena, CA.

Fullmer, W. D., Lee, S. Y., & Lopez de Bertodano, M. A. (2014). An artificial viscosity for the ill-posed one-dimensional incompressible two-fluid model. *Nuclear Technology, 185*, 296–308.

Fullmer, W. D., & Lopez de Bertodano, M. A. (2015). An assessment of the virtual mass force in RELAP5/MOD3.3 for the bubbly flow regime. *Nuclear Technology, 191*(2), 185–192.

Fullmer, W. D., Lopez de Bertodano, M. A., & Zhang, X. (2013). Verification of a higher-order finite difference scheme for the one-dimensional two fluid model. *Journal of Computational Multiphase Flows, 5*, 139–155.

Gidaspow, D. (1974). Round table discussion (RT-1-2): Modeling of two-phase flow. In *Proceedings of the 5th International Heat Transfer Conference, Tokyo, Japan, September 3–7*.

Holmås, H., Sira, T., Nordsveen, M., Langtangen, H. P., & Schulkes, R. (2008). Analysis of a 1D incompressible two fluid model including artificial diffusion. *IMA Journal of Applied Mathematics, 73*, 651–667.

ISL, Information Systems Laboratories. (2003). *RELAP5/MOD3.3 code manual, Vol. 1: Code structure, system models, and solution methods*. NUREG/CR-5535/Rev P3-Vol I.

Ishii, M. (1977). *One-dimensional drift-flux model and constitutive equations for relative motion between phases in various two-phase flow regimes* (ANL-77-47). Argonne National Laboratory.

Ishii, M., & Hibiki, T. (2006). *Thermo-fluid dynamics of two-phase flow*. New York: Springer.

Kataoka, I., & Ishii, M. (1987). Drift flux model for large diameter pipe and new correlation for pool void fraction. *International Journal of Heat and Mass Transfer, 30*, 1927.

Kocamustafaogullari, G. (1985). Two-fluid modeling in analyzing the interfacial stability of liquid film flows. *International Journal of Multiphase Flow, 11*, 63–89.

Krishnamurthy, R., & Ransom, V. H. (1992). *A non-linear stability study of the RELAP5/MOD3 two-phase model*. Paper presented at Japan-U.S. Seminar Two-Phase Flow Dynamics, Berkeley, California, July 5–11.

Lafferty, N., Ransom, V. H., & Lopez De Bertodano, M. A. (2010). RELAP5 analysis of two-phase decompression and rarefaction wave propagation under a temperature gradient. *Nuclear Technology, 169*, 34.

Park, J.-W., Drew, D. A., & Lahey, R. T., Jr. (1998). The analysis of void wave propagation in adiabatic monodispersed bubbly two-phase flows using an ensemble-averaged two-fluid model. *International Journal of Multiphase Flow, 24*, 1205.

Park, J.-W., Drew, D. A., Lahey, R. T., Jr., & Clausse, A. (1990). Void wave dispersion in bubbly flows. *Nuclear Engineering and Design, 121*, 1.

Pauchon, C., & Banerjee, S. (1986). Interphase momentum interaction effects in the averaged multifield model. Part I: Void propagation in bubbly flows. *International Journal of Multiphase Flow, 12*, 559.

Pokharna, H., Mori, M., & Ransom, V. H. (1997). Regularization of two-phase flow models: A comparison of numerical and differential approaches. *Journal of Computational Physics, 87*, 282.

Richtmeyer, R. D., & Morton, K. W. (1967). *Difference methods for initial-value problems* (2nd ed.). New York: Interscience.

Stuhmiller, J. H. (1977). The influence of interfacial pressure forces on the character of two-phase flow model equations. *International Journal of Multiphase Flow, 3*, 551.

U.S. Nuclear Regulatory Commission. (2008). *TRACE V5.0: Theory manual.*

Vreman, A. W. (2011). Stabilization of the Eulerian model for incompressible multiphaseflow by artificial diffusion. *Journal of Computational Physics, 230*, 1639–1651.

Zuber, N. (1964). On the dispersed two-phase flow in the laminar flow regime. *Chemical Engineering Science, 19*, 897.

Zuber, N., & Findlay, J. (1965). Average volumetric concentrations in two-phase flow systems. *Journal of Heat Transfer, 87*, 453.

Chapter 9
Two-Fluid Model CFD

Abstract In this chapter we first address TFM CFD simulations for several cases of bubbly flow that are of engineering interest. We then consider the application of the two-phase 1D TFM stability results of previous chapters to TFM CFD, specifically to the numerical convergence of an ill-posed model.

Multidimensional TFM CFD usually implies turbulence modeling and this chapter addresses both a stable Reynolds Averaged (RANS) TFM and an unstable RANS (URANS) model.

The first topic is the derivation of a k–ε model for a RANS TFM and several applications are discussed. Next a special near-wall TFM averaging is performed to match the two-phase logarithmic law of the wall of Marie et al. (International Journal of Multiphase Flow, 23: 227–247, 1997) to the two-phase k–ε model so that convergence is preserved. Finally the simulation of a chaotic bubble plume with a URANS model is considered. A subscale Smagorinsky model stabilizes the nonlinear small scale turbulence but is not sufficient for linear stability of the model. Additional interfacial forces are needed and the stabilizing effect of the interfacial pressure difference and the collision force is investigated, following the analysis of Chap. 5. These forces prevent ill-posed short wavelength oscillations and allow convergence of the numerical model. The convergence of the turbulence spectrum of the well-posed model is demonstrated using the statistical approach of Chap. 4 for chaotic flow.

9.1 Introduction

The derivation of a stable Reynolds Averaged (RANS) TFM with a two-phase k–ε model is presented first. For those cases where the Reynolds stresses stabilize the TFM the RANS TFM offers a stable solution, only time-averaged void fractions and velocities are obtained. This type of model is of practical engineering interest and it is validated for several prototypical cases of turbulent bubbly flow: decay of grid generated turbulence, jet flow, boundary layer flow, and pipe flow.

Next a special near-wall TFM averaging based on the geometry of the bubbles is performed. This near-wall TFM matches the two-phase logarithmic law of the wall of Marie et al. (1997) to the k–ε model so that near-wall grid convergence is

M.L. de Bertodano et al., *Two-Fluid Model Stability, Simulation and Chaos*,
DOI 10.1007/978-3-319-44968-5_9

preserved. Numerical convergence and validation in the near-wall region are assessed for three cases of vertical bubbly flow: a boundary layer and pipe up and down flows.

Finally the effect of the ill-posed condition on a CFD TFM is addressed. A multidimensional TFM inherently incorporates more physics than a 1D model, in particular turbulence and the associated stresses. While these make the model more stable they do not resolve the linear stability ill-posed issue, which is that the KH instability is embedded in the TFM averaging whether the problem is 1D or 3D. Therefore, the 1D linear stability analyses of previous chapters are pertinent to multidimensional models. To illustrate this, the stability of URANS TFM for a bubble plume is addressed. The subscale Smagorinsky model stabilizes the small scale turbulence of the TFM in a similar way to the viscous stresses of the 1D TFM of Chap. 3, i.e., subscale stresses are necessary for nonlinear stability but not sufficient for linear stability, so additional short wave physics is needed. In particular, the interfacial pressure difference and collision forces discussed in Chap. 5 are employed to obtain a well-posed TFM that preserves the plume dynamics. The statistical convergence of the solution of the well-posed TFM is explored using the ideas of chaotic flow of Chap. 4.

9.2 Incompressible Multidimensional TFM

9.2.1 Model Equations

The multidimensional TFM of Ishii and Hibiki (2006) applied to vertical bubbly two-phase flows is,

$$\boxed{\frac{\partial}{\partial t}\alpha_k\rho_k + \nabla\cdot\alpha_k\rho_k\bar{\boldsymbol{u}}_k = \Gamma_k,} \tag{9.1}$$

$$\boxed{\begin{aligned}\frac{\partial}{\partial t}\alpha_k\rho_k\bar{\boldsymbol{u}}_k + \nabla\cdot\alpha_k\rho_k\bar{\boldsymbol{u}}_k\bar{\boldsymbol{u}}_k = &-\alpha_k\nabla p_k + \nabla\cdot\alpha_k\left(\bar{\boldsymbol{\tau}}_k + \bar{\boldsymbol{\tau}}_k^T\right) + \alpha_k\rho_k\boldsymbol{g} + \boldsymbol{M}_{\mathrm{k i}}\\ &+ (p_{\mathrm{k i}} - p_k)\nabla\alpha_k - \bar{\boldsymbol{\tau}}_{\mathrm{k i}}\cdot\nabla\alpha_k + \bar{\boldsymbol{u}}_{\mathrm{k i}}\Gamma_k,\end{aligned}} \tag{9.2}$$

where, $k=1$ for liquid and $k=2$ for gas phase. α_k, ρ_k, $\bar{\boldsymbol{u}}_k$ are the void fraction, density, and time-averaged velocity vector corresponding to phase k. Since the research presented here is restricted to the case of adiabatic flows, the interface mass transfer rate is zero, $\Gamma_k = 0$. Similarly, the momentum transfer due to mass transfer is zero. Furthermore, $\bar{\boldsymbol{\tau}}_{\mathrm{k i}}$ is neglected for bubbly flows.

The term $\boldsymbol{M}_{\mathrm{k i}}$ represents the averaged contribution from the net momentum transfer occurring at the interface. The interfacial forces should be recognized by now to be of great significance to TFM stability and they will be a key aspect of this chapter. Finally, the jump condition for adiabatic flow, neglecting surface tension, is simply

$$\sum_{k=1}^{2} \boldsymbol{M}_{k\mathrm{i}} = 0. \tag{9.3}$$

9.2.2 Interfacial Momentum Transfer

The term $\boldsymbol{M}_{k\mathrm{i}}$ in Eq. (9.3) denotes the averaged momentum transfer occurring between the two phases, and it is important in determining their distribution and relative velocity. For bubbly flow we decompose it into

$$\boldsymbol{M}_{k\mathrm{i}} = \boldsymbol{M}_{k\mathrm{i}}^{\mathrm{D}} + \boldsymbol{M}_{k\mathrm{i}}^{\mathrm{L}} + \boldsymbol{M}_{k\mathrm{i}}^{\mathrm{VM}} + \boldsymbol{M}_{k\mathrm{i}}^{\mathrm{coll}}. \tag{9.4}$$

The terms on the RHS from left to right represent the contributions from drag, lift, virtual mass, and collision, respectively. In Chap. 5 the virtual mass force, the collision force, and the interfacial pressure were shown to play an important role in the stabilization of the 1D TFM applied to vertical bubbly flows. For the CFD calculations the virtual mass force already discussed in Sect. 5.2.2 is used. Likewise the collision force model of Sect. 5.4.2 is also used. The difference in pressure between the interface and the continuous phase also influences the stability and the phase distribution. Following Bernoulli's principle, for an inviscid flow past a sphere, the interfacial pressure difference is given by Eq. (5.10) with $C_{\mathrm{p}} = 0.25$. Drew and Passman (1999) conclude that for dispersed flow the pressure in the dispersed phase is almost equal to the pressure at the interface, and hence $p_2 \cong p_{2\mathrm{i}}$. Other forces will be discussed next in more detail.

9.2.3 Drag Force

The drag force has already been addressed for 1D in Sect. 5.4.4. For 3D Eq. (5.45) may be written as,

$$\boldsymbol{M}_{2\mathrm{i}}^{\mathrm{D}} = -\frac{3}{4}\alpha_2\rho_1\frac{C_{\mathrm{D}}}{d_{\mathrm{B}}}|\overline{\boldsymbol{u}}_{\mathrm{R}}|\overline{\boldsymbol{u}}_{\mathrm{R}}. \tag{9.5}$$

For small spherical bubbles in water, i.e., 1 mm, the drag coefficient, C_{D}, of Tomiyama et al. (1998) is,

$$C_{\mathrm{D}} = \frac{24}{Re_{\mathrm{D}}}\left(1 + 0.15\, Re_{\mathrm{D}}^{0.687}\right), \tag{9.6}$$

where Re_D is the bubble Reynolds number based on the bubble diameter, d_B, and the relative velocity, $\bar{\boldsymbol{u}}_r = \bar{\boldsymbol{u}}_2 - \bar{\boldsymbol{u}}_1$. We note that for the bubble size range of interest in this chapter, the Eötvös number correction introduced by Tomiyama et al. (1998) is not applicable.

For larger distorted bubbles in tap water, i.e., 3–5 mm, the drag coefficient according to Ishii and Chawla (1979) is given by

$$C_D = \frac{2}{3} d_B \sqrt{\frac{g\Delta\rho}{\sigma}} \left(\frac{1 + 17.67 f(\alpha_2)^{6/7}}{18.67 f(\alpha_2)} \right)^2 \tag{9.7}$$

where

$$f(\alpha_2) = (1 - \alpha_2)^{1.5} \tag{9.8}$$

As was discussed in Chaps. 2 and 5, algebraic expressions like the drag force do not play a major direct role in the stability of the TFM.

9.2.4 *Lift Force*

The lift force accounts for the transverse migration of the bubbles in the presence of velocity gradients and has not been discussed in previous chapters since it is not applicable in 1D. This force has been the object of many studies and we employ the correlation developed by Auton (1987) given by,

$$\boldsymbol{M}_{2i}^L = -\alpha_2 \rho_1 C_L \bar{\boldsymbol{u}}_r \times (\nabla \times \bar{\boldsymbol{u}}_1). \tag{9.9}$$

For a sphere in inviscid shear flow Auton obtained $C_L = 0.5$. Legendre and Magnaudet (1998) performed a full CFD simulation for viscous shear flow around a spherical bubble with slip at the surface and obtained a similar result to Eq. (9.9) for $10 < Re_D < 1000$, though C_L varies a little. In particular, $C_L = 0.45$ for $Re_D = 500$, i.e., a 1 mm bubble, and a dimensionless vorticity, $a = d_B \nabla \times \frac{u_1}{u_r} = 0.2$. However, bubbles flowing in tap water behave like rigid spheres and, furthermore, they may rotate as they move through the shear flow. Recently, Kurose and Komori (1999) solved the problem of a rotating sphere moving in linear shear flow with a full CFD simulation. In this case not only a, but also the dimensionless sphere rotation, $\Omega = \Omega_2/u_r$, was considered. It turns out that assuming $\Omega = a/2$ and $Re_D = 100$ the prediction of Kurose and Komori can be reproduced very closely with Eq. (9.9) and $C_L = 0.28$ for $0 < a < 0.4$. However, if $\Omega = 0$ for solid spheres at moderate Re_D in shear flow, as proposed by Bagchi and Balachandar (2003), the model of Kurose and Komori predicts $C_L = -0.07$. So the prediction of the lift coefficient has substantial variations. Experimental results for bubbly flow are in good agreement with the result of Kurose and Komori and the assumption that $\Omega = a/2$. For example, the experiment of Naciri (1992) for a bubble in a vortex showed that $C_L = 0.28$ for $10 < Re_D < 120$ and $a = 0.25$. Tomiyama's experimental data (2002) for a bubble in Couette flow fit Eq. (9.9) with $C_L = 0.288$ for a range of vorticities. Unfortunately the lift force on large distorted bubbles in the turbulent near-wall region is not well understood. There are two such cases assessed in this chapter, the bubbly boundary layer experiment of Marie et al. (1997) and the bubbly pipe flow experiment of Serizawa et al. (1986). $C_L = 0.1$ is found to fit the data satisfactorily for both cases. $C_L = 0.15$ produces a better fit for the downflow data of Wang et al. (1987).

The stability of the lift force was analyzed by Lucas et al. (2005). They found that a positive lift coefficient is stabilizing whereas a negative value destabilizes the flow. In this chapter, we will only consider cases with positive coefficients.

9.2.5 Wall Force

In addition to the interfacial forces in Eq. (9.4) there is a wall force that acts on the bubbles. $\boldsymbol{M}_{2\mathrm{i}}^{\mathrm{W}}$ represents the interphase momentum transfer due to the wall force. The wall force, apart from being significant on its own merit, balances the lift force in the near-wall region. The wall lubrication force derived by Antal et al. (1991) is a hydrodynamic force that prevents the bubble from touching the wall given by the following correlation,

$$\boldsymbol{M}_{2\mathrm{i}}^{\mathrm{W}} = -\alpha_2 \rho_1 C_{\mathrm{wall}} |\overline{\boldsymbol{u}}_{\mathrm{R}}|^2 \boldsymbol{n}, \tag{9.10}$$

where $\underline{n}$ is the wall outward normal vector and

$$C_{\mathrm{wall}} = \min\left\{0, -\left(\frac{c_{\mathrm{w1}}}{d_{\mathrm{B}}} + \frac{c_{\mathrm{w2}}}{y_{\mathrm{wall}}}\right)\right\}. \tag{9.11}$$

The coefficients are given by Antal et al. (1991) as,

$$c_{\mathrm{w1}} = -0.06|\overline{u}_{\mathrm{R}}| - 0.104, \quad c_{\mathrm{w2}} = 0.147. \tag{9.12}$$

These coefficients were selected based on the laminar bubbly flow data of Nakoryakov et al. (1986) with 1 mm spherical bubbles. For the cases of turbulent bubbly two-phase flows with distorted bubbles addressed later in this chapter, the coefficient values used are $c_{\mathrm{w1}} = -0.01$ and $c_{\mathrm{w2}} = 0.05$ following Frank et al. (2004). These coefficients are arbitrary and vary from author to author. A less uncertain treatment of the TFM in the near-wall region that is based on the geometry and size of the bubbles instead of the wall force will be presented in Sect. 9.5.

9.2.6 Laminar Pipe Flow

The wall-bounded vertical bubbly flows have two important aspects which need to be considered, viz., physically appropriate boundary conditions and the void fraction prediction in the near-wall region. Before proceeding to the complicated scenario of turbulent two-phase flows with distorted bubbles, a simpler case of laminar two-phase flows with small spherical bubbles is analyzed where the boundary conditions are straightforward. The vertical air–water two-phase flow experiments of Nakoryakov et al. (1986) were conducted in a pipe having a diameter of 1.5 cm. The bubbles generated in the experiments were spherical, having $d_B = 0.87$ mm. The numerical calculations reported here are carried out using an in-house code (Vaidheeswaran et al. 2017b). The governing equations, following the work of Antal et al. (1991), are given by,

$$(1-\alpha_2)\frac{dp}{dz} = \frac{1}{r}\frac{d}{dr}\left[r(1-\alpha_2)\mu_1\frac{du_1}{dr}\right] + \frac{3}{4}\alpha_2\rho_1\frac{C_L}{d_B}u_R^2 + (1-\alpha_2)\rho_1 g + \mu_1\frac{du_1}{dr}\frac{d\alpha_2}{dr}, \tag{9.13}$$

$$\begin{aligned}\alpha_2\frac{d}{dr}\left[\frac{3}{20}\alpha_2(1-\alpha_2)\rho_1 u_R^2\right] - C_L\rho_1\alpha_2 u_R\frac{du_1}{dr} - \frac{1}{2}\alpha_2(1-\alpha_2)\rho_1 u_R^2\frac{d\alpha_2}{dr} \\ - \frac{\alpha_2\rho_1 u_r^2}{\frac{1}{2}d_B}\left(C_{W1} + C_{W2}\frac{d_B}{2y_0}\right) + \frac{3}{20}\alpha_2\rho_1 u_R^2\frac{d\alpha_2}{dr} = 0.\end{aligned} \tag{9.14}$$

Equations (9.13) and (9.14) can be reduced to,

$$\mu_1\frac{d^2u_1}{dr^2} + \frac{\mu_1}{r}\frac{du_1}{dr} = \frac{dp}{dz} - \frac{3}{8}\frac{\alpha_2}{(1-\alpha_2)}\rho_1\frac{C_D}{d_B}u_R^2 - \rho_1 g, \tag{9.15}$$

$$\frac{1}{5}\alpha_2(1-\alpha_2)\rho_1 u_r^2\frac{d\alpha_2}{dr} = -C_L\alpha_2\rho_1 u_r\frac{du_1}{dr} - \frac{\alpha_2\rho_1 u_R^2}{\frac{1}{2}d_B}\left(C_{W1} + C_{W2}\frac{d_B}{2y_0}\right). \tag{9.16}$$

Equations (9.15) and (9.16) are solved iteratively starting from an initial uniform distribution with a tolerance for the numerical uncertainty of 10^{-6} for the variables u_1 and α_2. The results obtained are similar to those reported by Antal et al. (1991). It can be seen that the numerical results are in good agreement with the data as shown in Fig. 9.1. The void fraction distribution has a distinct peak near the wall with a width of d_B approximately. Furthermore, it can be seen that when the grid is refined, both the void fraction and liquid velocity distributions do not change, indicating that the solution have already converged. Hence it can be concluded that the TFM with the wall force model of Antal et al. (1991) can be verified and validated for small bubbles and low Reynolds number two-phase flow data.

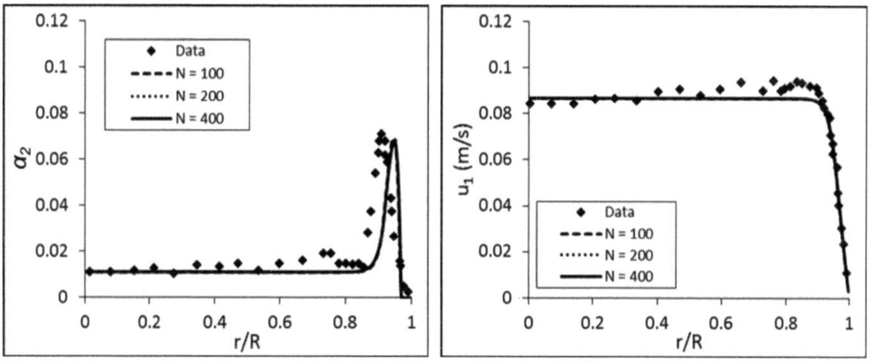

Fig. 9.1 Standard TFM convergence of volume fraction and liquid velocity profiles for Nakoryakov et al. (1986)

9.3 RANS Two-Fluid Model

9.3.1 Reynolds Stress Stabilization

The Reynolds-averaged Navier–Stokes equations are widely used in engineering. The main advantage is that the Reynolds stresses provide physical stabilization to the momentum equations. Starting with the Prandtl mixing length model, many options now exist to constitute the Reynolds stresses. We adopt the standard k–ε model of Launder and Spalding (1974) which is widely used in CFD.

The first TFM averaged k–ε model for particle flows was developed by Elghobashi and Abou-Arab (1983). Kataoka and Serizawa (1989) derived a similar set of equations for bubbly flows. Lopez de Bertodano et al. (1994b) partitioned this set into two kinetic energy equations, one for the shear-induced and another for the bubble-induced components and proposed appropriate closures for bubbly flow. More recently, Chahed et al. (2003) separated the bubble-induced turbulence into the pseudo-turbulence or bubble displacement part, which is nondissipative, and the turbulent dissipative part which occurs at the bubble wake.

9.3.2 Single-Phase k–ε *Model*

For single-phase flow the conservation equations for the Reynolds stresses of the Reynolds-Averaged Navier–Stokes (RANS) model may be derived by time averaging the dot product of the Navier–Stokes equations and the fluctuating component of the liquid velocity, $\boldsymbol{u}'$ Taking the trace of these equations yields the conservation equation for the turbulent kinetic energy (Tennekes and Lumley 1974),

$$\frac{Dk}{Dt} = -\nabla \cdot \overline{\boldsymbol{u}'\left(k + \frac{p'}{\rho}\right)} - \overline{\boldsymbol{u}'\boldsymbol{u}'} : \nabla \overline{\boldsymbol{u}} - \nu \overline{\nabla \boldsymbol{u}' : (\nabla \boldsymbol{u}')^T}, \tag{9.17}$$

where the turbulent kinetic energy is given by

$$k = \frac{1}{2}\overline{\boldsymbol{u}' \cdot \boldsymbol{u}'}. \tag{9.18}$$

and p' the fluctuating pressure of the fluid. The first term on the right-hand side of Eq. (9.17) is a turbulent transport term. The second term is the production of the turbulent kinetic energy, which may be shown to be identical to the loss of kinetic energy of the mean flow. Finally, the last term represents the viscous dissipation which is positive definite.

The single-phase k–ε model, which was developed by Launder and Spalding (1974), is based on a set of constitutive relations for the right-hand side of Eq. (9.17). The Reynolds stress tensor in the production term is modeled according to the Boussinesq relation,

$$\overline{\boldsymbol{u}'\boldsymbol{u}'} = -\nu_t\left(\nabla \overline{\boldsymbol{u}} + \nabla \overline{\boldsymbol{u}}^T\right), \tag{9.19}$$

where ν_t is the turbulent kinematic viscosity. Similarly, the turbulent transport term is modeled as a diffusion term,

$$\overline{\boldsymbol{u}'\left(k + \frac{p'}{\rho}\right)} \cong \overline{\boldsymbol{u}'k} = -\nu_t \nabla k. \tag{9.20}$$

The remaining term is the turbulence dissipation rate,

$$\varepsilon \triangleq \nu \overline{\nabla \boldsymbol{u}' : (\nabla \boldsymbol{u}')^T}. \tag{9.21}$$

Thus, the turbulence kinetic energy transport equation may be written as

$$\frac{Dk}{Dt} = \nabla \cdot \nu_t \nabla k + \nu_t\left(\nabla \overline{\boldsymbol{u}} + \nabla \overline{\boldsymbol{u}}^T\right) : \nabla \overline{\boldsymbol{u}} - \varepsilon. \tag{9.22}$$

The turbulence dissipation rate, which is the key of the k–ε model is constituted with another transport equation proposed by Hanjalic and Launder (1972),

$$\frac{D\varepsilon}{Dt} = \frac{1}{\sigma_\varepsilon} \nabla \cdot \nu_t \nabla \varepsilon + \frac{1}{k/\varepsilon}\left[c_{\varepsilon 1}\nu_t\left(\nabla \overline{\boldsymbol{u}} + \nabla \overline{\boldsymbol{u}}^T\right) : \nabla \overline{\boldsymbol{u}} - c_{\varepsilon 2}\varepsilon\right], \tag{9.23}$$

where (Launder and Spalding 1974), $\sigma_\varepsilon = 1.3$, $c_{\varepsilon 1} = 1.44$, and $c_{\varepsilon 2} = 1.92$. This equation possesses production and dissipation terms that are similar to those in the kinetic energy transport Eq. (9.22) divided by the turbulence time constant, $\tau_t = k/\varepsilon$.

Finally, the correlation for the turbulent viscosity is derived from Prandtl's mixing length theory, assuming that the turbulence time constant is proportional to the time that it takes for a fluid particle to move half way around a large eddy. The resulting equation proposed by Launder and Spalding (1974) is

$$\nu_{\mathrm{t}} = c_{\mu}\frac{k^2}{\varepsilon} \tag{9.24}$$

with $c_{\mu} = 0.09$. Once the total kinetic energy is calculated the Reynolds stresses may be computed using an extension of Eq. (9.22) that includes the normal stresses,

$$\overline{\boldsymbol{u}'\boldsymbol{u}'} = -\nu_{\mathrm{t}}\left(\nabla\overline{\boldsymbol{u}} + \nabla\overline{\boldsymbol{u}}^T\right) + \frac{2}{3}\boldsymbol{A}k, \tag{9.25}$$

where $\boldsymbol{A}$ is the turbulence anisotropy tensor which may be calculated by one of the various algebraic stress models available, e.g., Naot and Rodi (1982). Naturally, for the case of isotropic turbulence $\boldsymbol{A} = \boldsymbol{I}$.

Let us next consider the homogeneous decay of turbulence, since it is a very simple case. Indeed, there are no turbulence or velocity gradients so there are no diffusion terms and no production terms, only dissipation. Then Eq. (9.22) becomes

$$\frac{Dk}{Dt} = -\varepsilon \tag{9.26}$$

and similarly Eq. (9.23) simplifies to

$$\frac{D\varepsilon}{Dt} = -c_{\varepsilon 2}\frac{\varepsilon}{k}\,\varepsilon \tag{9.27}$$

thus combining Eqs. (9.26) and (9.27) and integrating yields

$$\frac{\varepsilon}{\varepsilon_0} = \left(\frac{k}{k_0}\right)^{c_{\varepsilon 2}}, \tag{9.28}$$

where k_0 and ε_0 are the initial conditions. Inserting this into Eq. (9.26) yields

$$\frac{D}{Dt}\left(\frac{k}{k_0}\right) = -\frac{\varepsilon_0}{k_0}\left(\frac{k}{k_0}\right)^{c_{\varepsilon 2}} \tag{9.29}$$

which has the solution

$$\frac{k}{k_0} = \left[\frac{1}{1 + (c_{\varepsilon 2} - 1)\frac{\varepsilon_0}{k_0}t}\right]^{\frac{1}{c_{\varepsilon 2}-1}}. \tag{9.30}$$

Since $c_{\varepsilon 2} = 1.92 \cong 2$, then Eq. (9.30) may be written approximately as

$$\frac{k}{k_0} = \left[\frac{1}{1 + \frac{\varepsilon_0}{k_0} t}\right] \cong \mathrm{e}^{-\frac{t}{k_0/\varepsilon_0}}, \tag{9.31}$$

so it becomes clear that the turbulence decays with time constant $\tau = k_0/\varepsilon_0$.

9.3.3 Two-Phase k–ε Model

Kataoka and Serizawa (1989) obtained an equation for the two-phase turbulent kinetic energy which is similar to Eq. (9.17) and was derived with the two-fluid time averaging method (Ishii 1975),

$$\alpha_1 \frac{Dk}{Dt} = -\nabla \cdot \overline{\alpha_1 \boldsymbol{u}_1' \left(k + \frac{p'}{\rho_1}\right)} - \alpha_1 \overline{\boldsymbol{u}_1' \boldsymbol{u}_1'} : \nabla \overline{\boldsymbol{u}_1'} - \alpha_1 \nu \overline{\nabla \boldsymbol{u}_1' : \nabla \boldsymbol{u}_1'^T} + S_{k\mathrm{i}} \tag{9.32}$$

where 1 is the liquid phase subscript, α_1 is the liquid fraction, k is the kinetic energy of the liquid phase, and $S_{k\mathrm{i}}$ is the interfacial source of turbulence, which is

$$S_{k\mathrm{i}} = \boldsymbol{M}_{1\mathrm{i}}^{\mathrm{D}} \overline{\boldsymbol{u}_{\mathrm{R}}}, \tag{9.33}$$

assuming no interfacial area change, where $\boldsymbol{M}_{1\mathrm{i}}^{\mathrm{D}}$ is the interfacial drag force, $\overline{\boldsymbol{u}_{\mathrm{R}}} = \overline{\boldsymbol{u}_2} - \overline{\boldsymbol{u}_1}$. This term represents the work done by the gas on the liquid phase which gets transformed into turbulent eddies. Constituting a kinetic energy transport equation from Eq. (9.32) is performed in a similar way to the single-phase flow case,

$$\alpha_1 \frac{Dk}{Dt} = \alpha_1 \frac{D}{Dt} \nabla \cdot [\alpha_1 \nu_{\mathrm{t}} \nabla k] + \alpha_1 \left[\nu_{\mathrm{t}1} \left(\nabla \overline{\boldsymbol{u}}_1 + \nabla \overline{\boldsymbol{u}}_1^T\right) : \nabla \overline{\boldsymbol{u}} - \varepsilon\right] + S_{k\mathrm{i}}. \tag{9.34}$$

The problem again is how to constitute the dissipation term, ε, for the case of bubbly flows. The straightforward way to model the two-phase dissipation equation is to proceed in an analogous way to the single-phase k–ε model. Then

$$\alpha_1 \frac{D\varepsilon}{Dt} = \nabla \cdot \left[\alpha_1 \frac{\nu_{\mathrm{t}}}{\sigma_\varepsilon} \nabla \varepsilon\right] + \alpha_1 \frac{1}{k/\varepsilon} \left[C_{\varepsilon 1} \nu_{\mathrm{t}1} \left(\nabla \overline{\boldsymbol{u}}_1 + \nabla \overline{\boldsymbol{u}}_1^T\right) : \nabla \overline{\boldsymbol{u}}_1 - C_{\varepsilon 2} \varepsilon\right] + S_{\varepsilon\mathrm{i}} \tag{9.35}$$

and the problem that remains is how to constitute the interfacial source of dissipation term, $S_{\varepsilon\mathrm{i}}$. For the case of bubbles rising in a still-water tank, assuming steady fully developed flow, Eqs. (9.34) and (9.35) become:

$$0 = -\alpha_1 \varepsilon + S_{ki} \tag{9.36}$$

$$0 = -\alpha_1 C_{\varepsilon 2} \frac{\varepsilon}{k} \varepsilon + S_{\varepsilon i} \tag{9.37}$$

So the interfacial production of dissipation becomes

$$S_{\varepsilon i} = C_{\varepsilon 2} \frac{\varepsilon}{k} S_{ki}. \tag{9.38}$$

9.3.4 *Decay of Grid Generated Turbulence*

9.3.4.1 Single Time-Constant Model

For two-phase homogenous decay of turbulence Eqs. (9.34) and (9.35) reduce to

$$\alpha_1 \frac{Dk}{Dt} = S_{ki} - \alpha_1 \varepsilon, \tag{9.39}$$

$$\alpha_1 \frac{D\varepsilon}{Dt} = S_{\varepsilon i} - \alpha_1 \varepsilon_\varepsilon, \tag{9.40}$$

where, for uniform and homogenous flow, the diffusion and production terms are negligible as stated earlier. Inserting Eq. (9.38) into Eq. (9.40) yields

$$\alpha_1 \frac{D\varepsilon}{Dt} = C_{\varepsilon 2} \frac{\varepsilon}{k} (S_{ki} - \alpha_1 \varepsilon) \tag{9.41}$$

and then, similarly to the single-phase case, Eqs. (9.39) and (9.41) may be combined to yield Eq. (9.31). Inserting this result into Eq. (9.42) gives

$$\frac{D}{Dt}\left(\frac{k}{k_0}\right) = \frac{S_{ki}}{\alpha_1 k_0} - \frac{\varepsilon_0}{k_0}\left(\frac{k}{k_0}\right)^{C_{\varepsilon 2}} \tag{9.42}$$

which may be solved by separation of variables. Defining $a = \left(\frac{S_{ki}}{\alpha_1 \varepsilon_0}\right)^{1/2}$, $\tau = \frac{k_0}{\varepsilon_0}$ and assuming $C_{\varepsilon 2} \cong 2$ the solution to Eq. (9.45) is

$$\frac{k}{k_0} = a \tanh\left[a\, t/\tau + \text{Arccosh}\left(-\frac{a}{\sqrt{-1 + a^2}}\right)\right]. \tag{9.43}$$

In the limit $a \to 0$ Eq. (9.46) reduces to Eq. (9.34). The asymptotic value of k/k_0 is a and the time constant for turbulence decay is τ/a. Since $a = \left(\frac{S_{ki}}{\alpha_1 \varepsilon_0}\right)^{1/2}$, we see that

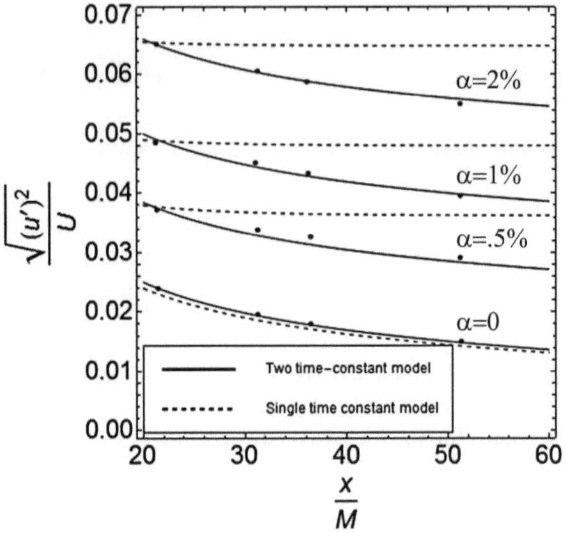

Fig. 9.2 Comparison of two-phase turbulence models with grid generated turbulence bubbly flow data (*bullet*) of Lance and Bataille (1991), inside a vertical 2 m long square channel (450 mm × 450 mm)

the asymptotic value depends on the initial dissipation rate, ε_0, though it should only depend on the bubble-induced turbulence. Moreover, Fig. 9.2 shows (Lance and Bataille 1991) that the time constant for the decay of bubbly flow grid generated turbulence, both with bubbles and without them, is approximately the same, i.e., $\tau = \left(\frac{k_0}{\varepsilon_0}\right)_{1\phi} \neq a\left(\frac{k_0}{\varepsilon_0}\right)_{2\phi}$. Therefore, the single time-constant model fails to reproduce both the asymptotic value and the time constant of homogeneous decay of bubbly flow turbulence, as shown in Fig. 9.2.

9.3.4.2 Two Time-Constant Model

It has been shown that the single time constant extension of the k–ε model to two-phase flow does not match the data of Lance and Bataille (1991). A two time constant model is now formulated in terms of physical arguments (Lopez de Bertodano et al. 1994a). It will be shown that it can be cast in a form consistent with the two-phase averaged equations for turbulence, i.e., Eqs. (9.34) and (9.35). The experimental data of Lance and Bataille (1991), Fig. 9.2, shows that for low void fractions the linear superposition of grid generated shear-induced (SI) turbulence and bubble-induced (BI) turbulence is applicable,

$$k = k_{SI} + k_{BI}. \tag{9.44}$$

The bubble-induced turbulence is primarily composed of the irrotational motion, i.e., pseudo-turbulence due to liquid displacement by the bubbles, for all but Taylor bubbles. However, there is a small rotational contribution due to the wakes.

The approach of linear superposition for an inherently nonlinear process such as turbulence should only be considered to be an approximation based on the assumption that the nonlinear coupling between shear-induced and bubble-induced fluctuations is weak. Even for the case of dilute flows, i.e., negligible bubble-to-bubble interactions, Lance et al. (1996) have identified three types of nonlinear couplings. First, there is the stretching of the shear-induced vortices in the potential flow around the bubbles. Second, there is the deformation of the bubbles by these vortices which changes the virtual volume coefficient of the bubbles and also changes the drag force. Third, there is the interaction between the small vortices in the wakes with the shear-induced vortices. Another type of nonlinear coupling has been identified by Squires and Eaton (1990) for particle-laden flows. They performed direct numerical simulation analysis of small particles in homogeneous turbulence and found that the drag of the particles on the eddies produced a damping effect on the turbulence field of the continuous phase. This effect was greater for higher particle loadings (i.e., higher volume fractions) and it was uniform across the whole energy spectrum. For the case of bubbles which are relatively large it would be expected that the low frequency part of the spectrum would be primarily affected. Another nonlinear *effect* could occur if the bubbles preferentially locate themselves in certain regions of the turbulence field. Anyhow all these effects are being neglected for this simple approach and the task that remains is to determine at which point the linear superposition approach is no longer valid.

In order to use the superposition model given by Eq. (9.44) it is necessary to constitute the bubble-induced turbulence. Arnold et al. (1989) performed cell averaging to obtain the pseudo-turbulence around a group of spheres in potential flow. His result is

$$\left(\overline{\boldsymbol{u}'\boldsymbol{u}'}\right)_{\mathrm{BI}} = \alpha_2 \left[\frac{1}{20}\overline{\boldsymbol{u}}_{\mathrm{R}}\overline{\boldsymbol{u}}_{\mathrm{R}} + \frac{3}{20}|\overline{\boldsymbol{u}}_{\mathrm{R}}|^2 \boldsymbol{I}\right], \tag{9.45}$$

which may be rewritten in matrix form as

$$\left(\overline{\boldsymbol{u}'\boldsymbol{u}'}\right)_{\mathrm{BI}} = \begin{pmatrix} 4/5 & 0 & 0 \\ 0 & 3/5 & 0 \\ 0 & 0 & 3/5 \end{pmatrix} \alpha_2 \frac{1}{2} C_{\mathrm{vm}} |\overline{\boldsymbol{u}}_{\mathrm{R}}|^2, \tag{9.46}$$

where for potential flow around a sphere, the virtual volume coefficient is $C_{\mathrm{vm}} = 1/2$. The bubble-induced turbulent kinetic energy is obtained by taking the trace of Eq. (9.46)

$$k_{\mathrm{BIa}} = \alpha_2 \frac{1}{2} C_{\mathrm{vm}} |\overline{\boldsymbol{u}}_{\mathrm{R}}|^2. \tag{9.47}$$

For completeness of the formulation the effects of inertia and diffusion may be included even though they are negligible. A first-order relaxation type transport equation for bubble-induced turbulence is thus obtained,

$$\alpha_1 \frac{Dk_{\mathrm{BI}}}{Dt} = \nabla \cdot [\alpha_1 \nu_{\mathrm{t}} \nabla k_{\mathrm{BI}}] + \frac{1}{\tau_{\mathrm{b}}}(k_{\mathrm{BIa}} - k_{\mathrm{BI}}), \tag{9.48}$$

where k_{BIa} is the asymptotic value given in Eq. (9.47) for the case when $\bar{\boldsymbol{u}}_{\mathrm{R}}$ reaches its terminal value, and τ_{b} is the time constant of the bubble-induced turbulence. Hence, a second time constant has been introduced. This time constant is based on the assumption that the majority of the bubble-induced turbulence is pseudo-turbulence due to the potential displacement of the liquid by the bubbles and is therefore nondissipative. The effect of the wakes is neglected.

The shear-induced turbulence of the liquid is modeled with the standard single-phase transport equation, Eq. (9.22), modified by the two-phase volumetric fraction,

$$\alpha_1 \frac{Dk_{\mathrm{SI}}}{Dt} = \nabla \cdot [\alpha_1 \nu_{\mathrm{t}} \nabla k_{\mathrm{SI}}] + \alpha_1 [P_{\mathrm{SI}} - \varepsilon_{\mathrm{SI}}] \tag{9.49}$$

where P_{SI} and $\varepsilon_{\mathrm{SI}}$ are the production and the dissipation of the shear-induced turbulence. P_{SI} is given by the second term on the right-hand side of Eq. (9.22). $\varepsilon_{\mathrm{SI}}$ is given by the standard $k - \varepsilon$ model transport equation modified for two-phase flow in an analogous way as Eq. (9.49). Following the assumption of linear superposition Eqs. (9.48) and (9.49) are then added,

$$\alpha_1 \frac{Dk}{Dt} = \nabla \cdot [\alpha_1 \nu_{\mathrm{t}} \nabla k_{\mathrm{l}}] + \alpha_1 [P_{\mathrm{SI}} - \varepsilon_{\mathrm{SI}}] + \frac{1}{\tau_{\mathrm{b}}}(k_{\mathrm{BIa}} - k_{\mathrm{BI}}). \tag{9.50}$$

Now Eqs. (9.32) and (9.50) are equivalent if the dissipation is the superposition of shear-induced and bubble-induced components,

$$\alpha_1 \varepsilon = \alpha_1 \varepsilon_{\mathrm{SI}} + \frac{k_{\mathrm{BI}}}{\tau_{\mathrm{b}}} \tag{9.51}$$

and

$$S_{k\mathrm{i}} = \frac{k_{\mathrm{BI_a}}}{\tau_{\mathrm{b}}} \tag{9.52}$$

Finally, using the well-known expression for interfacial drag

$$\boldsymbol{M}_{1\mathrm{i}}^{\mathrm{D}} = \alpha_2 \frac{3}{4} \frac{C_{\mathrm{D}}}{d_{\mathrm{B}}} \rho_1 \boldsymbol{u}_{\mathrm{R}}^2 \tag{9.53}$$

and inserting Eqs. (9.33) and (9.47) into Eq. (9.52) yields the requirement that the bubble-induced relaxation time constant must be

$$\tau_{\mathrm{b}} = \left(\frac{2}{3}\frac{C_{\mathrm{vm}}}{C_{\mathrm{D}}}\right)\frac{d_{\mathrm{B}}}{u_{\mathrm{R}}} \tag{9.54}$$

for this model to be consistent with Kataoka's two-phase-averaged turbulent kinetic energy equation. This time constant is proportional to the residence time of a bubble, which is an appropriate value for pseudo-turbulence, i.e., arising from potential flow. This time scale is usually very short compared to the time constant of the shear-induced turbulence and may be neglected for most practical cases. Hence, Eq. (9.48) reduces to:

$$k_{\mathrm{BI}} = k_{\mathrm{BI_a}}. \tag{9.55}$$

Applying this model to the problem of bubbles rising in a still water tank yields the asymptotic solution

$$k_{\mathrm{SI}} = 0, \quad \varepsilon_{\mathrm{SI}} = 0, \quad k = k_{\mathrm{BI}} = k_{\mathrm{BI_a}}, \tag{9.56}$$

which is consistent with the model. For the case of homogenous decay of turbulence the inertial terms are added,

$$\frac{Dk_{\mathrm{SI}}}{Dt} = -\varepsilon_{\mathrm{SI}}, \quad k_{\mathrm{BI}} = k_{\mathrm{BI_a}} \tag{9.57}$$

$$\frac{D\varepsilon_{\mathrm{SI}}}{Dt} = -\frac{C_{\varepsilon 2}}{k_{\mathrm{SI}}/\varepsilon_{\mathrm{SI}}}\varepsilon_{\mathrm{SI}} \tag{9.58}$$

So the shear-induced turbulence decays as in the case of single-phase flow, given by Eq. (9.30), and the bubble-induced turbulence, Eq. (9.55), is simply added to it. A comparison between the single time constant model with the two time constant model vs. the data of Lance and Bataille for grid generated bubbly flow turbulence is shown in Fig. 9.2. It can be seen that the two time constant model is in close agreement with the data.

9.3.5 Turbulent Pipe Flow

The two time constant k–ε model is now incorporated into the TFM Eqs. (9.1)–(9.3) to perform CFD calculations for bubbly flow in pipes. The Reynolds stresses are constituted using linear superposition:

$$\overline{\boldsymbol{u}_1'\boldsymbol{u}_1'} = \overline{\boldsymbol{u}_1'\boldsymbol{u}_1'}_{\mathrm{SI}} + \overline{\boldsymbol{u}_1'\boldsymbol{u}_1'}_{\mathrm{BI}} \tag{9.59}$$

where the shear-induced (SI) component is given by Eq. (9.25) and the bubble-induced component is given by Eq. (9.46). The principle of superposition is also used for the two-phase turbulent viscosity. The turbulence in the continuous phase is modeled following the linear superposition model proposed by Sato and Sekoguchi (1975) for bubbly flow,

$$\nu_{1t} = \nu_{1tSI} + \nu_{1tBI}, \tag{9.60}$$

where the terms on the RHS stand for the shear-induced eddy viscosity, and the bubble-induced eddy viscosity, respectively. The shear-induced turbulent viscosity is now calculated according to the standard k–ε model,

$$\nu_{1tSI} = c_\mu \frac{k_{SI}^2}{\varepsilon_{SI}}. \tag{9.61}$$

The bubble-induced component is modeled using the correlation of Sato and Sekoguchi (1975) given by,

$$\nu_{1tBI} = C_{\mu b}\alpha_2 \frac{d_B}{2} |\bar{\boldsymbol{u}}_R| \tag{9.62}$$

where the recommended value of $C_{\mu b} = 1.2$ is used.

The TFM with the k–ε model is validated with the pipe data of Serizawa et al. (1986) with particular emphasis on the turbulence model. However, the boundary conditions and convergence will be deferred to Sect. 9.4.

Serizawa et al. (1986) conducted air–water experiments over a wide range of liquid and gas superficial velocities at atmospheric conditions in a vertical pipe 2.58 m long having a diameter of 60 mm. Liquid velocity and air void fraction were measured at $L/D = 43$. For the current analysis, the case corresponding to $j_2 = 0.077$ m/s, $j_1 = 1.36$ m/s is chosen for benchmarking. The mean bubble diameter was 3 mm. A comparison with this set of data by Serizawa et al. (1986) for bubbly upflow in a pipe is shown in Figs. 9.3, 9.4, 9.5, and 9.6.

Figure 9.3 shows that the void fraction profile calculated with $C_L = 0.1$ results in a satisfactory void distribution. The correct distribution of the buoyancy in the flow is the prerequisite to the assessment of the velocity and turbulence fields which follows. First, Fig. 9.4 shows that the velocity profile becomes flatter due to the "chimney" effect caused by the bubble peak near the wall. Interestingly, Fig. 9.5 then shows that the liquid phase turbulence levels for bubbly flow are lower than for single-phase flow, which is counter intuitive. Concurrently, the shear stress, i.e., $\overline{u'v'} = \nu_t du/dr$, which is also lower, is shown in Fig. 9.6. This follows primarily from the flatter velocity profile. The reduction in the normal stresses may be explained now in terms of Figs. 9.4 and 9.6; for two-phase flow the velocity profile is flatter and the Reynolds shear stress is lower, except near the wall, so the production of shear-induced turbulence:

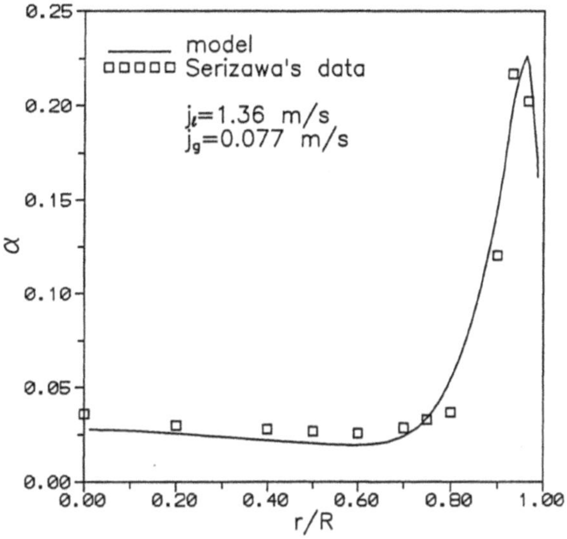

Fig. 9.3 α_2 validation, Serizawa et al. (1986). Reprinted from Lopez de Bertodano et al., JFE (1994), with permission from ASME

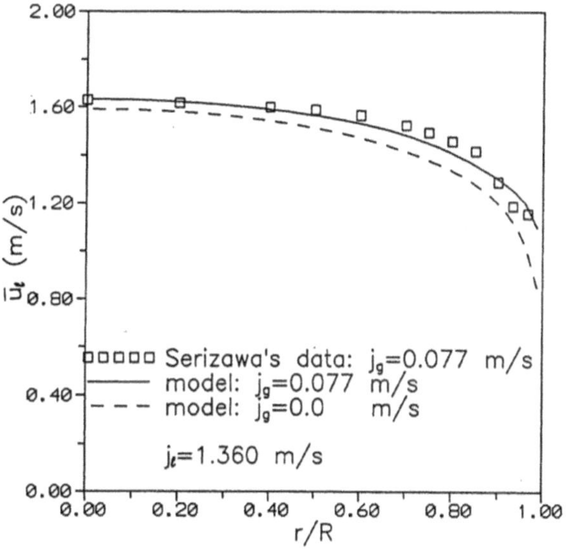

Fig. 9.4 $\bar{u}_1$ validation, Serizawa et al. (1986). Reprinted from Lopez de Bertodano et al., JFE (1994), with permission from ASME

$$P_{\mathrm{SI}} = \overline{u'v'\frac{du}{dr}} \tag{9.63}$$

is reduced significantly and the contribution from bubble-induced turbulence cannot make up that difference. Therefore, the turbulence suppression is a direct result of the reduction of shear-induced production. It is not caused directly by some additional bubble-induced dissipation, as one may speculate a priori.

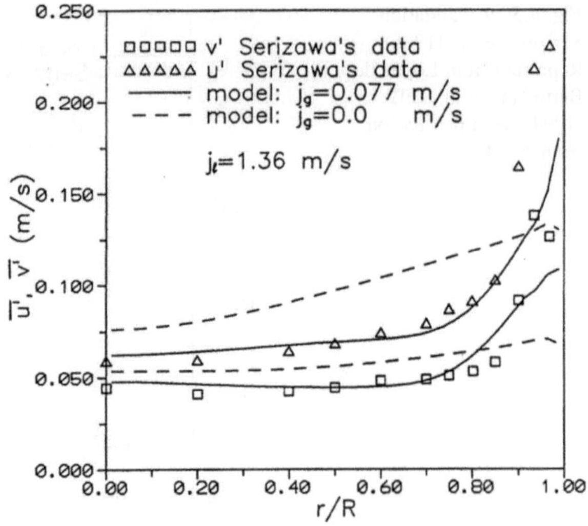

Fig. 9.5 $\overline{u'}$, $\overline{v'}$ validations, Serizawa et al. (1986). Reprinted from Lopez de Bertodano et al., JFE (1994), with permission from ASME

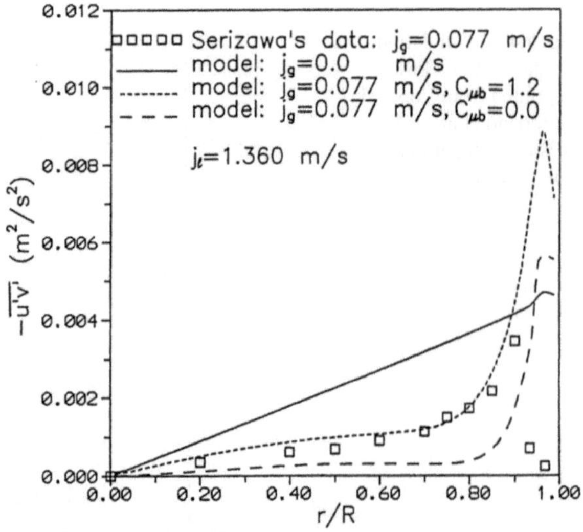

Fig. 9.6 $\overline{u'v'}$ validations, Serizawa et al. (1986). Reprinted from Lopez de Bertodano et al., JFE (1994), with permission from ASME

The bubble-induced turbulence, which is the dominant component of the turbulence at the centerline of the pipe, is calculated using $C_{\mathrm{vm}} = 2$. This value, which is much higher than for spheres moving in a straight line in potential flow, is appropriate for oblate distorted bubbles moving in helical paths. Lance and Bataille (1991) developed an equation for the kinetic energy of the flow around such bubbles starting with the potential flow equations of Saffman (1956). Then they proceeded to measure the shape and trajectory of 5 mm bubbles rising in a tank of still water. From their equations and measurements, $1.2 < C_{\mathrm{vm}} < 3.4$, thus the value chosen for the pipe computations is reasonable.

Finally, the results for the shear stress shown in Fig. 9.6, with $(C_{\mu b} = 1.2)$ and without $(C_{\mu b} = 0)$ bubble-induced turbulent viscosity, indicate that Sato's model, Eq. (9.62), matches the data satisfactorily.

9.3.6 Turbulent Diffusion Force

Eulerian turbulent transport models for particles based on a convection diffusion equation have been pursued for many years, e.g., Batchelor and Townsend (1956). However, these models cannot be called a TFM in the strict sense. Another option is the Lagrangian approach where the motions of discrete particles are tracked through the flow field using Lagrangian equations while an Eulerian formulation is used to solve the conservation equations for the continuous phase. The effect of the particles on the continuous phase is taken into account by introducing appropriate source terms in the continuous phase conservation equations. The Lagrangian approach has been applied by Shuen et al. (1983) and Mostafa and Mongia (1988) among many others. While the Lagrangian method is compatible with the TFM we are strictly interested in the Eulerian approach. Reeks (1991, 1992) derived an Eulerian TFM starting from a probability distribution function equation (i.e., a kinetic equation). This fundamental approach yields the net fluctuating interphase force as a function of local gradients, which permits the development of a constitutive relation for the turbulent diffusion of particles as a force. Another similar approach is the Favre-averaged turbulent diffusion force of Burns et al. (2004).

Physically, turbulent diffusion is the result of the fluctuating component of the forces acting on the bubbles. In the simplest case, the turbulent diffusion force at a point is the ensemble average of the fluctuating component of the drag force of all the bubbles whose trajectories intersect that point. The kinetic equation obtained by Reeks (1991, 1992) describes the evolution of the probability density function (PDF) for the bubbles in phase space, $(\boldsymbol{x}, \boldsymbol{u}_2, t)$, and so it has the information of the bubble trajectories in it. The Eulerian TFM momentum equation for the bubbles is obtained from the ensemble averaged of the first moment of the kinetic equation. The result of this analysis is that the effect of turbulent transport of the particles appears as a force. For the case of homogeneous turbulence, the diffusion force model obtained by Reeks is, including the added mass of the bubbles:

$$\boldsymbol{M}_2^{\mathrm{TD}} = -(\rho_2 + C_{\mathrm{VM}}\rho_1)\boldsymbol{\Lambda}^T \cdot \nabla\alpha, \tag{9.64}$$

where

$$\boldsymbol{\Lambda}^T = \overline{\Delta\boldsymbol{x}(\boldsymbol{x}, t)\boldsymbol{f}(\boldsymbol{x}, t)}.$$

$\boldsymbol{f}(\boldsymbol{x}, t) = \frac{1}{\tau_b}\boldsymbol{u}_1'(\boldsymbol{x}, t)$ is a "driving force" obtained from the fluctuating component of the instantaneous equation of motion for a particle:

$$\frac{d\boldsymbol{u}_2'(\boldsymbol{x},t)}{dt} = -\frac{1}{\tau_b}\boldsymbol{u}_2'(\boldsymbol{x},t) + \boldsymbol{f}(\boldsymbol{x},t).$$

$\Delta\boldsymbol{x}(\boldsymbol{x},t)$ is the displacement of a particle from its initial position at time zero to position $\underline{x}$ at time t. This displacement may be obtained integrating the equation of motion along a particle's trajectory. The result is that:

$$\boldsymbol{\Lambda} = \tau_b\int_0^t\left(1 - e^{\frac{t-s}{\tau_b}}\right)\overline{\boldsymbol{f}(\boldsymbol{x} - \boldsymbol{u}_2(t-s),s)\boldsymbol{f}(\boldsymbol{x},t)}ds$$

To simplify the notation $(\boldsymbol{x},s)$ will stand for a point along a particle's trajectory at time s that reaches position $\boldsymbol{x}$ at time t. The average inside the integral then is:

$$\overline{\boldsymbol{f}(\boldsymbol{x},s)\boldsymbol{f}(\boldsymbol{x},t)} = \frac{1}{\tau_b^2}\overline{\boldsymbol{u}_1'(\boldsymbol{x},s)\boldsymbol{u}_1'(\boldsymbol{x},t)}$$

It is assumed that the turbulence is homogeneous (though not necessarily isotropic) and that the turbulence autocorrelation follows the usual Markov law,

$$\overline{\boldsymbol{u}_1'(x,s)\boldsymbol{u}_1'(x,t)} = \overline{\boldsymbol{u}_1'\boldsymbol{u}_1'}\,e^{-\frac{t-s}{\tau_1}}.$$

where τ_1 is the Lagrangian turbulence time constant along a particle's trajectory. Combining the previous three equations and taking the limit as $t \to \infty$ yields

$$\boldsymbol{\Lambda} = \frac{\tau_1}{\tau_b}\left(\frac{\tau_1}{\tau_1 + \tau_b}\right)\overline{\boldsymbol{u}_1'\boldsymbol{u}_1'},$$

so the turbulence diffusion force, Eq. (9.64), becomes

$$\boldsymbol{M}_2^{\mathrm{TD}} = -(\rho_2 + C_{\mathrm{VM}}\rho_1)\frac{\tau_1}{\tau_b}\left(\frac{\tau_1}{\tau_1 + \tau_b}\right)\overline{\boldsymbol{u}_1'\boldsymbol{u}_1'}^T \cdot \nabla\alpha, \tag{9.65}$$

where the time constant of small bubbles is derived using Eqs. (9.54) and (9.5),

$$\tau_b = \frac{1}{18(1 + 0.15Re^{0.687})}\frac{C_{\mathrm{VM}}\rho_1 d_b^2}{\mu_1}. \tag{9.66}$$

Final closure is provided by the k–ε model which is used to calculate τ_1 and $\overline{\boldsymbol{u}_1'\boldsymbol{u}_1'}$. The eddy relaxation time based on the k–ε model is,

$$\tau_1 = C_\mu^{3/4}\frac{k}{\varepsilon}, \tag{9.67}$$

where $C_\mu = 0.09$. Another considerable effect influencing dispersion is eddy cross over. It is characterized by the time scale $\tau_R = \lambda/u_R$, Lopez de Bertodano (1992), where λ is the Eulerian length scale of the eddies. Estimating this length scale with the $k - \varepsilon$ model mixing length it is concluded that $\tau_R \ll \tau_2$, which implies that eddy cross over has a negligible effect on the dispersion of the small bubbles. The normal Reynolds stress components may be obtained from the last term in Eq. (9.25). In particular, for the case of the two-phase jet analyzed in the next section the transverse components are approximately one half the value of the axial component. Specifically, Lopez de Bertodano (1992) shows that in the limit of very small bubbles it is possible to reduce the two-fluid conservation of mass and momentum equations of the dispersed phase to a single "convection-diffusion" conservation equation with the diffusivity given by the tensor,

$$\nu 2t_{ij} = \tau_1 \overline{u'_{1i} u'_{1j}} = \tau_1 c_{ij} k \tag{9.68}$$

where the bubble-induced turbulence has been neglected. The diagonal components of diffusivity are obtained combining Eqs. (9.67) and (9.68),

$$\nu 2t_{ij} = 0.165 \frac{c_{ij} k^2}{\varepsilon}. \tag{9.69}$$

Comparing this to the $k - \varepsilon$ model definition of turbulent diffusivity,

$$\nu_t = \frac{C_\mu k^2}{\varepsilon}, \tag{9.70}$$

one obtains $c_{ij} = 0.545\,\boldsymbol{I}$. Thus

$$\overline{\boldsymbol{u}'_1 \boldsymbol{u}'_1} = 0.545\, k\boldsymbol{I}, \tag{9.71}$$

so that, according to this model, the diffusivity of very small bubbles matches the diffusivity of momentum in the k–ε model.

The effect of the turbulence diffusion force on the TFM is stabilizing, like any diffusive mechanism. Since it is proportional to ∇ the gradient of the void fraction, it stabilizes the TFM in a similar way to other mechanisms that were addressed in the book, i.e., the hydrostatic force and the interfacial pressure term. Drew and Passman (1999) quantified this effect on their stability analysis of bubbly flow and obtained real characteristics for higher void fractions compared to those obtained without the effect. Lucas et al. (2005) compared the stability of turbulent diffusion and lift and found that the destabilizing effect of negative lift may overcome the turbulent diffusion force stabilization.

9.3.7 *Bubbly Jet*

The bubbly jet Case I of Sun (1985) was selected to test the TFM implementation of the turbulent diffusion force. In addition, the interfacial forces presented in Sect. 9.2 are used, including the lift force of Tomiyama et al. (2002).

The bubbly jet was injected vertically upward from a 5.08 mm nozzle into a still water tank. The bubbles had a diameter of 1 mm ± 0.11 mm. The inlet jet velocity was 1.65 m/s and the void fraction was 2.4 %. Mean and fluctuating properties of both phases were measured using LDA while bubble concentration was measured using flash photography. Data were measured at $x/D = 24$, 40, 60. This is the case with the lowest bubble flow rate so the turbulence modulation by the bubbles is negligible.

An adjustment to the turbulence model was needed because of the well-known fact that the standard k–ε model does not fit the data for single-phase axisymmetric jets very closely. It is common practice to modify one of the constants in the model, $c_{\varepsilon 2}$, from 1.92 to 1.87. This correction has been applied to the simulation (Lopez de Bertodano 1992).

To solve the problem, it is assumed that the mean flow is steady, axisymmetric, incompressible, and isothermal, and that both phases have constant physical properties. The equations were discretized on a uniform cylindrical grid, 80 jet diameters long and 20 jet diameters in diameter, using a finite volume procedure. The boundary condition on the wall of the cylinder is the free stream condition (i.e., the gradients of the velocities are zero). At the top of the cylinder, the pressure is atmospheric and the boundary conditions at the inlet are specified from the known inflow conditions.

Figures 9.7 and 9.8 show the liquid velocity and turbulent kinetic energy profiles at three axial positions. These results are essentially the same as the single-phase profiles because the void fraction is so low. Therefore, they do not reflect any effect in the liquid flow due to changes in the bubble interfacial forces. On the other hand, they have a strong effect on these forces, i.e., the lift force is proportional to the vorticity of the liquid flow and the turbulence diffusion force is proportional to the

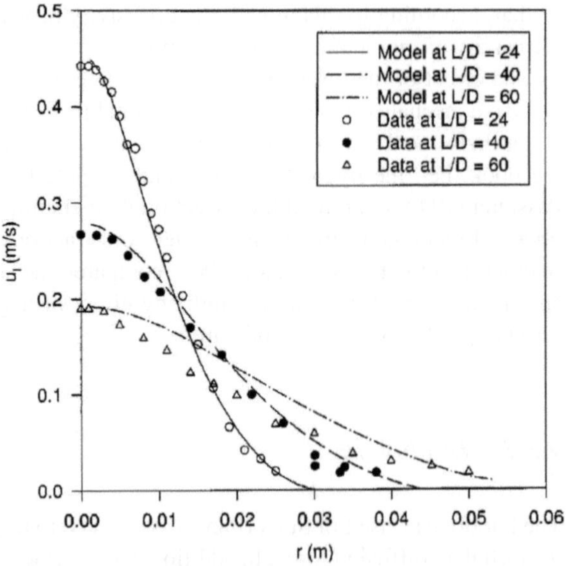

Fig. 9.7 $\overline{u_l}$ validation, Sun (1985)

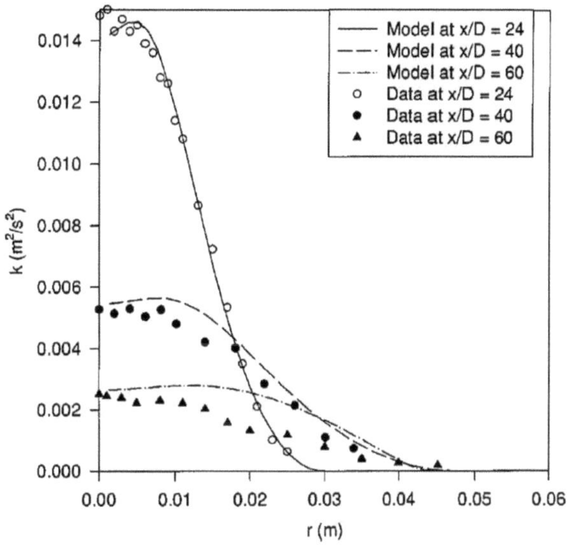

Fig. 9.8 *k* validation, Sun (1985)

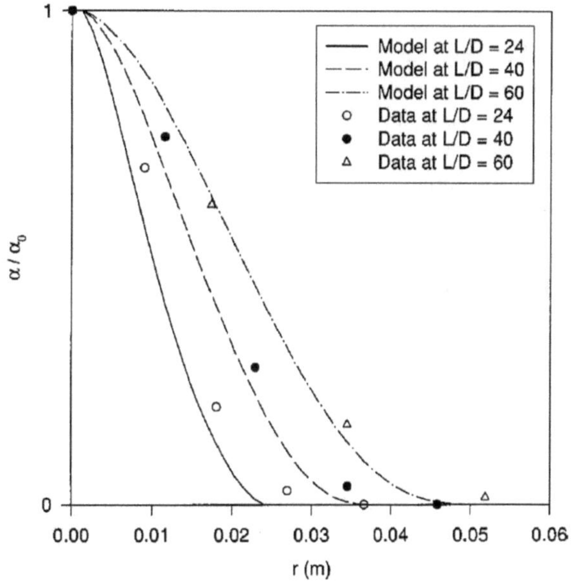

Fig. 9.9 α_2 validation, Sun (1985)

turbulence intensity. Furthermore, since there is good agreement between the simulation and the liquid flow data the interfacial forces may be assessed accurately.

Figure 9.9 shows the bubble distributions at three axial positions. The agreement is acceptable. Initially the model produced too much lateral migration of the bubbles. After inspection of the solution it was found that the drag in the model, i.e., Eqs. (9.5) and (9.6), was too low compared to the data, Fig. 9.10, close to the jet nozzle where the turbulence is intense. It is known that particles in a turbulent fluid have significantly lower relative velocities compared to a bubbles rising in a still fluid. Brucato et al. (1998) were able to correlate their reduced relative velocity data, as well as other's, with

$$C_D = C_{D\infty}\left[1 + 8.76 \times 10^{-4}\left(\frac{d_B}{\lambda_T}\right)^3\right], \tag{9.72}$$

where the Taylor length scale of the turbulence is defined as

$$\lambda_T = \left(\frac{\nu^3}{\varepsilon}\right)^{1/4}. \tag{9.73}$$

The void fraction predictions are in good agreement with the data once this correction is performed. However, Fig. 9.10 shows that even though the relative velocity is closer to the data near the jet centerline, the relative velocity is now under predicted in the periphery. As a consequence the lateral migration of the bubbles is also slightly under predicted.

The rigorously derived models for lift and turbulent diffusion were applied without any arbitrary constants, except for a few simplifying assumptions, e.g., homogeneous turbulence, and since the results did not vary significantly in the range $0 < C_L < 0.288$, this case provides an effective validation of the turbulent diffusion force.

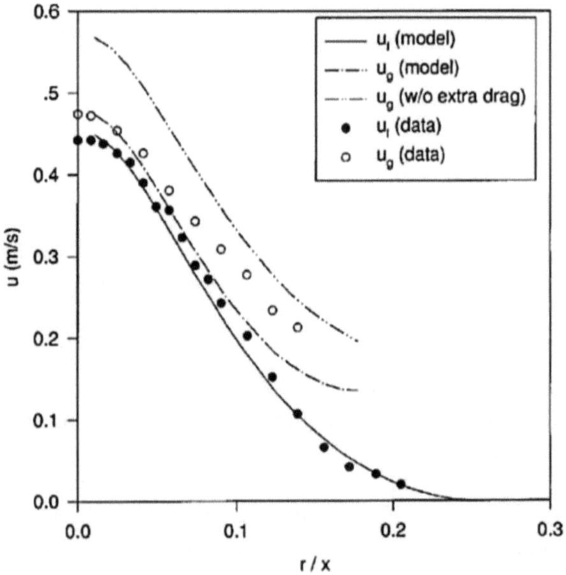

Fig. 9.10 $\overline{u_1}$, $\overline{u_2}$ validations, Sun (1985)

9.4 Near-Wall Two-Fluid Model

9.4.1 Wall Boundary Conditions

The logarithmic wall function approach of Launder and Spalding (1974) for single-phase turbulent flow CFD with the k–ε turbulence model is used to specify the boundary conditions for the velocity, turbulence kinetic energy, and dissipation at the wall adjacent node. This allows matching of the near-wall turbulence region represented by the logarithmic law with the free stream CFD model. For Two-Fluid Model CFD matching, the standard TFM is inconsistent with either the single-phase logarithmic law, which is the usual approach, or other two-phase logarithmic laws, so the solution does not converge in the near-wall region. This section presents a near-wall TFM averaging approach that is consistent with the two-phase logarithmic law of Marie et al. (1997) and that allows convergence.

9.4.2 Two-Phase Logarithmic Law of the Wall of Marie et al. (1997)

The boundary conditions for the liquid phase velocity, turbulence kinetic energy, and turbulence eddy dissipation are usually prescribed using the logarithmic law of the wall approach, which is based on the mixing length theory of Prandtl (1925). For two-phase flows, the presence of bubble buoyancy in the near-wall region affects the liquid phase shear stress distribution. Hence the conventional wall function approach may not be applicable, and it needs to be reformulated. Marie et al. (1997) proposed a modified logarithmic law of the wall including the buoyancy effects in the near-wall region. For the derivation, the wall-peaked void fraction profile is approximated with a double step function distribution as shown in Fig. 9.11. Upon integrating the momentum balance equation, and simplifying, it reduces to,

$$-\overline{u'v'} = u_*'^2 - g\left(\alpha_\mathrm{p} - \alpha_\infty\right) d_\mathrm{B} = u_*^{x2}. \tag{9.74}$$

The modified velocity scale in the boundary layer is given by u_*^x, which can be represented in terms of the original scaling velocity as

$$u_*^{x2} = \beta^2 u_*'^2, \tag{9.75}$$

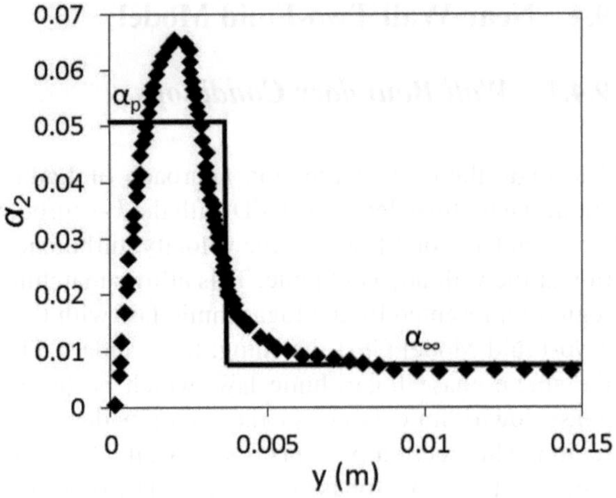

Fig. 9.11 Double-step function approximation of Marie et al. (1997)

where

$$\begin{aligned} \beta^2 &= \left(1 - \frac{F_R \alpha_x}{t^2}\right), \quad F_R = \frac{g d_B}{u_*^2}, \\ t^2 &= \left(\frac{u'_*}{u_*}\right), \quad \alpha_x = \alpha_p - \alpha_\infty. \end{aligned} \tag{9.76}$$

Using the modified velocity scale, the logarithmic relation in two-phase flow becomes

$$u^{+x} = \frac{1}{\kappa} \ln y^{+x} + C^x. \tag{9.77}$$

It can be seen from the experimental data in Fig. 9.12 that the slope remains the same, i.e., the logarithmic relation is preserved, but the y intercept, i.e., C^x in Eq. (9.77) varies with the dispersed phase concentration and is related to the constant appearing in the single-phase logarithmic law of the wall as

$$C^x = C + y_0^+ \left(\frac{1}{\beta} - 1\right) - \frac{1}{\kappa} \ln \beta \tag{9.78}$$

where $y_0^+ = 11$, $C = 5$, and $k = 0.41$.

It follows that the two-phase k_{SI} and ε_{SI} boundary conditions are given by

$$k_{SI} = \frac{u_*^{x2}}{\sqrt{C_\mu}}, \quad \varepsilon_{SI} = \frac{u_*^{x3}}{\kappa y}. \tag{9.79}$$

The void fraction profile obtained with the standard TFM used in the previous section results in a near-wall peaked void fraction distribution that is inconsistent with the boundary condition of Marie et al. (1997) because it is not based on the

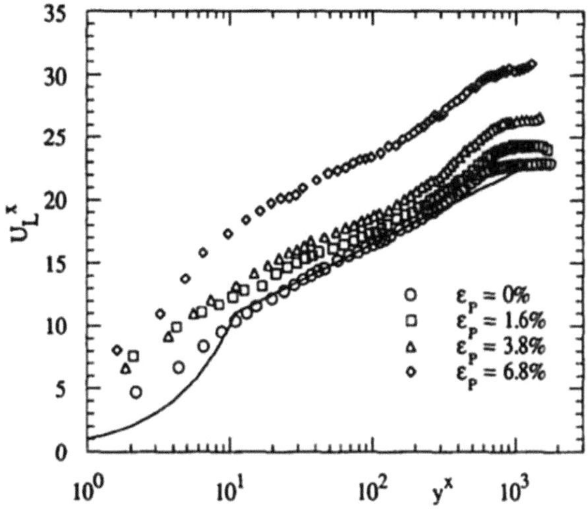

Fig. 9.12 Modified logarithmic law of the wall. Reprinted from Marie et al. (1997), with permission from Elsevier

double step void function approximation. The problem calls for a consistent TFM near-wall averaging approach.

9.4.3 Near-Wall Averaging

The behavior of the TFM near a wall has been a topic of concern for many years. Larreteguy et al. (2002) and Moraga et al. (2006) developed a bubble-center-averaged TFM that solves the problem of the discrete nature of the bubble wall force when it is controlled by surface tension. Vaidheeswaran et al. (2017b) proposed the TFM near-wall averaging addressed herein, based on the geometry of the bubbles touching the wall instead. While the wall-force model preserves the dynamics of the bubbles near the wall, the geometric method assumes a static equilibrium which is adequate for RANS. Furthermore the geometric method does not rely on coefficients that depend on experimental data that are very scarce and it is compatible with Marie's two-phase logarithmic law, as shall be demonstrated in Sect. 9.4.5.

The basic idea is to obtain a void distribution that is consistent with the two-step void distribution of the two-phase logarithmic law of Marie et al. (1997) near the wall by removing the inter-phase momentum transfer terms from the near-wall TFM up to a specific distance from the wall (see Fig. 9.13). The actual void distribution is then reconstructed based on the assumed shape of the bubble. For the cases considered in the current study, the bubbles are assumed to be spherical and nondeformable. Experiments of Tran-Cong et al. (2008) have demonstrated that when the bubbles are small, they have a tendency to be trapped in the near-wall region after colliding and do not bounce back. Hence a simple geometric picture of a spherical bubble touching the wall is proposed. Given that the measured void fraction is proportional to the chord length, the void fraction profile could be represented as,

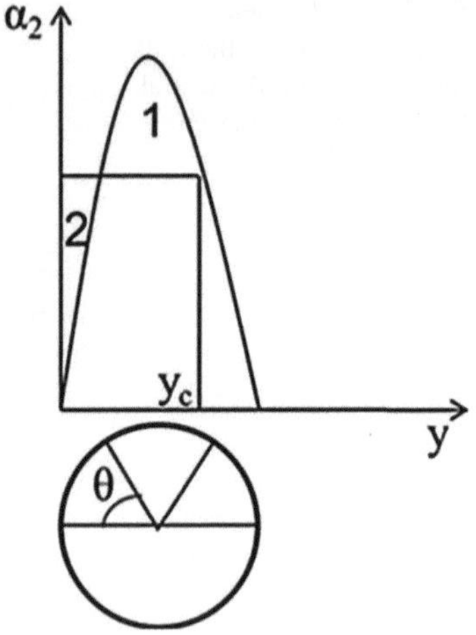

Fig. 9.13 Schematic of geometric model, location of y_c. Reprinted from Vaidheeswaran et al. (2016) with permission from ASME

$$\alpha_2 = \frac{\alpha_p \theta_c}{1 - \cos\theta_c} \sin\theta \tag{9.80}$$

y_c is determined such that the resulting integral void fraction in the wall layer has the same value before and after averaging. This constraint could be replaced by a mathematical statement that the area represented by region 1 equals that of region 2 in Fig. 9.13. This results in the following expression,

$$\theta_c \sin\theta_c + \cos\theta_c - 1 = 0, \quad \theta_c = 133.5°, \quad y_c = 1.67 r_b. \tag{9.81}$$

Thus, the momentum transfer due to the lift force is truncated at y_c, and the wall force is removed completely. In fact, for the turbulent case of Marie et al. (1997) it can be seen that the approximated void fraction profile is in good agreement with the measured values as shown in Fig. 9.14.

The proposed geometric modeling approach eliminates the need for a wall force model, since the geometric condition alone is sufficient to predict the void fraction distribution in the near-wall region. The actual void fraction distribution up to y_c from the wall may then be reconstructed using Eq. (9.80). It is worth mentioning at this point that the wall force model of Antal et al. (1991) that is commonly used for CFD analysis was proposed first based on the laminar flow experiments of Nakoryakov et al. (1986) presented in Sect. 9.2.6 that were conducted at very low flow rates ($u_1 = 0.1$ m/s). Also, the bubbles in the experiments had a mean diameter of 0.87 mm whereas typical air–water turbulent bubbly flows commonly observed in the laboratory, or the fish tank, have larger bubble sizes (3–8 mm). Thus, the applicability of the wall force model of Antal et al. (1991) for turbulent air–water bubbly flows becomes questionable since the physics of the flow around the bubbles in the near-wall region may be significantly different. The more complete models for the wall force have been proposed by Larreteguy et al. (2002) and Moraga et al. (2006) based on bubble center averaging consider the effect of bubble contact

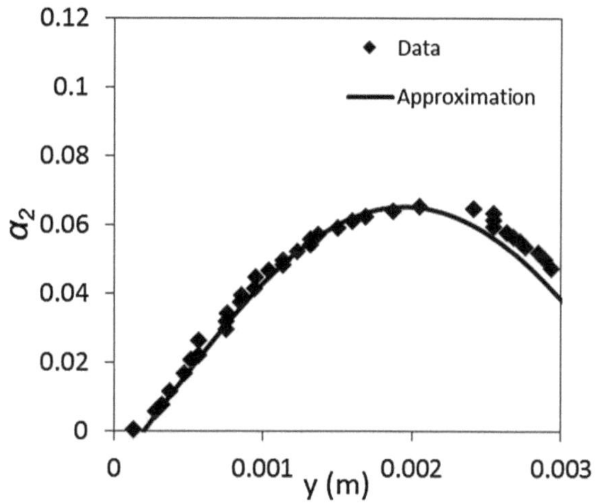

Fig. 9.14 Comparison of geometric model with data of Marie et al. (1997). Reprinted from Vaidheeswaran et al. (2016) with permission from ASME

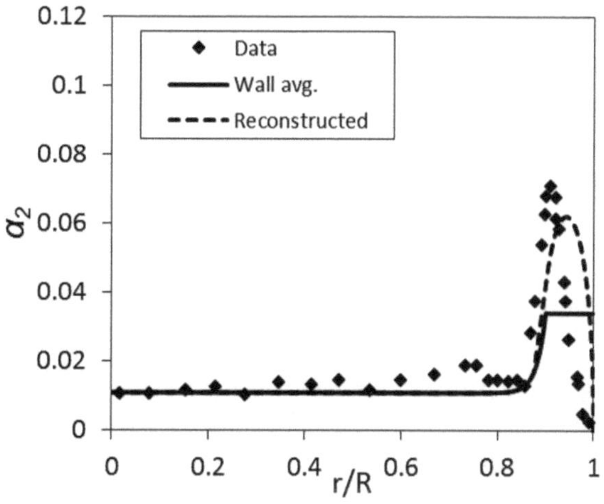

Fig. 9.15 Comparison of near-wall-averaged and reconstructed volume fraction profiles for Nakoryakov et al. (1986). Reprinted from Vaidheeswaran et al. (2016) with permission from ASME

with the wall, bubble deformation, and surface tension. However, it is challenging to incorporate these models in the existing CFD commercial codes.

9.4.4 Laminar Pipe Flow Revisited

The near-wall averaging technique is first applied to the laminar two-phase flow case of Nakoryakov et al. (1986) to demonstrate that it is consistent even for a case when a two-step void distribution approach is not required, cf. Sect. 9.2.6. The near-wall-averaged TFM results in a step function void fraction distribution in the near-wall region as shown in Fig. 9.15. The actual void fraction distribution is then reconstructed with Eq. (9.80) from the averaged uniform value in the near-wall

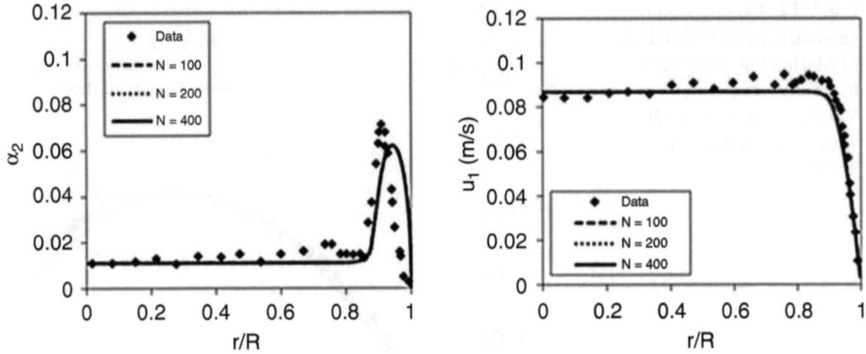

Fig. 9.16 Near-wall-averaged TFM convergence of volume fraction and liquid velocity profiles for Nakoryakov et al. (1986). Reprinted from Vaidheeswaran et al. (2016) with permission from ASME

region as shown in Fig. 9.15. Physically, the reconstructed curve in the near-wall region represents the shape of the bubbles. Then, in the far-wall region up to a distance of y_c from the wall, the averaged and the reconstructed curves overlap and the solution converges. The void fraction and the liquid velocity predictions are shown to be in reasonable agreement with the data of Nakoryakov et al. (1986) and are grid independent as shown in Fig. 9.16.

9.4.5 Turbulent Bubbly Boundary Layer

Having performed the numerical analysis of laminar bubbly flow in a pipe, CFD calculations are now carried out for turbulent two-phase flows using ANSYS CFX 15.0. The air–water data of Marie et al. (1997), Serizawa et al. (1986), and Wang et al. (1987) are chosen to compare the applicability of the single-phase logarithmic boundary condition with the near-wall TFM in combination with the two-phase logarithmic law of Marie et al. (1997).

The bubbly flow experiments of Marie et al. (1997) were conducted over a vertical flat plate inside a vertical 2 m long square channel (450 mm × 450 mm) to study the two-phase flow boundary layer development. The measured mean bubble diameter was 3.5 mm. The liquid velocity and air void fraction at the inlet were uniform and 1 m/s and 1.5 %, respectively.

We first present the results of the standard TFM with the single-phase logarithmic law of the wall. The computational domain considered is 1.5 m high, 50 mm wide, 4 mm deep. Two grid sizes 1 mm and 0.5 mm are used to perform the grid convergence test. The restrictions regarding spatial discretization are worth pointing out. When, the size of the mesh is more than 1 mm, it would result in having less than 2 grid points in the near-wall peak region which is not sufficient to resolve the void fraction distribution. When the mesh size is less than 0.5 mm, the wall $y^+ < 30$ which makes the application of the modified wall function approach inappropriate.

The air void fraction results are compared with the data at 1 m downstream the leading edge of the flat plate. It can be seen that the void fraction predictions using a coarse mesh, i.e., $\Delta y = 1\,\text{mm}$, agree well with the data as shown in Fig. 9.17 (left). However, the liquid velocity profile is underpredicted in the near-wall region, Fig. 9.17 (right). This may result from the use of the single-phase wall function developed for single-phase flows which is inadequate to describe the physics of turbulent bubbly flows. Also, when a grid refinement test is performed, it is seen that the void fraction profile does not converge.

The CFD calculations are performed again using the near-wall-averaged TFM with the two-phase wall function approach implemented via user Fortran routines in ANSYS CFX 15.0. The reconstructed void fraction profile is seen to be in good agreement with the data as shown in Fig. 9.18 (left). When the modified wall function is used, the boundary conditions now account for the change in shear stress in the near-wall region due to the presence of the bubbles and the velocity predictions become more accurate as shown in Fig. 9.18 (right) compared to the results obtained

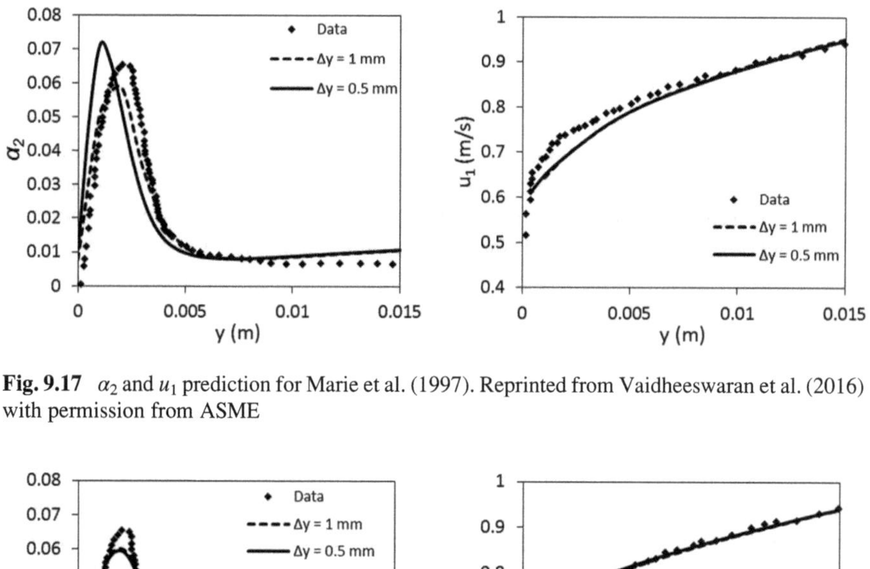

Fig. 9.17 α_2 and u_1 prediction for Marie et al. (1997). Reprinted from Vaidheeswaran et al. (2016) with permission from ASME

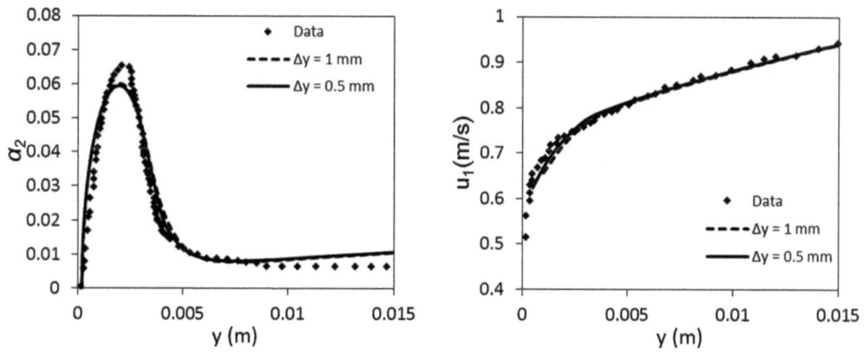

Fig. 9.18 Convergence of volume fraction and velocity distributions—Marie et al. (1997). Reprinted from Vaidheeswaran et al. (2016) with permission from ASME

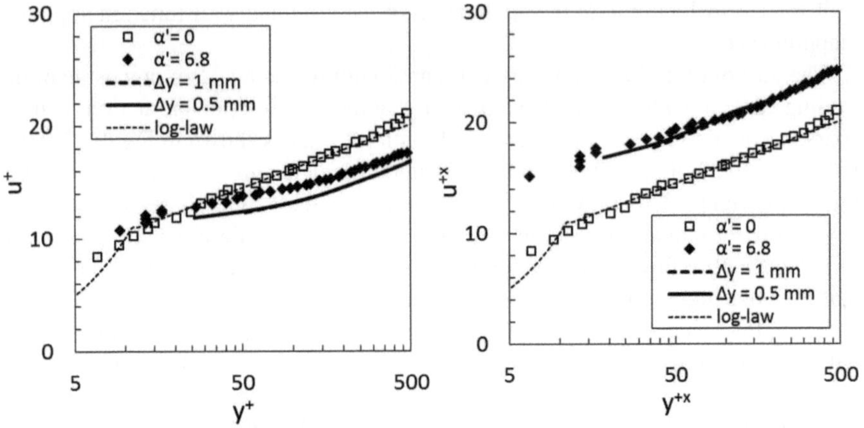

Fig. 9.19 Dimensionless velocity profiles using standard logarithmic law (*left*) and modified scaling velocity with new near-wall averaged TFM (*right*). Reprinted from Vaidheeswaran et al. (2016) with permission from ASME

with the default CFD TFM approach. Furthermore when the grid is refined, it can be seen that both the void fraction and the liquid velocity distributions converge.

Figure 9.19 shows a more revealing picture obtained by plotting the velocities of Figs. 9.17 and 9.18 in the well-known dimension less logarithmic scale. Both pictures include the asymptotic case of the single-phase logarithmic law, by including Marie's single-phase data, for comparison. The left graphic compares the standard TFM with single-phase logarithmic law with the experimental data, showing a significant discrepancy. The right picture shows the two-phase approach which accounts for buoyancy, showing better agreement with the experiment.

9.4.6 *Turbulent Pipe Flow Revisited*

A similar improvement is observed for the case of Serizawa et al. (1986). The convergence test of the near-wall-averaged TFM was performed with 1 mm, 0.5 mm, and 0.25 mm grids. However, the void fraction profile as shown in Fig. 9.20 (left) does not converge completely because of the grid restriction, i.e., $y^+ > 30$, i.e., a grid size of 1 mm is not sufficient to resolve the void fraction profile, while grid sizes less than 0.25 mm are too fine to apply the modified wall function. On the other hand, the CFD velocity predictions converge. Nevertheless, this is a significant improvement over the single-phase log law approach with a wall force, and furthermore it makes the application of the wall force with arbitrary coefficients unnecessary.

Finally, the proposed near-wall averaging method is extended to the case of downward bubbly flows. The data of Wang et al. (1987) shown in Fig. 9.21 were obtained for air–water two-phase flows in a pipe having an inner diameter of

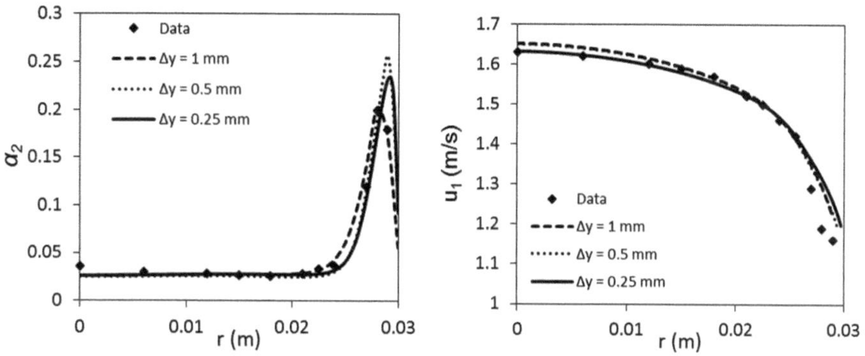

Fig. 9.20 $\alpha_2$and u_1 predictions compared to upflow data of Serizawa et al. (1986). Reprinted from Vaidheeswaran et al. (2016) with permission from ASME

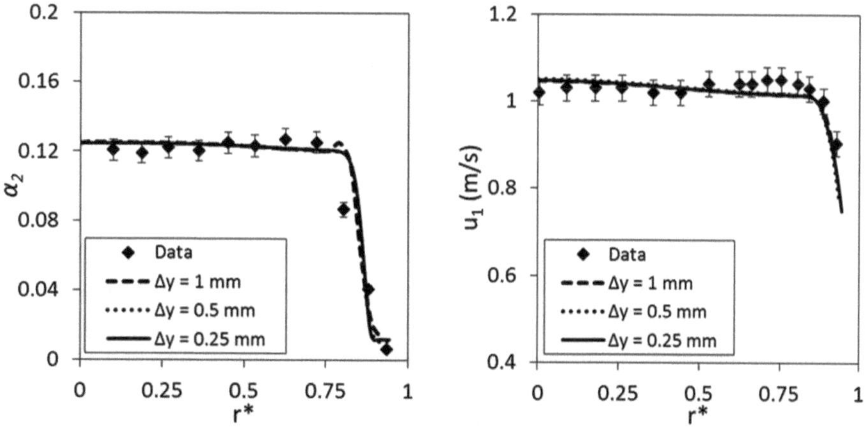

Fig. 9.21 α_2 and u_1 predictions compared to downflow data of Wang et al. (1987). Reprinted from Vaidheeswaran et al. (2016) with permission from ASME

57.15 mm. The measurements are reported at $L/D = 35$. The void fraction distribution obtained for downflow is different from that seen typically for upward bubbly flows and it can be considered as a limiting case for the application of the near-wall Two-Fluid Model averaging. The void coring phenomenon leads to the absence of bubbles in the near-wall region which results in $\beta = 1.0$, and the modified log-law reduces to the single-phase universal log-law. In the current study, CFD calculations were performed for $j_g = 0.1$ m/s and $j_f = 0.94$ m/s. The grid convergence test was performed for 1 mm, 0.5 mm, and 0.25 mm grids. The CFD predictions with $C_L = 0.15$ are in good agreement with the data. Furthermore, Fig. 9.21 shows that the proposed technique yields grid convergence. The near-wall TFM matches the results of Lopez de Bertodano et al. (1994b), obtained with the single-phase logarithmic law for a case where the single-phase logarithmic is applicable.

The conditions considered in the current analyses are restricted to bubbly flows having low to moderate Reynolds numbers where the bubble-induced turbulence and buoyancy are comparable to the shear-induced turbulence effects. At the other limit, when the flow rates are sufficiently high that the shear-induced effects are dominant over the buoyancy effects in the near-wall region, the theory of Marie et al. (1997) results in $\beta \approx 1.0$ and the single-phase logarithmic law is again recovered as it should. This limit applies to many cases of industrial interest, in particular convective boiling at high flow rates.

9.5 URANS Two-Fluid Model

9.5.1 Stability

In this section we address the stability of an URANS TFM for a plane bubble plume. This type of model is typically referred to as an LES model in the literature. The LES TFM approach was introduced by Deen et al. (2001) who observed that such a model is required to capture the transient behavior of a bubble plume. The LES TFM for bubble plumes and shear layers was later improved by several other researchers, for example, Lakehal et al. (2002), Zhang et al. (2006), Niceno et al. (2008), and Ma et al. (2016).

The multidimensional stability analysis of the plane bubble plume has been performed computationally by various researchers using simplifications of the TFM, e.g., Caballina et al. (2003). However, no Orr–Sommerfeld analysis appears to exist to the present. The closest Orr–Sommerfeld analysis was performed by Narayanan and Lakehal (2002) for the shear layer instability of a particle 2D TFM. They included viscosity and avoided the KH instability by setting $u_2 = u_1$, under which condition the shear layer instability is well-posed.

We do not intend to perform a multidimensional stability analysis here. Instead we will focus on the effect of the 1D ill-posed condition on a multidimensional TFM, specifically on the stability effects of the interfacial pressure difference and the collision force. It is very common to neglect these terms because their contribution to the fidelity of the model is small. However, we have seen in Sects. 5.4.1 and 5.4.2 that they allow the formulation of a well-posed TFM. It will be demonstrated that this is also true for a CFD TFM and that without terms like these a grid convergence test is not feasible since the ill-posed oscillations interfere with the solution as the mesh is refined. The common remedy is to use relatively coarse meshes, which is similar to the standard practice with 1D TFM codes discussed in Chap. 8. However, the TFM is a continuum set of partial differential equations and the mesh size of the computational model should not be restricted arbitrarily. Thus, it seems preferable to have a physically well-posed model that prevents the KH instability while preserving the plume meandering instability.

9.5.2 Constitutive Relations

We now proceed to simulate a turbulent bubbly flow case that is unstable, i.e., a bubble plume, with a TFM that captures the large-scale dynamics of the flow. This case illustrates a nonlinear behavior of the TFM similar to that analyzed in Chap. 4. The advantage of CFD TFM formulations is that the Reynolds stresses are generally considered so the models are Lyapunov stable as long as they are well-posed, since linear stability remains a concern. The 1D stability analysis of Sect. 5.5.1 is extended to a CFD TFM to illustrate the applicability of the 1D stability results in a multidimensional situation. In particular, the collision force and the interfacial pressure difference are included to stabilize the otherwise ill-posed model. The other interfacial forces that are used in this TFM are the virtual mass force of Sect. 5.2.2 with $C_{\mathrm{VM}} = 0.5$, the drag force of Ishii and Chawla (1979), and the lift force of Naciri (1992) with $C_{\mathrm{L}} = 0.25$, which plays an important role in the bubble distribution.

We adopt a URANS TFM to capture the plume dynamics. While the TFM equations (9.1)–(9.4) are still applicable, the RANS TFM of section 9.3 that averages out all turbulent structures is not. First the interfacial turbulent diffusion interfacial force presented in Sect. 9.3.6 becomes unnecessaryunnecessary since the eddy scales that transport the bubbles are resolved. Second, the shear-induced eddy viscosity in Eq. (9.60) is closed using the subgrid scale viscosity model of Smagorinsky (1963) given by,

$$\nu_{1\mathrm{tSI}} = (C_{\mathrm{s}}\Delta)^2|S_1| \tag{9.82}$$

where $C_{\mathrm{S}} = 0.1$ and the filter size is given by Lakehal et al. (2002) as $\Delta = 1.5\, d_{\mathrm{B}}$. This is a relatively simple model, and more elaborate models exist, e.g., the model of Lakehal et al. (2002) validated with the two-phase shear layer measurements of Roig et al. (1997), or the model of Niceno et al. (2008). However, our interest is the stability of the URANS TFM rather than the fidelity of the constitutive model of turbulence.

9.5.3 Plane Bubble Plume

The bubble plume simulations of Vaidheeswaran et al. (2017a) were performed with ANSYS FLUENT 15.0 (ANSYS CFX 15.0 2013) for the conditions of the bubble plume experiment of Reddy Vanga (2004).

In TFM CFD it is common practice to use a grid size restriction based on the bubble diameter. However, to impose this criterion for the small-scale bubble column of Reddy Vanga (2004), where for a typical bubble size of 3 mm the minimum grid size criterion is approximately 5 mm, is too restrictive because the test section is only 2 cm in the shortest dimension. Therefore, in this section the bubble size criterion is removed. Furthermore, and more important, a rigorous convergence test will be performed.

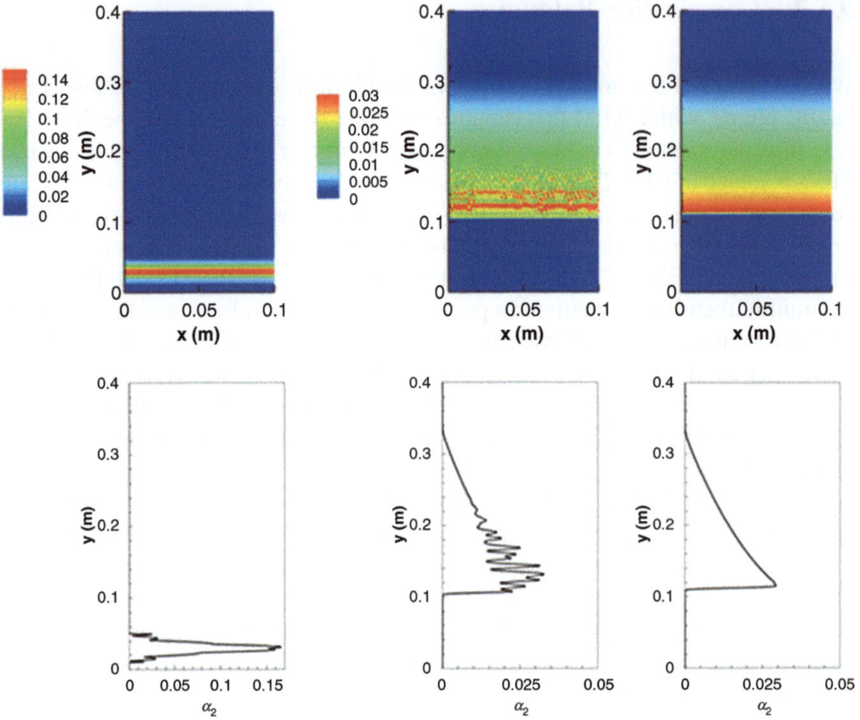

Fig. 9.22 Void fraction contours (*left*) initial condition, (*center*) ill-posed URANS TFM, and (*right*) well-posed URANS TFM, $\Delta x = 1.25$ mm

Before performing the turbulent bubble plume CFD simulations, a simulation of a void wave travelling within a rectangular domain 0.1 m by 0.5 m is carried out, very similar to the "Gedankenexperiment" presented in Sect. 5.5.1 for the nonlinear 1D TFM analysis. Free-slip boundary conditions are applied to the wall in order to have a quasi 1D two-phase flow problem, in order to address the issue of ill-posedness separate from turbulence or wall boundary conditions. The initial void fraction contour shown in Fig. 9.22 (left) consists of a Gaussian with a superposed small high frequency perturbation. An implicit bounded second order temporal scheme is used along with the MUSCL scheme for spatial discretization. The grid size and time step size used are 1.25 mm and 2 ms, respectively, and the calculations are run for 2 s. Figure 9.22 (center and right) is a comparison of the simulation results with the ill-posed and well-posed CFD TFMs. The void fraction distribution obtained with the ill-posed TFM shows short wavelength void fraction lumps which are nonphysical. This typical behavior is similar to the high frequency oscillations obtained with the ill-posed 1D TFM shown in Fig. 9.7. When the interfacial pressure difference term is included in the TFM formulation, the well-posed model exhibits no such instabilities.

Having extended the findings of 1D linear stability analysis to a CFD TFM, the bubble plume experiment of Reddy Vanga (2004) is simulated. Reddy Vanga (2004) performed experiments using a narrow rectangular vertical channel 10 cm

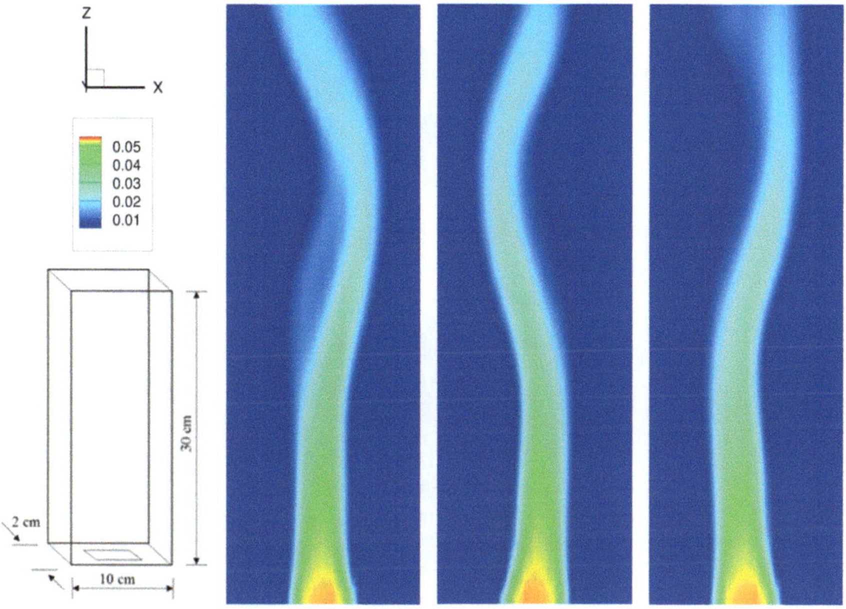

Fig. 9.23 Instantaneous void fraction contours, $\Delta x = 5$ mm

by 2 cm in cross section filled with stagnant water. Air was injected into the test section through a sparger at the bottom and the top was open to the atmosphere. The mean size of the bubbles was 3 mm. Void fraction and bubble velocities were measured 8 cm above the bottom via mesh wire sensors. The sensors were 2.4 mm apart and the wire meshes were 3.03 mm × 2.22 mm. Two distinct two-phase flow regimes were observed: the plume meandered near the inlet, and farther downstream, the bubbles became uniformly dispersed, and appeared to have a steady-state velocity. The section close to the inlet is the focus of the current study, with the objective to capture the plume dynamics and to perform a convergence study. The computational domain shown in Fig. 9.23 (left) corresponding to the lower part of the experimental test section is 30 cm m tall. The inlet has an area of 2 cm by 1 cm to simulate the bubble sparger. Low and a high void fraction cases, $j_g = 2$ mm/s and $j_g = 6$ mm/s, respectively, are simulated to test the interfacial pressure difference and the collision force stabilizations separately, cf. Fig. 5.2. The grid sizes used are 5 mm, 2.5 mm, and 1.25 mm. The time step sizes chosen are 8 ms, 4 ms, and 2 ms, respectively, for $j_g = 2$ mm/s, and 4 ms, 2 ms, and 1 ms, respectively, for $j_g = 6$ mm/s. The statistics are obtained running the simulations for 100 s.

The CFD simulations for $j_g = 2$ mm/s demonstrate the effect of the interfacial pressure difference. When the default CFD TFM is used with the 5 mm mesh, the dynamics of the plume are shown in Fig. 9.23 for three successive instantaneous void fraction contours. Initially the TFM without the interfacial pressure difference was used for the simulations. When the mesh is refined to 1.25 mm, the solutions start exhibiting high void fraction lumps, Fig. 9.24 (left), similar to those previously obtained for the ill-posed traveling void wave problem. Hence, regardless of the

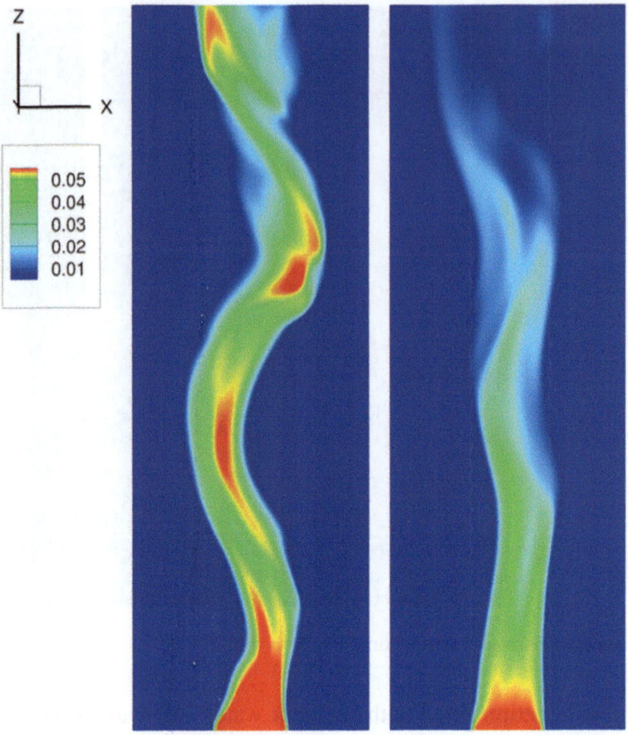

Fig. 9.24 Void fraction contours for turbulent bubble plume with ill-posed (*left*) and well-posed (*right*) URANS TFMs, $\Delta x = 1.25$ mm, $j_g = 2$ mm/s

dimensionality of the problem, the inherent ill-posedness manifests itself as nonphysical oscillations in the solution. When the interfacial pressure difference term is added, the CFD TFM becomes well-posed and the fine-mesh solution is free from nonphysical void fraction lumps as shown in Fig. 9.24 (right).

The analysis of the statistics obtained from the CFD simulations is performed next. Surprisingly for the 5 mm mesh, the time series of void fraction at the channel center indicates a periodic behavior as shown in Fig. 9.25 (left), given that the turbulent bubble plume is expected to be chaotic. The FFT spectrum shown in Fig. 9.25 (right) indicate distinct and discrete frequencies corresponding to a limit cycle, which is comparable to the spectrum of the KY limit cycle shown in Fig. 4.12, and is not in accordance with the known physics of turbulent bubble plumes. This changes abruptly when the mesh is refined to $\Delta x = 2.5$ mm and 1.25 mm, as seen when the spectra in Fig. 9.27 are compared with the spectrum in Fig. 9.25. The FFT spectra of the times series shown in Fig. 9.26, obtained with the two fine grids become continuous in the frequency domain, which indicates the expected chaotic behavior. It is remarkable that a correction motivated by the linear stability behavior of the TFM may have a significant impact on the nonlinear simulations. This result highlights the importance of short wave physics that may

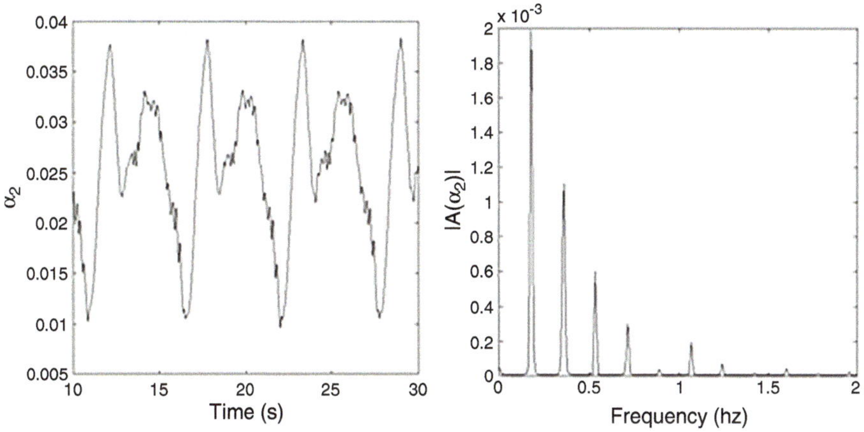

Fig. 9.25 $j_g = 2$ mm/s (*left*) Time series, (*right*) FFT of α_2, $\Delta x = 5$ mm

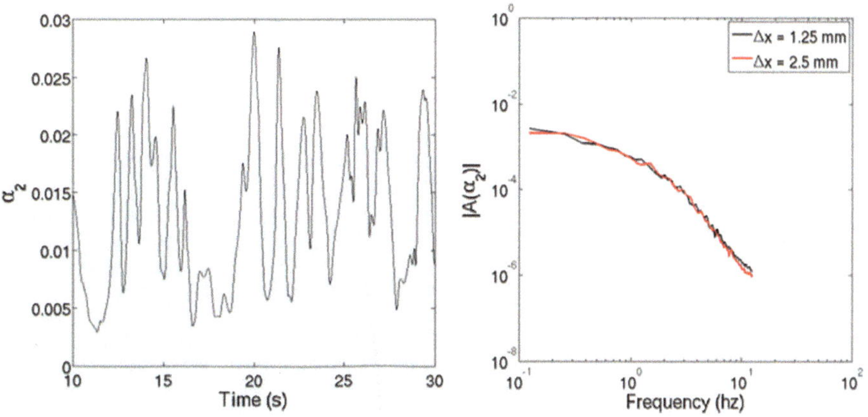

Fig. 9.26 Time series of α_2 for $j_g = 2$ mm/s, (*left*) $\Delta x = 2.5$ mm, (*right*) FFT, $\Delta x = 2.5$ mm and 1.25 mm

appear to have a negligible effect to the fidelity of the model but are significant to the stability.

It can also be seen in Fig. 9.26 that numerical convergence is achieved for this case when comparing the spectra of the two fine meshes, an idea explored also in Chap. 4 for chaotic behavior. In addition, the time-averaged void fraction in Fig. 9.27 shows agreement with the experimental void fraction data of obtained with a wire mesh. While this validation is not as detailed as other works (Niceno et al. 2008, Zhang et al. 2006, Ma et al. 2016) since no turbulence measurements were obtained by Reddy Vanga (2004), it demonstrates that the model is sufficiently accurate to justify the stability analysis.

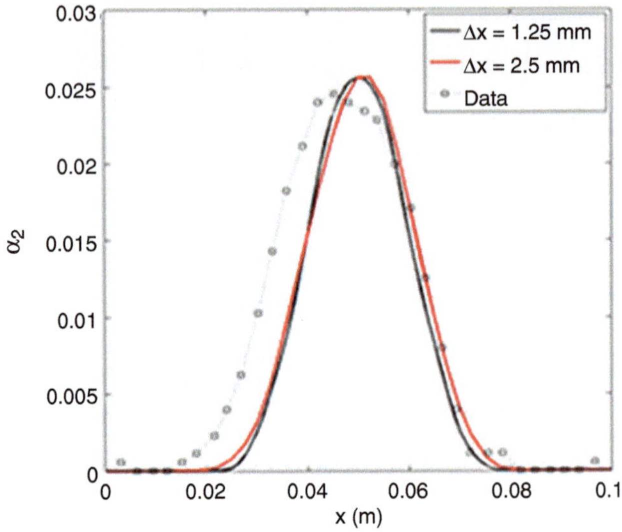

Fig. 9.27 Comparison with the data (Reddy Vanga 2004) for $j_g = 2$ mm/s

Fig. 9.28 Void fraction contours, URANS TFMs without collision (*left*) and with collision (*right*), $\Delta x = 1.25$ mm, $j_g = 6$ mm/s

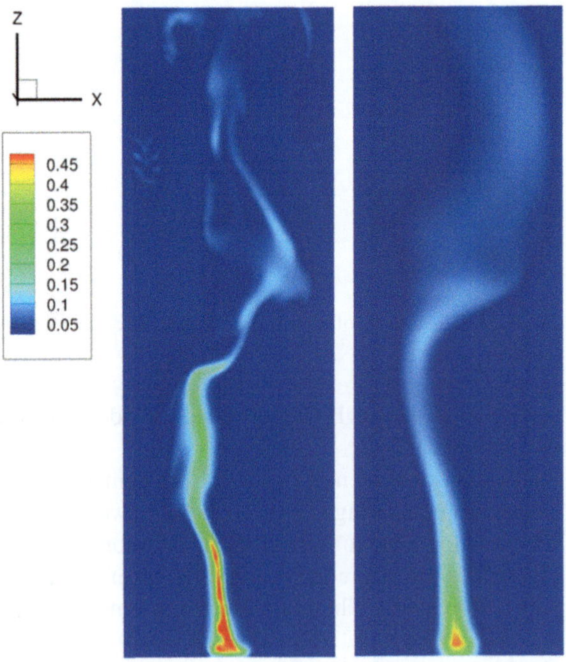

The CFD simulations for $j_g = 6$ mm/s are performed next to demonstrate the effect of the collision term on the TFM stabilization, cf. Fig. 5.2. When the fine mesh URANS TFM simulation with the interfacial pressure difference term alone is performed, it can be seen that a lumpy void fraction distribution occurs in the region near the inlet, Fig. 9.28 (left). This can be explained by the 1D linear stability analyses of the Sect. 5.4.2, since the cutoff void fraction for the TFM well-posedness with the interfacial pressure difference term alone is 26 % and simulated the void fraction near the sparger exceeds 40 %. Hence the model without the collision force is ill-posed as an initial value problem. Adding the collision term makes the model well-posed which results in a smooth void fraction distribution near the inlet as seen in Fig. 9.28 (right).

Thus, the interfacial pressure and the collision mechanisms of Chap. 5 allow the formulation of a bubbly flow TFM that is unconditionally well-posed but captures the bubble plume instability. The short wave stabilization achieved by adding these physical, though not necessarily unique, mechanisms was demonstrated through linear and nonlinear stability analyses in Chap. 5. These physical terms were chosen over others because they stabilize the model, but it is not claimed that they make the model complete. Nevertheless, the point is that the TFM becomes well-posed while the dominant physics, i.e., drag and turbulence, are represented correctly and are not tampered by artificial means. The current results also show the benefit that a well-posed TFM obtained from a 1D linear stability analysis has on the nonlinear performance of a multidimensional model. Furthermore, the spectra of Fig. 9.26 for bubbly vertical flow resemble those of Fig. 4.21 for horizontal flow, which suggest Lyapunov stability, although the calculation of the largest Lyapunov Exponent has not been performed for this case.

Summarizing, it is demonstrated that a well-posed URANS TFM that includes well-posed short wave physics results in well-behaved turbulent simulations when a fine mesh is employed. This then enables convergence in a statistical sense. On the other hand, adopting the bubble size grid size restriction to regularize an ill-posed TFM might result in an unrealistic outcome, e.g., a limit cycle which is an inappropriate representation of a turbulent bubbly plume.

9.6 Summary and Discussion

In this chapter TFM turbulence is considered in some detail, because of its significance to the engineering CFD cases of interest. However, a well-posed multidimensional TFM is also necessary to guarantee model stability and convergence, so several stability lessons learned in previous chapters are applicable.

First, a well-posed stable Reynolds averaged TFM model with a two-phase k–ε model was derived for dilute bubbly flows. This well-posed and stable model was validated with experimental air–water data for homogeneous decay of turbulence and turbulent flow in a boundary layer, a pipe, and a jet.

Then, a special near-wall TFM averaging was performed to warrant the matching of the two-phase logarithmic law of the wall of Marie et al. (1997) to the k–ε model so that convergence is preserved. Numerical convergence and validation in the near-wall region were assessed for three cases of vertical bubbly flow: a boundary layer and pipe up and down flows. An added advantage of the technique is that the Reynolds-averaged TFM is made independent of the wall force with its uncertain coefficients, thus making it less arbitrary for turbulent flows with distorted bubbles.

Finally, a well-posed and Lyapunov stable URANS TFM was obtained by adding the interfacial pressure and collision forces, and was demonstrated to avoid high frequency nonphysical components in the solution. The simulations of a bubble plume also demonstrated that a well-posed TFM is necessary to obtain a converged fine mesh turbulent solution. It was further observed that a coarse grid causes the transition from turbulence to a limit cycle for one particular case. Thus, coarse meshes employed to stabilize an ill-posed model may have unwanted consequences.

Three additional transverse forces, that are relevant to a CFD TFM vs. a 1D TFM, were incorporated along the chapter: the lift force, the wall force, and the turbulent diffusion force. We make no pretense that these comprise a complete set but they are the most significant in an engineering practice and add nuances of their own to the stability of the TFM.

In closing, TFM turbulence modeling was considered in some detail in this chapter but the TFM interfacial structure modeling was left out for expediency. It deserves better consideration and the interested reader is referred to the interfacial area transport models of Ishii and Hibiki (2006). Interfacial area dynamics is a promising avenue where to apply the tools and findings presented in the book. In principle the role of interfacial structure modeling to the two-phase stability of the TFM should be comparable to the role of turbulence modeling to the stability of the single-phase Navier–Stokes equations. From an engineering perspective further research is needed to apply the lessons learned about the stability of single-phase RANS turbulent flow models to the stability of two-phase Reynolds-averaged TFMs. In particular, the development of Reynolds-averaged models without regularization for those flow regimes that have large interfacial structures like horizontal wavy flow or vertical churn turbulent flow represents a challenge.

References

ANSYS CFX 15.0. (2013). *CFX solver theory guide.*

Antal, S. P., Lahey, R. T., & Flaherty, J. E. (1991). Analysis of phase distribution in fully developed laminar bubbly two-phase flow. *International Journal of Multiphase Flow, 17*, 635–652.

Arnold, G. S., Drew, D. A., & Lahey, R. T. (1989). Derivation of constitutive-equations for interfacial force and Reynolds stress for a suspension of spheres using ensemble cell averaging. *Chemical Engineering Communications, 86*, 43–54.

Auton, T. R. (1987). The lift force on a spherical body in a rotational flow. *Journal of Fluid Mechanics, 183*, 199–218.

Bagchi, P., & Balachandar, S. (2003). Effect of turbulence on the drag and lift of a particle. *Physics of Fluids, 15*, 3496–3513.
Batchelor, G. K., & Townsend, A. A. (1956). Turbulent diffusion. In G. K. Batchelor & R. M. Davies (Eds.), *Surveys in mechanics: G. I. Taylor anniversary volume*. Cambridge: Cambridge University Press.
Brucato, A., Grisafi, F., & Montante, G. (1998). Particle drag coefficients in turbulent fluids. *Chemical Engineering Science, 53*(18), 3295–3314.
Burns, A. D., Frank, T., Hamill, I., & Shi, J. M. (2004). The Favre averaged drag model for turbulent dispersion in Eulerian multiphase flows. In *5th International Conference on Multiphase Flow, Yokohama, Japan, May 30–June 4*.
Caballina, O., Climent, E., & Dusek, J. (2003). Two-way coupling simulations of instabilities in a plane bubble plume. *Physics of Fluids, 15*(6), 1535–1544.
Chahed, J., Roig, V., & Masbernat, L. (2003). Eulerian-Eulerian two-fluid model for turbulent gas-liquid bubbly flows. *International Journal of Multiphase Flow, 29*(1), 23–49.
Deen, N. G., Solberg, T., & Hjertager, B. H. (2001). Large eddy simulation of the gas-liquid flow in a square cross-sectioned bubble column. *Chemical Engineering Science, 56*, 6341–6349.
Drew, D. A., & Passman, S. L. (1999). *Theory of multicomponent fluids* (Applied Mathematical Sciences). Berlin: Springer.
Elghobashi, S. E., & Abou-Arab, T. W. (1983). A two-equation turbulence model for two-phase flows. *Physics of Fluids, 26*, 931.
Frank, T., Shi, J. M., & Burns, A. D. (2004). Validation of Eulerian multiphase flow models for nuclear safety applications. In *Third International Symposium on Two-Phase Flow Modeling and Experimentation, Pisa*.
Hanjalic, K., & Launder, B. E. (1972). A Reynolds stress model of turbulence and its application to thin shear flows. *Journal of Fluid Mechanics, 98*, 58–69.
Ishii, M. (1975) Thermo-Fluid Dynamic Theory of Two-Phase Flow. Eyrolles, Paris.
Ishii, M., & Chawla, T. C. (1979). *Local drag laws in dispersed two-phase flow*. Argonne: ANL.
Ishii, M., & Hibiki, T. (2006). *Thermo-fluid dynamics of two-phase flow*. New York: Springer.
Kataoka, I., & Serizawa, A. (1989). Basic equations of turbulence in gas-liquid 2-phase flow. *International Journal of Multiphase Flow, 15*(5), 843–855.
Kurose, R., & Komori, S. (1999). Drag and lift forces on a rotating sphere in laminar shear flows. *Journal of Fluid Mechanics, 384*, 183–206.
Lakehal, D. D., Smith, B. L., & Milelli, M. (2002). Large eddy simulation of bubbly turbulent shear flows. *Journal of Turbulence, 3*.
Lance, M., & Bataille, J. (1991). Turbulence in the liquid phase of a uniform bubbly air-water flow. *Journal of Fluid Mechanics, 222*, 95–118.
Lance, M., & Lopez de Bertodano, M. (1996). Phase distribution phenomena and wall effects in bubbly two-phase flows. In *Multiphase science and technology* (Vol. 8, Ch. 2, pp. 69–123). Begell House, Inc.
Larreteguy, A. E., Drew, D. A., & Lahey, Jr., R. T. (2002). A particle center-averaged two-fluid model for wall-bounded bubbly flows. In *ASME Fluid Engineering Division Summer Meeting, Montreal*.
Launder, B. E., & Spalding, D. B. (1974). The numerical computation of turbulent flows. *Computer Methods in Applied Mechanics and Engineering, 3*, 269–289.
Legendre, D., & Magnaudet, J. (1998). The lift force on a spherical bubble in viscous linear flow. *Journal of Fluid Mechanics, 368*, 81–126.
Lopez de Bertodano, M. (1992). *Turbulent bubbly two-phase flow in a triangular duct*. Troy, NY.
Lopez de Bertodano, M., Lahey, R. T., Jr., & Jones, O. C. (1994a). Development of a k-epsilon model for bubbly two-phase flow. *Journal of Fluids Engineering, 116*, 128–134.
Lopez de Bertodano, M., Lahey, R. T., Jr., & Jones, O. C. (1994b). Phase distribution in bubbly two-phase flow in vertical ducts. *International Journal of Multiphase Flow, 20*(5), 805–818.
Lucas, D., Prasser, H.-M., & Manera, A. (2005). Influence of the lift force on the stability of a bubble column. *Chemical Engineering Science, 60*(13), 3609–3619.

Ma, T., Ziegenhein, T., Lucas, D., & Frohlich, J. (2016). Large eddy simulations of the gas-liquid flow in a rectangular bubble column. *Nuclear Engineering and Design, 299*, 146–153.

Marie, J. L., Moursali, E., & Tran-Cong, S. (1997). Similarity law and turbulence intensity profiles in a bubbly boundary layer. *International Journal of Multiphase Flow, 23*, 227–247.

Moraga, F. J., Larreteguy, A. E., Drew, D. A., & Lahey, R. T., Jr. (2006). A center-averaged two-fluid model for wall-bounded bubbly flows. *Computers and Fluids, 35*, 429–461.

Mostafa, A. A., & Mongia, H. C. (1988). On the interaction of particles and turbulent fluid flow. *International Journal of Heat and Mass Transfer, 31*(10), 2063–2075.

Naciri, M. A. (1992). *Contribution a l'etude des forces exercees por un liquide sur une bulle de gaz, masse ajoutee et interactions hydrodynamiques*. Ph. D. thesis, L'Ecole Central de Lyon, Lyon, France.

Nakoryakov, V. E., Kashinsky, O. N., Kozmenko, B. K., & Gorelik, R. S. (1986). Study of upward bubbly flow at low liquid velocities. *Izvestija Sibirskogo Otdelenija Akademii Nauk SSSR, 16*, 15–20.

Naot, D., & Rodi, W. (1982). Calculation of secondary currents in channel flow. *Proceedings of the American Society of Civil Engineers, 108*(HY8), 948–968.

Narayanan, C. & Lakehal, D. (2002) Temporal instabilities of a mixing layer with uniform and nonuniform particle loadings, Physics of Fluids, 14(11):3775–3789.

Niceno, B., Dhotre, M. T., & Deen, N. G. (2008). One-equation sub-grid scale (SGS) modelling for Euler-Euler large eddy simulation (EELES) of dispersed bubbly flow. *Chemical Engineering Science, 63*, 3923–3931.

Prandtl, L. (1925). A report on testing for built-up turbulence. *ZAMM Journal of Applied Mathematics and Mechanics, 5*.

Reddy Vanga, B. N. (2004). *Experimental investigation and two fluid model large eddy simulations of re-circulating turbulent flow in bubble columns*. West Lafayette, IN: Purdue University.

Reeks, M. W. (1991). On a kinetic equation for the transport of bubbles in turbulent flows. *Physics of Fluids A, 3*(3), 446–456.

Reeks, M. W. (1992). On the continuum equations for dispersed bubbles in non-uniform flows. *Physics of Fluids A, 4*(6), 1290–1302.

Roig, V., Suzanne, C., & Masbernat, L. (1997). Experimental investigation of a turbulent bubbly mixing layer. *International Journal of Multiphase Flow, 24*, 35–54.

Saffman, P. G. (1956). On the rise of small air bubbles in water. *Journal of Fluid Mechanics, 1*, 249–275.

Sato, Y., & Sekoguchi, K. (1975). Liquid velocity distribution in two-phase bubble flow. *International Journal of Multiphase Flow, 2*, 79–95.

Serizawa, A., Kataoka, I., & Michiyoshi, I. (1986). Phase distribution in bubbly flow, data set no. 24. In *The Second International Workshop on Two-Phase Flow Fundamentals, Troy*.

Shuen, J. S., Solomon, A. S. P., Zhang, Q. F., & Faeth, G. M. (1983). *A theoretical and experimental study of turbulent particle-laden lets* (Technical Report 168293, NASA-CR).

Smagorinsky, J. (1963). General circulation experiments with the primitive equations. I. The basic experiment. *Monthly Weather Review, 91*, 99–165.

Squires, K. D., & Eaton, J. K. (1990). Particle response and turbulence modification in isotropic turbulence. *Physics of Fluids A: Fluid Dynamics, 2*(7), 1191–1203.

Sun, T.-Y. (1985). *A theoretical and experimental study on noncondensible turbulent bubbly jets*. Ph.D. Dissertation, The Pennsylvania State University, University Park, PA.

Tennekes, H. & Lumley, J. (1974). A First Course in Turbulence. The MIT Press.

Tomiyama, A., Kataoka, I., Zun, I., & Sakaguchi, T. (1998). Drag coefficients of single bubbles under normal and micro gravity conditions. *JSME International Journal Series B Fluids and Thermal Engineering, 41*(2), 472–479.

Tomiyama, A., Tamai, H., Zun, I., & Hosokawa, S. (2002). Transverse migration of single bubbles in simple shear flows. *Chemical Engineering Science, 57*(11), 1849–1858.

Tran-Cong, S., Marie, J. L., & Perkins, R. J. (2008). Bubble migration in a turbulent boundary layer. *International Journal of Multiphase Flow, 34*, 786–807.

Vaidheeswaran, A., Fullmer, W. D., & Lopez de Bertodano, M. (2017). Effect of collision force on well-posedness and stability of the two-fluid model for vertical bubbly flows. *Nuclear Science and Engineering* (Accepted for publication).

Vaidheeswaran, A., Prabhudharwadkar, D., Guilbert, P., Buchanan, Jr., J. R., & Lopez de Bertodano, M. (2017). New two-fluid model near-wall averaging and consistent matching for turbulent bubbly flows. *Journal of Fluids Engineering*, 139(1), 011302-1, 011302-11.

Wang, S. K., Lee, S. J., Jones, O. C., Jr., & Lahey, R. T., Jr. (1987). 3-D turbulence structure and phase distribution measurements in bubbly two-phase flows. *International Journal of Multiphase Flow, 13*, 327–343.

Zhang, D., N. G. Deen and J. A. Kuipers (2006) "Numerical Simulation of the Dynamic Flow Behavior in a Bubble Column: A study of Closures for Turbulence and Interface Forces," Chemical Engineering Science, *61*:7593–7608.

Appendix A: One-Dimensional Two-Fluid Model

A.1 1D Two-Fluid Model Derivation

To derive the 1D TFM, the full 3D TFM, obtained by proper temporal or statistical averaging, must be area averaged. The 3D TFM of Ishii and Hibiki (2006) will serve as the starting point:

$$\frac{\partial}{\partial t}\alpha_k\rho_k + \nabla\cdot\alpha_k\rho_k\bar{\boldsymbol{u}}_k = \Gamma_k, \tag{A.1}$$

$$\begin{aligned}\frac{\partial}{\partial t}\alpha_k\rho_k\bar{\boldsymbol{u}}_k + \nabla\cdot\alpha_k\rho_k\bar{\boldsymbol{u}}_k\bar{\boldsymbol{u}}_k &= -\alpha_k\nabla p_k + \nabla\cdot\alpha_k\left(\bar{\boldsymbol{\tau}}_k + \bar{\boldsymbol{\tau}}_k^T\right) + \alpha_k\rho_k\boldsymbol{g} + \boldsymbol{M}_{\mathrm{ki}} \\ &\quad + \left(p_{\mathrm{ki}} - p_k\right)\nabla\alpha_k - \bar{\boldsymbol{\tau}}_{\mathrm{ki}}\cdot\nabla\alpha_k + \bar{\boldsymbol{u}}_{\mathrm{ki}}\Gamma_k,\end{aligned} \tag{A.2}$$

where α_k, ρ_k, $\bar{\boldsymbol{u}}_k$, and p_k are the volume fraction, density, velocity field, and pressure of the k phase. The additional source terms Γ, $\bar{\boldsymbol{\tau}}_k$, $\bar{\boldsymbol{\tau}}_k^T$, $\boldsymbol{g}$, $\bar{\boldsymbol{u}}_{\mathrm{ki}}$, $\bar{\boldsymbol{\tau}}_{\mathrm{ki}}$, p_{ki}, and $\boldsymbol{M}_{\mathrm{ki}}$ are the mass transfer at the interface, shear stress tensor, turbulent Reynolds stress, body force, interfacial mass transfer velocity, interfacial shear stress, interfacial pressure, and generalized drag, respectively. The enthalpy or energy equations have been neglected on the basis of the incompressible and isothermal assumptions. Additionally, the incompressible assumption,

$$\frac{D_k\rho_k}{Dt} = \frac{\partial\rho_k}{\partial t} + \bar{\boldsymbol{u}}_k\cdot\nabla\rho_k = 0, \tag{A.3}$$

will be used to simplify the equations before averaging.

The fluids are assumed to be in thermal equilibrium so that $\Gamma_k = 0$ and is eliminated from Eqs. (A.1) to (A.2). Then applying Eq. (A.3) to Eq. (A.1) the simplified continuity equation becomes:

M.L. de Bertodano et al., *Two-Fluid Model Stability, Simulation and Chaos*,
DOI 10.1007/978-3-319-44968-5

$$\frac{\partial \alpha_k}{\partial t} + \nabla \cdot (\alpha_k \overline{\boldsymbol{u}}_k) = 0 \tag{A.4}$$

Additionally, the current work will only be concerned with stratified flows so that $\boldsymbol{M}_{ki} = 0$. The resulting equations will only be valid for cases without dispersed phase structures. The turbulent viscosity hypothesis will be used to constitute the Reynolds stress so that a single effective stress can be used in place of the sum of the viscous and turbulent stresses. Lastly, the phasic pressure terms will be grouped into conservative form. The resulting simplified momentum equations become:

$$\rho_k \left(\frac{\partial}{\partial t} \alpha_k \overline{\boldsymbol{u}}_k + \nabla \cdot \alpha \overline{\boldsymbol{u}}_k \overline{\boldsymbol{u}}_k \right) = -\nabla \alpha_k p_k + \nabla \cdot \alpha_k \overline{\boldsymbol{\tau}}_k^{\mathrm{eff}} - \nabla \alpha_k \cdot \overline{\boldsymbol{\tau}}_{ki} + p_{ki} \nabla \alpha_k + \alpha_k \rho_k \boldsymbol{g} \tag{A.5}$$

Now in order to obtain an appropriate 1D TFM, the 3D field equations need to be area averaged over the channel cross-section, shown schematically in Fig. A.1. The channel is assumed to be rectangular so that an area averaged value is defined by

$$\langle \phi \rangle = \int_0^W \int_0^H \phi dydz, \tag{A.6}$$

where W and H are the channel width and height. The void-weighted mean value given by

$$\langle\langle \phi \rangle\rangle = \frac{\langle \alpha \phi \rangle}{\langle \alpha \rangle} \tag{A.7}$$

will also be utilized. It should be noted that the applicability of a 1D model for wavy-stratified two-phase flow poses the restrictive long-wavelength assumption, as the transverse component of the flow will be lost. While there have been some attempts at constituting a pseudo-2D model, e.g., Ransom and Hicks (1984), the equations derived later have remained as general as possible in accordance with standard thermal-hydraulics models that are needed to analyze the full spectrum of flow conditions. While the traditional assumption of $v = w = 0$ is applied, the transverse component of the pressure field will be retained. It is essential that the transverse pressure component be included and in a manner that does not affect the final form of the 1D TFM, i.e., it is desired to construct a single-pressure model, the form used in the vast majority of 1D TFM codes.

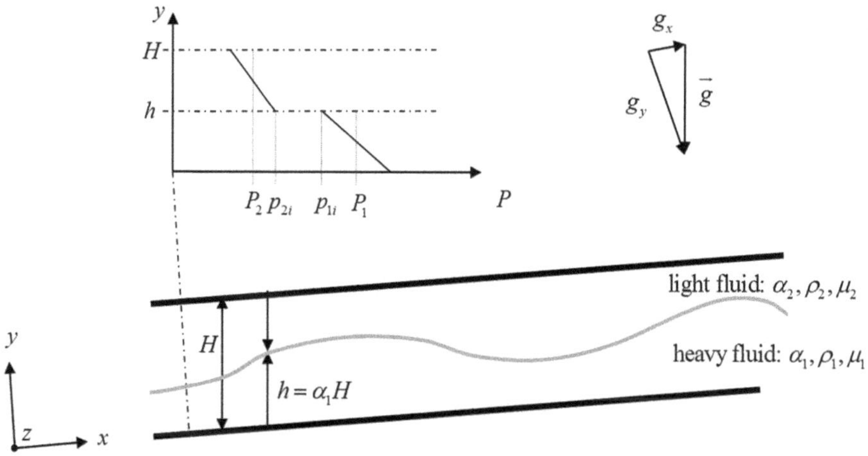

Fig. A.1 Geometry and transverse pressure distribution for stratified two-phase flow in an inclined, rectangular channel, reprinted from Lopez de Bertodano et al. (2013), with permission from Begell House.

By area averaging Eqs. (A.4) and (A.5) the 1D TFM equations are obtained:

$$\frac{\partial\langle\alpha_k\rangle}{\partial t}+\frac{\partial}{\partial x}\langle\alpha\rangle\langle\langle u_k\rangle\rangle=0 \tag{A.8}$$

$$\rho_k\left[\frac{\partial}{\partial t}\langle\alpha_k\rangle\langle\langle u_k\rangle\rangle+\frac{\partial}{\partial x}C_{uk}\langle\alpha_k\rangle\langle\langle u_k\rangle\rangle^2\right]=-\langle\alpha_k\rangle\frac{\partial\langle\langle p_k\rangle\rangle}{\partial x}+\left(p_{ki}-\langle\langle p_k\rangle\rangle\right)\frac{\partial\langle\alpha_k\rangle}{\partial x}$$
$$\frac{\partial}{\partial x}\langle\alpha_k\rangle\langle\langle\tau_{kxx}^{\mathrm{eff}}\rangle\rangle-\frac{4\alpha_{kw}}{D_H}\tau_{kw}-\frac{1}{H}\tau_{ki}+\langle\alpha_k\rangle\rho_k g_x \tag{A.9}$$

$$0=-\langle\alpha_k\rangle\frac{\partial\langle\langle p_k\rangle\rangle}{\partial y}+\langle\alpha_k\rangle\rho_k g_y \tag{A.10}$$

where the variables C_{uk}, α_{kw}, D_{H}, τ_{kw}, and τ_{ki} are the momentum flux distribution parameter, mean wall void fraction, hydraulic diameter, wall shear, and interfacial shear, respectively. After area averaging in the conservative form, the pressure-void fraction terms have again been rearranged into the nonconservative form. The additional wall term in the 1D momentum equation is an effect of averaging the in-plane shear stress components across the channel area. This form includes the approximation that the wall shear in each direction is roughly equivalent. The spanwise momentum equation becomes trivial under the assumption that gravity will not act in the z direction. The distribution parameters are shape factors, i.e., they account for the velocity profiles across the averaging plane, and cannot be predicted as part of a 1D TFM and, hence, require closure. In this work, all flux shape factor terms are neglected by assuming a value of unity—an appropriate assumption if both phases are in the turbulent regime (Ishii and Hibiki 2006). More complete models—particular for laminar flows—can be constructed from experimental data or analytical solutions of transverse velocity profiles. Interested readers should consult the classic work of Kocamustafaogullari (1985) or more recent studies by Biberg (2007) or Picchi et al. (2014), among others. Several terms require additional closure because the four field equations can only be used to solve four independent variables, specifically a void fraction, each velocity, and a pressure. For simplicity, the averaging brackets will be dropped at this point. In order to make the model fit into the framework of a general model, applicable to all types of flow topography, a single pressure model is required. A first-order approximation is to simply assume a uniform pressure, $p_1 = p_2 = p_\mathrm{i}$, where subscripts 1 and 2 indicate the heavier and lighter fluids, respectively. While this may be applicable for vertical flows, when the gravity vector acts in the transverse direction the uniform pressure assumption is physically incorrect. Therefore, Eq. (A.10), which gives the phasic pressures a hydrostatic pressure distribution in the transverse direction, may be used to prescribe the average phasic pressures. The difference between the phase-side interface pressure and the average phase pressure is

$$p_k = p_{k\mathrm{i}} \mp \frac{1}{2}\rho_k g_y h_k, \tag{A.11}$$

where $h_k = \alpha_k H$ is the phase height and assumed to be a product of the void fraction and the channel height. The $\mp$ of the hydrostatic effect assumes that gravity acts in the negative y direction and that the lighter phase, $k = 2$, rests on top. Equation (A.11) can be directly inserted into the interfacial pressure difference term in Eq. (A.9) gives:

$$(p_{k\mathrm{i}} - p_k)\frac{\partial \alpha_k}{\partial x} = \pm\frac{1}{2}\alpha_k \rho_k g_y H \frac{\partial \alpha_k}{\partial x}. \tag{A.12}$$

Additionally, the axial pressure gradient can also be expressed with the phase-side interface pressure as the reference pressure:

$$-\alpha_k \frac{\partial p_k}{\partial x} = -\alpha_k \frac{\partial}{\partial x}\left(p_{k\mathrm{i}} \mp \frac{1}{2}\alpha_k \rho_k g_y H\right) = -\alpha_k \frac{\partial p_{k\mathrm{i}}}{\partial x} \pm \frac{1}{2}\rho_k g_y H \frac{\partial \alpha_k}{\partial x}. \tag{A.13}$$

The second term of Eq. (A.13) is added to the identical interfacial pressure term in Eq. (A.12) which changes the coefficient to unity. Now the interfacial pressure jump condition can be related by the surface tension force which is approximated by Ramshaw and Trapp (1978):

$$p_{2\mathrm{i}} - p_{1\mathrm{i}} = \sigma \frac{\partial^2 h_1}{\partial x^2} \tag{A.14}$$

Using the (lighter) phase-2 side interface pressure as the reference, the phase-1 pressure gradient becomes:

$$-\alpha_1 \frac{\partial p_{1\mathrm{i}}}{\partial x} = -\alpha_1 \frac{\partial}{\partial x}\left(p_{2\mathrm{i}} - \sigma H \frac{\partial^2 \alpha_1}{\partial x^2}\right) = -\alpha_1 \frac{\partial p_{2\mathrm{i}}}{\partial x} + \alpha_1 \sigma H \frac{\partial^3 \alpha_1}{\partial x^3}. \tag{A.15}$$

The combined total pressure effect for each phase is

$$\begin{aligned} -\alpha_2 \frac{\partial p_2}{\partial x} + (p_{2\mathrm{i}} - p_2)\frac{\partial \alpha_2}{\partial x} &= -\alpha_2 \frac{\partial p_{2\mathrm{i}}}{\partial x} + \alpha_2 \rho_2 g_y H \frac{\partial \alpha_2}{\partial x} \\ -\alpha_1 \frac{\partial p_1}{\partial x} + (p_{1\mathrm{i}} - p_1)\frac{\partial \alpha_1}{\partial x} &= -\alpha_1 \frac{\partial p_{2\mathrm{i}}}{\partial x} - \alpha_1 \rho_1 g_y H \frac{\partial \alpha_1}{\partial x} + \alpha_1 \sigma H \frac{\partial^3 \alpha_1}{\partial x^3} \end{aligned} \tag{A.16}$$

The total or effective viscous stress is given by Drew and Passman (1998):

$$\tau_{kxx}^{\mathrm{eff}} = \rho_k \nu_k^{\mathrm{eff}} \frac{\partial u}{\partial x} \tag{A.17}$$

The effective viscosity of Fullmer et al. (2011) will be used

$$\nu_k^{\mathrm{eff}} = C_\varepsilon \left(\nu_k + \nu_k^t\right) \tag{A.18}$$

where ν_k^t is the turbulent viscosity and the C_ε coefficient is an adjustment factor that scales the shear stress so that the 1D dissipation is roughly equivalent to its multidimensional counterpart. Fullmer et al. (2011) performed a CFD simulation of the Thorpe (1969) experiment and calculated, as a first-order approximation, $\nu_k^t = 1.3 \times 10^{-5}\,\mathrm{m^2/s}$ and $C_\varepsilon = 8.1$, which is used for both phases. In the equations used for linear stability analysis only the second derivative of the velocity is considered since the product of the first derivatives of the void fraction and the velocity are nonlinear products of a perturbation. However, for nonlinear analysis with TFIT code, the conservative form of the viscous term is implemented:

$$\frac{\rho_k \nu_k^{\mathrm{eff}}}{\Delta x}\left[(\alpha_k)_{j+\frac{1}{2}}^n \frac{(u_k)_{j+1}^n - (u_k)_j^n}{\Delta x} - (\alpha_k)_{j-\frac{1}{2}}^n \frac{(u_k)_j^n + (u_k)_{j-1}^n}{\Delta x}\right], \tag{A.19}$$

which replaces the linearized form shown in Eqs. (A.17) and (A.18).

The interfacial and wall shear terms will be closed with a simple Darcy model,

$$\tau_{k\mathrm{i}} = \frac{1}{2} f_{\mathrm{i}} \rho_2 |u_{\mathrm{R}}| (u_k - u_{nk}), \tag{A.20}$$

and

$$\tau_{k\mathrm{w}} = \frac{1}{2} f_k \rho_k u_k^2, \tag{A.21}$$

where $u_\mathrm{R} \equiv u_2 - u_1$ is the relative velocity and u_o is the velocity of the other phase. The coefficients f_i and f_k are the interface and wall friction factors. While they are generally modeled as functions of the phasic Reynolds numbers, see e.g. (Andritsos and Hanratty 1987; Kowalski 1987; Hurlburt and Hanratty 2002; Ullmann and Brauner 2006; Biberg 2007), here they will be treated as constants. The average void fraction at the wall is given by the fraction of the wetted perimeter occupied by phase k, $\alpha_{\mathrm{w}k} = P_{\mathrm{w}k}/P_\mathrm{w}$. While $\alpha_{\mathrm{w}k} \neq \alpha_k$ in general, for the present case of purely stratified flow it is given by $\alpha_{\mathrm{w}k} = (W + 2\alpha_k H)/P_\mathrm{w}$.

Finally, by collecting the terms in Eqs. (A.12)–(A.21), the resulting incompressible, isothermal, 1D TFM equations for horizontal stratified flow in a rectangular channel become: for horizontal stratified flow in a rectangular channel become:

$$\boxed{\frac{D_1\alpha_1}{Dt} + \alpha_1 \frac{\partial u_1}{\partial x} = 0,} \tag{A.22}$$

$$\boxed{\frac{D_2\alpha_2}{Dt} + \alpha_2 \frac{\partial u_2}{\partial x} = 0,} \tag{A.23}$$

$$\boxed{\begin{aligned} \rho_1 \frac{D_1 u_1}{Dt} = &-\frac{\partial p_{2\mathrm{i}}}{\partial x} - \rho_1 g_y H \frac{\partial \alpha_1}{\partial x} + \sigma H \frac{\partial^3 \alpha_1}{\partial x^3} + \frac{\rho_1}{\alpha_1}\frac{\partial}{\partial x}\left(\alpha_1 \nu_1^{\mathrm{eff}} \frac{\partial u_1}{\partial x}\right) \\ &+ \rho_1 g_x - \frac{W + 2\alpha_1 H}{\alpha_1 A}\frac{f_1}{2}\rho_1 |u_1| u_1 - \frac{1}{\alpha_1 H}\frac{f_\mathrm{i}}{2}\rho_2 |u_2 - u_1|(u_1 - u_2), \end{aligned}} \tag{A.24}$$

$$\boxed{\begin{aligned} \rho_2 \frac{D_2 u_2}{Dt} = &-\frac{\partial p_{2\mathrm{i}}}{\partial x} + \rho_2 g_y H \frac{\partial \alpha_2}{\partial x} + \frac{\rho_2}{\alpha_2}\frac{\partial}{\partial x}\left(\alpha_2 \nu_2^{\mathrm{eff}} \frac{\partial u_2}{\partial x}\right) + \rho_2 g_x \\ &- \frac{W + 2\alpha_2 H}{\alpha_2 A}\frac{f_2}{2}\rho_2 |u_2| u_2 - \frac{1}{\alpha_2 H}\frac{f_\mathrm{i}}{2}\rho_2 |u_2 - u_1|(u_2 - u_1). \end{aligned}} \tag{A.25}$$

Appendix B: Mathematical Background

B.1 Introduction

The topics of fluid wave analyses presented in this appendix provide an introduction to the mathematical methods that are used in the book for the analysis of the TFM stability. Unfortunately most of these topics are rarely encountered in introductory texts of fluid mechanics or undergraduate mathematics so we have collected them here.

The 1D TFM linear stability analysis is based on wave mechanics (i.e., Fourier analysis). Therefore, it is logical to start with the simplest linear mathematical wave model which is the one-way wave equation. The advantage of starting at such a basic model is that this equation has an analytic solution, whereas the TFM generally does not. Furthermore, its linear stability analysis is straightforward so that the characteristic and dispersion analyses used throughout the book may be understood more easily. The one-way wave equation also allows the mathematical definition of an ill-posed model using the dispersion relation. This alleged aspect of the TFM is one of the driving ideas of the book. At this point we will only define it as a mathematical property of the model, i.e., as a singularity in the growth rate of a perturbation as the wavelength approaches zero. A more physical rationale for this behavior is explored throughout the book. In the end, the book advocates that the TFM need not be ill-posed under any circumstance, but it is the pursuit of the question, rather than the outcome, that leads to a better understanding of TFM stability. In this sense the ill-posed question is very valuable.

Another key endeavor is nonlinear wave analysis and so the most basic place to start is Burgers' equation, which is a well-understood equation of modern mathematics, and again it has an analytic solution. The Drift-Flux void propagation equation follows immediately because of its similarity to Burgers' equation. Anybody familiar with two-phase flow analysis will appreciate the introduction of the Drift-Flux Model, which is one of the most successful theories of two-phase flow. What may not be so familiar is the demonstration that the similarity of the void

M.L. de Bertodano et al., *Two-Fluid Model Stability, Simulation and Chaos*,
DOI 10.1007/978-3-319-44968-5

propagation equation to Burgers' equation implies that material shocks and expansion waves are ubiquitous in the material wave behavior of the 1D TFM.

We are seldom able to obtain an analytical solution to the TFM, and numerical methods play a central role in two-phase flow modeling, so the third topic that is covered is the numerical stability of finite difference schemes. Since it is important to separate the numerical stability from the stability of the model, the von Neumann stability analysis is of key importance. The von Neumann analysis is defined in terms of a growth matrix which is different to the growth rates obtained from the dispersion analysis of the differential model, but fortunately it turns out that they are simply related. Various explicit and implicit first-order finite difference methods used in industrial codes are analyzed. Then some well-known second-order methods that are better suited for unsteady wavy flows are also analyzed. The compromises that are always made, consciously or not, in numerical mathematics are highlighted.

The Shallow Water Theory (SWT) of Whitham (1974) is then introduced and the dispersion analysis is performed to identify the kinematic instability. This model is of great significance to us because it is closely related to the TFM material wave linear stability and nonlinear simulations performed in various chapters. Chapter 2 illustrates how close this relation is, where the incompressible TFM is reduced exactly to SWT using the fixed flux assumption. The stability of SWT has been extensively researched in the past and these important results may now be applied directly to the TFM.

Finally, there is a very brief and incomplete introduction to chaos theory. This fascinating subject has made great strides in the past decades and the reader is encouraged to pursue a more extensive and authoritative introduction, e.g., Strogatz (1994).

B.2 Linear Stability

B.2.1 One-Way Wave Equation

The one-way wave equation is the simplest of the hyperbolic equations, or initial value PDEs,

$$\frac{\partial u}{\partial t} + c\frac{\partial u}{\partial x} = 0 \tag{B.1}$$

where u is, for now, an unspecified dimensionless variable. The solution, obtained by the method of characteristics, is $u(x,t) = u_0(x - ct)$, so the amplitude remains constant along characteristic lines $x - ct = x_0$. For example, Fig. B.1 shows the case for $c = 1\,\mathrm{m/s}$ and $u_0(x) = \frac{1}{\sqrt{2\pi}}\mathrm{e}^{-\frac{1}{2}x^2}$. In this case the normal distribution function propagates along the characteristic lines unchanged.

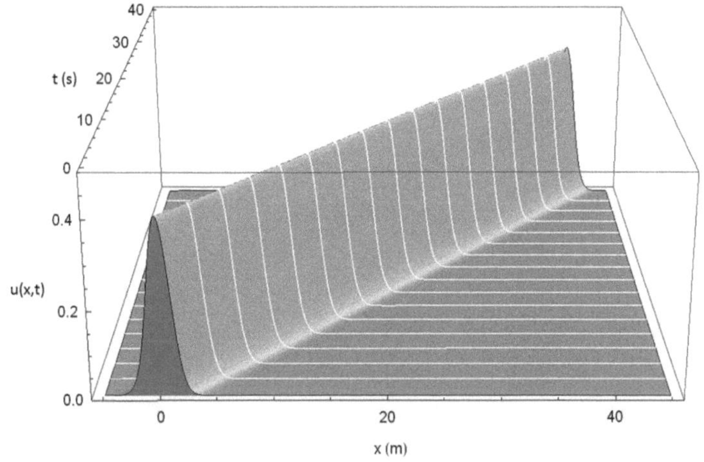

Fig. B.1 One-way wave equation wave

Adding a source term, the wave equation becomes

$$\frac{\partial u}{\partial t} + c\frac{\partial u}{\partial x} = au, \tag{B.2}$$

where a is constant. In this case the solution is

$$u(x,t) = u_0(x - ct)\mathrm{e}^{at}, \tag{B.3}$$

which is shown in Fig. B.2 for the case $a = 0.02$. The amplitude of the wave increases exponentially but the width remains constant. This case leads to the definition of instability that will be carried throughout this text: the equation is unstable if $a > 0$ and stable otherwise.

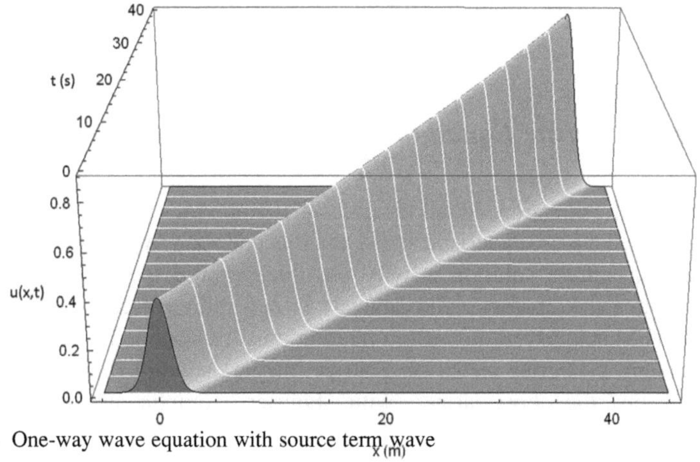

Fig. B.2 One-way wave equation with source term wave

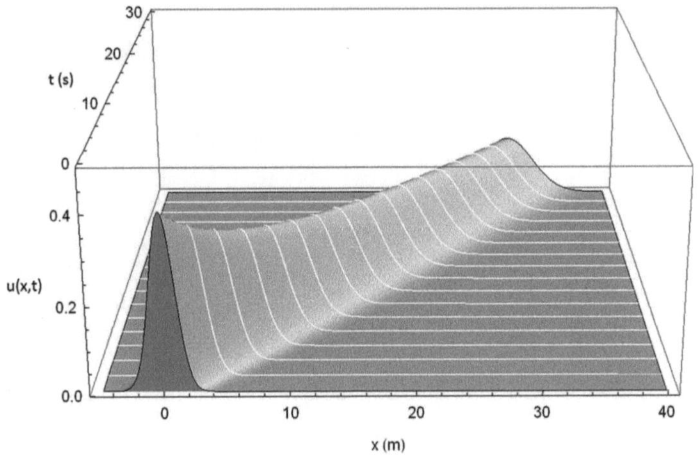

Fig. B.3 One-way wave equation with viscous diffusivity wave

The one-way wave equation may include a diffusion term instead to obtain the well-known convection–diffusion equation,

$$\frac{\partial u}{\partial t} + c\frac{\partial u}{\partial x} = \nu\frac{\partial^2 u}{\partial x^2}. \tag{B.4}$$

The equation is now parabolic, not hyperbolic, and has a well-known solution (Strang 2007),

$$u(x,t) = \frac{1}{\sqrt{4\pi t}}\int_{-\infty}^{\infty} u_0(s)e^{-\frac{(x-s-ct)^2}{4\nu t}}ds. \tag{B.5}$$

The solution for $\nu = 0.01$ is shown in Fig. B.3. In this case the amplitude of the wave decreases and the width increases as the wave travels along the characteristic.

B.2.2 Characteristic Equation and Dispersion Relation

The characteristic equation, for all the cases in the previous section, is $x - ct = x_0$, where c is called the characteristic speed and it is real for a hyperbolic equation. Of course such a simple solution is only possible because the equation is linear. However, since the characteristic analysis is not sufficient for some important aspects of TFM stability, a dispersion relation must be obtained too. For an initial value problem a Fourier series or Fourier integral solution may be proposed, depending on the boundaries, with Fourier components of the form $u = u_0 e^{i(kx-\omega t)}$ where k is the wave number related to the wavelength by $k = \frac{2\pi}{\lambda}$.

Inserting this into Eq. (B.4) the dispersion relation that relates the frequency to the wave number is obtained

$$\omega = ck - i\nu k^2. \tag{B.6}$$

which is in fact part of the solution. The other part is the amplitude of the Fourier components to match the initial condition. Two quantities derived from the dispersion relation are the wave speed and the group speed. The wave speed is given by $c_{\text{wave}} = \frac{\omega}{k}$,

$$c_{\text{wave}} = c - i\nu k. \tag{B.7}$$

In this case the wave speed has a negative imaginary component proportional to the wave number, which means that the short wave components are being dissipated. The imaginary component of the wave speed is related to the speed at which the wave decays due to viscous dissipation and not to the translational speed of the wave. An important aspect of the wave speed obtained from the dispersion relation is that in the limit as $k \to \infty$ the real part becomes the characteristic speed and the equation is well-posed if the imaginary part is equal or less than zero. This may be generalized to state that the characteristic speed is obtained from the dispersion analysis in the limit of infinitesimal wavelengths.

The group speed is given by $c_{\text{group}} = \frac{d\omega}{dk}$,

$$c_{\text{group}} = c - 2i\nu k. \tag{B.8}$$

The dotted line in Figure B.4 shows the dispersion relation in terms of the wavelength, $\lambda = \frac{2\pi}{k}$. When $\nu > 0$ Eq. (B.6) determines that $\mathrm{e}^{-\omega_i t} = \mathrm{e}^{-\nu k^2 t}$ so the wave growth is negative and the wave amplitude decreases by viscous dissipation as shown in Fig. B.3. This is a desirable feature of the convection–diffusion equation, i.e., a parabolic problem, and the model is called "well-posed." When $\nu = 0$, the equation becomes hyperbolic and the growth rate is zero for all wavelengths as shown by the dashed line. Now let's suppose that the dependent variable has an imaginary component, $u = u_r + iu_i$, which is common practice in electromagnetic wave theory, and also $c = c_r + ic_i$ where c_i represents the artificially prescribed wave growth speed. Then $\mathrm{e}^{-\omega_i t} = \mathrm{e}^{\left(c_i k - \nu k^2\right)t}$. When $\nu = 0$ and $c_i > 0$ the wave growth rate goes to infinity as $\lambda \to 0$. This is called an "ill-posed" problem and of course it makes no physical sense. Similar ill-posed behavior emerges in incomplete TFMs in a more subtle way due to the Kelvin–Helmholtz instability. Finally consider the case when $\nu > 0$ and $c_i > 0$. Now Fig. B.4 shows a cutoff wavelength below which the wave growth rate becomes negative. This is called a "well-posed" problem but it is unstable for wavelengths larger than the cutoff. It is important to distinguish the difference between an unstable problem and an "ill-posed" one, i.e., while unstable flows exhibit wave growth beyond a cutoff wavelength, ill-posed problems result in unbounded growth as $\lambda \to 0$.

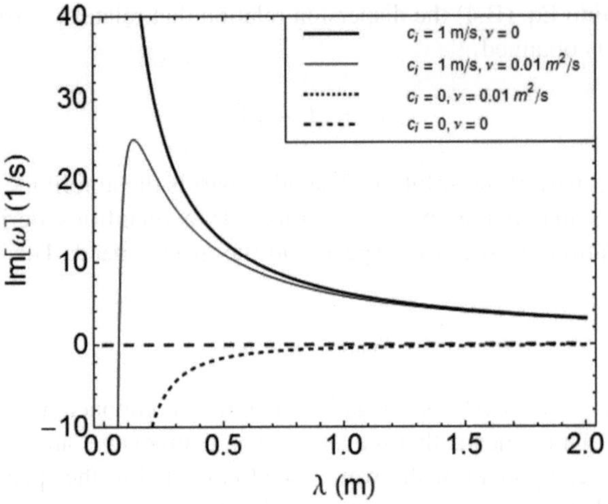

Fig. B.4 Dispersion relation for one-way wave equation with artificially imposed wave growth

Of course the main advantage of the dispersion relation is that it predicts the wave growth for all wavelengths, but it does more than that. Consider the linearized Korteweg–de Vries equation,

$$\frac{\partial u}{\partial t} + c\frac{\partial u}{\partial x} = -\sigma\frac{\partial^3 u}{\partial x^3} \quad \rightarrow \quad \omega = ck - \sigma k^3. \tag{B.9}$$

The real wave speed $c_{\text{wave}} = c - \sigma k^2$ is not constant but depends on the wavelength so that waves of different wavelengths "disperse." For instance, it is well known that surface tension causes dispersion in ripples.

B.3 Nonlinear Models

B.3.1 Burgers' Equation

The formation of kinematic shocks is a significant phenomenon of the 1D TFM. Kinematic shocks are well known; they are described by Wallis (1969) for two-phase flows and by Whitham (1974) for surface nonlinear waves of the type

$$\frac{\partial u}{\partial t} + c(u)\frac{\partial u}{\partial x} = 0. \tag{B.10}$$

A simple example of relevance to the TFM is Burgers' equation,

$$\frac{\partial u}{\partial t} + u\frac{\partial u}{\partial x} = \nu\frac{\partial^2 u}{\partial x^2} \tag{B.11}$$

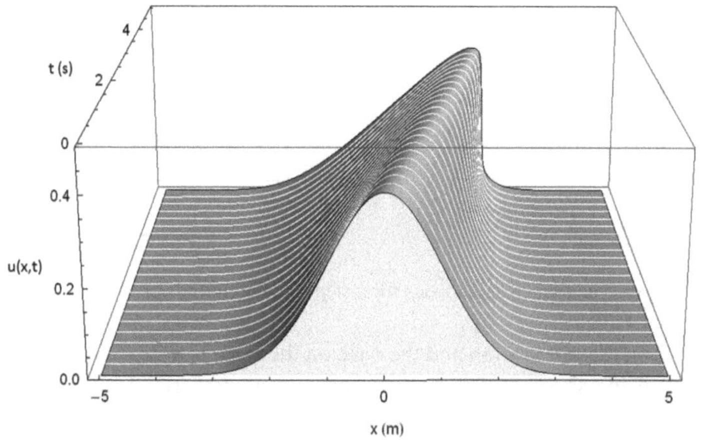

Fig. B.5 Burgers equation wave

A solution to this equation with $\nu = 0$ is shown in Fig. B.5. The initial condition is a normal (Gaussian) wave, $u(x,0) = \frac{1}{\sqrt{2\pi}\sigma_s} e^{-(x-\mu)^2/2\sigma_s^2}$ with center at $\mu = 0$ and standard deviation $\sigma_s = 1$. It is possible to see the two nonlinear features of interest: a shock and a fan that develop simultaneously.

The general solution to Burgers' equation can be obtained using the method of characteristics, described by Whitham (1974). The solution propagates unchanged along characteristic lines $x - ut = x_0$ such that

$$\frac{dx}{dt} = u(x,t), \qquad \frac{du}{dt} = 0. \tag{B.12}$$

and the solution to this pair of ode's is $x(t) = x_0 + u(x_0,0)t$, $u(x,t) = u(x_0,0)$ or

$$u(x_0,t) = u(x_0 + u(x_0,0)t) \tag{B.13}$$

This is an implicit solution that is valid as long as the characteristic lines do not intersect. However, if $\frac{du(x_0,0)}{dx_0} < 0$, the characteristics intersect and form a shock. This is illustrated by the Riemann problem, a fundamental solution of Burgers' equation for the simple case of a step function initial condition. For example, if $u_0(x) = 0,\ 1$ for $x < 0$ and $u_0(x) = 1, 0$ for $x \geq 0$, Fig. B.6 shows the two different sets of characteristics along which the solution propagates.

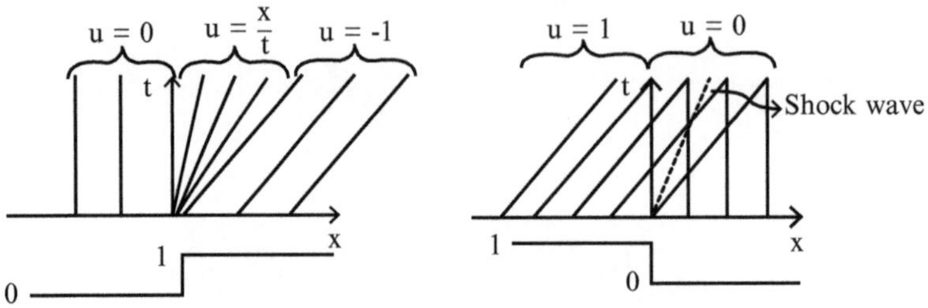

Fig. B.6 Burgers equation characteristics for a step function initial condition

The case on the left is a fan and the case on the right is a shock. The left and right fields beyond the shock or fan remain constant as prescribed by Eq. (B.11). The speed of the shock is obtained by integrating the conservative form of Burgers' equation in a fixed interval $[x_L, x_R]$,

$$\int_{x_L}^{x_R} \left(\frac{\partial}{\partial t} u + \frac{\partial}{\partial x} \frac{u^2}{2} \right) dx = 0 \tag{B.14}$$

or

$$\frac{\partial}{\partial t} \int_{x_L}^{x_R} u \, dx + \frac{u_R^2}{2} - \frac{u_L^2}{2} = 0. \tag{B.15}$$

If the position of the shock is $X(t)$ then the integral Eq. (B.15) may be performed by parts

$$\frac{\partial}{\partial t} [(X - x_L) u_L + (x_R - X) u_R] + \frac{u_R^2}{2} - \frac{u_L^2}{2} = 0 \tag{B.16}$$

where x_L, x_R, u_L, u_R are constant. Then

$$\frac{dX}{dt} (u_L - u_R) + \frac{u_R^2}{2} - \frac{u_L^2}{2} = 0 \tag{B.17}$$

which has the simple solution

$$\frac{dX}{dt} = \frac{1}{2} (u_R + u_L) \tag{B.18}$$

Burgers' equation with $\nu > 0$ may be solved using the Cole–Hopf transformation which converts it into the heat equation

$$u = -2\nu \frac{1}{\phi} \frac{\partial \phi}{\partial x} \quad \rightarrow \quad \frac{\partial \phi}{\partial t} = \nu \frac{\partial^2 \phi}{\partial x^2} \tag{B.19}$$

with solution

$$u(x,t) = -2\nu \frac{\partial}{\partial x} \ln \left(\frac{1}{\sqrt{4\pi \nu t}} \int_{-\infty}^{\infty} e^{\left(-\frac{(x-s)^2}{4\nu t} - \frac{1}{2\nu} \int_0^s u(\xi, 0) d\xi \right)} ds \right) \tag{B.20}$$

Consider next Burgers' equation with additional source term

$$\frac{\partial u}{\partial t} + u\frac{\partial u}{\partial x} = au + \nu\frac{\partial^2 u}{\partial x^2} \tag{B.21}$$

There is a significant difference when the nonlinear behavior of Burgers' equation is compared to the linear one-way wave equation. Nonlinear stability analysis may be performed analytically by various methods like perturbation theory (Whitham 1974). Such analyses require mathematical tools that are not employed in this book. However, we will use numerical simulations quite often to demonstrate nonlinear effects.

To illustrate the difference between linear and nonlinear waves, Figs. B.7 and B.8 show the evolution of a solitary wave for the linear wave equation and for Burgers' equation with $c = 1/4$, $a = 0.02$ and $\nu = 0.01$. In the linear case the wave grows exponentially but in the nonlinear case it does so only during a short initial time interval. Afterward the amplitude decreases toward an asymptotic value because the wave growth generated by the source term au is dissipated at the shock by the viscous diffusivity and a quasi-equilibrium is reached. From a Fourier analysis perspective this may be seen as a transfer of energy from long wavelengths where the energy is generated to short ones where it is dissipated, i.e., a large amount of dissipation occurs at the shock. This is strictly a nonlinear phenomenon and is somewhat similar to the well-known turbulent eddy cascade mechanism, except that in 1D a shock occurs instead.

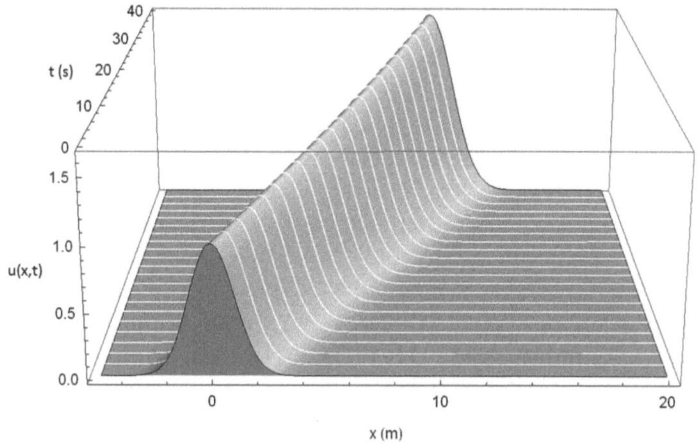

Fig. B.7 One-way wave equation with $c = 1/4$, $a = 0.02$, and $\nu = 0.01$

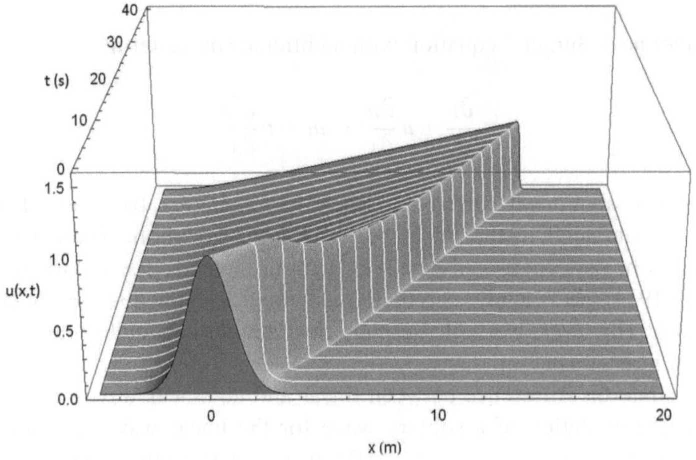

Fig. B.8 Burgers equation with $c = 1/4,\ a = 0.02$, and $\nu = 0.01$

B.3.2 Drift-Flux Void Propagation Equation

In Chap. 6 we obtain the Drift-Flux void propagation equation from a severe simplification of the TFM for bubbly flows:

$$\frac{\partial \alpha}{\partial t} + \left(C_0 j + V_{gj} + \alpha \frac{dV_{gj}}{d\alpha_2} \right) \frac{\partial \alpha}{\partial x} = 0 \tag{B.22}$$

where α is the gas volume fraction, or void fraction, and j is the total volumetric flux. The distribution parameter, C_0, and the drift velocity, V_{gj}, first defined by Zuber and Findlay (1965), are discussed in Chap. 6. Let us choose arbitrarily $C_0 = 1$, and $V_{gj} = V_{gj0}(1-\alpha)^n$. The important thing here is that we have a one-way wave equation model derived from the TFM that exhibits nonlinear wave propagation behavior. Let us consider a simple void solitary wave propagating along a bubble column with a uniform void fraction of 10 %. We select the drift velocity correlation $V_{gj} = 0.162(1-\alpha)^{1.75}\,\mathrm{m/s}$ for upward bubbly flow at atmospheric conditions corresponding to Eq. (6.32). The wave propagates upward along the channel as shown in Fig. B.9 and evolves nonlinearly into a backward facing shock. Comparing with Fig. B.5 this is the mirror image of Burger's solution and results from the fact that the nonlinearity in V_{gj} is a decreasing function of α. This model is stable in a local sense because all unstable behavior has been removed, as will be explained in Chap. 6.

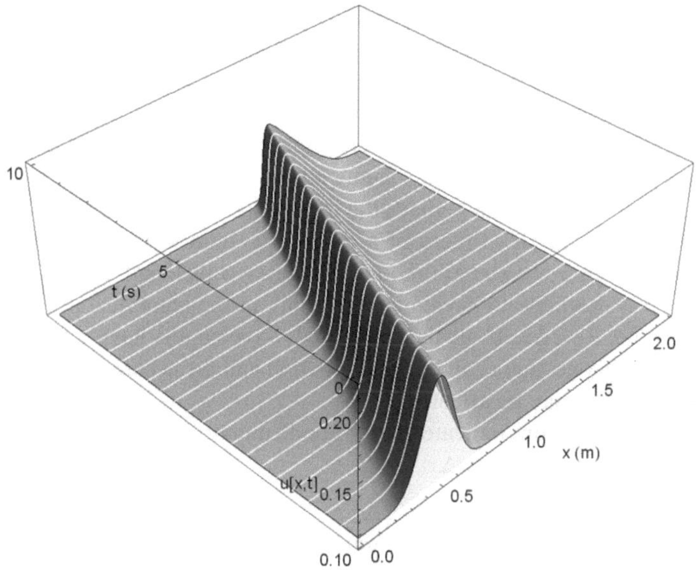

Fig. B.9 Void propagation equation wave

B.4 Numerical Stability

We now consider numerical stability since the TFM is seldom solved analytically. In any numerical solution procedure it is necessary to choose a spatial nodalization which is consistent with the shortest wavelength phenomena of interest. A solution component wavelength twice the spatial node spacing is the shortest wavelength that can be represented in the numerical scheme.

This cutoff point and associated node spacing should be established as a result of some added knowledge about the initial conditions and system behavior (Richtmeyer and Morton 1967). While the shortest wavelength that can be represented is $\lambda = 2\Delta x$, most numerical schemes seriously affect longer wavelengths as well. In formulating a numerical scheme and judging its characteristics, it is first necessary to decide upon the physical range of interest and establish a wavelength, λ_o, below which representation is not required. This limit should be chosen sufficiently small that the neglected components do not seriously affect the longer wavelength phenomena of interest. The cutoff wavelength would generally be chosen below a point where the differential equation system has an amplification factor less than unity.

Once a cutoff wavelength has been established, a numerical scheme must be constructed to give a satisfactory approximation to the differential system response for solution components having $\lambda > \lambda_o$. Since any numerical scheme has some distortion at wavelengths of order $2\Delta x$, the cutoff point, λ_o, should be at a wavelength longer than $2\Delta x$.

Even though asymptotic stability only requires a numerical scheme to have an amplification factor of order unity, there are practical reasons to require an amplification factor less than unity at the shortest wavelength, $\lambda = 2\Delta x$. First, the wavelengths shorter than λ_o are assumed to be unimportant and are not represented accurately. Thus, to keep these wavelengths from growing and affecting the solution of interest, the numerical scheme amplification factor for $\lambda = 2\Delta x$ should be less than one. Second, and perhaps the more important reason, in order to obtain an accurate numerical description of a physical system in which energy is transferred from the long wavelength modes to shorter wavelength modes, where it is dissipated, it is desirable to have an energy sink in the numerical scheme at short wavelength, i.e. $\lambda \sim 2\Delta x$. If all the numerical wavelengths have an amplification factor larger than one, there can be no dissipation and the accuracy of even the long wavelength modes may be affected.

Experience and theory indicate that a "good" numerical scheme should have the following properties for finite Δx:

1. Bounded growth rates for all $\lambda > 2\Delta x$,
2. Dissipation at the shortest wavelength, i.e., $|G(2\Delta x)|_{\max} < 1$,
3. As accurate as possible for all wavelengths $\lambda > \lambda_o$.

B.4.1 First-Order Explicit Upwind Scheme

The simplest scheme to solve the one-way wave equation numerically is the explicit first-order upwind scheme (FOU), i.e., Euler's scheme,

$$\frac{u_j^{n+1} - u}{\Delta t} + c\frac{u_j^n - u_{j-1}^n}{\Delta x} = 0, \tag{B.23}$$

where u is positive. The truncation error of this scheme is first order with respect to the discretization intervals, i.e., $\frac{1}{2}\Delta t\frac{\partial^2 u}{\partial t^2} + \frac{1}{2}c\Delta x\frac{\partial^2 u}{\partial x^2}$. We now proceed with the von Neumann analysis. Defining the Courant number $Co = \frac{c\Delta t}{\Delta x}$ Eq. (B.23) may be written as

$$u_j^{n+1} = (1 - Co)u_j^n + Co\, u_{j-1}^n. \tag{B.24}$$

Inserting $u_j^n = u_0 \mathrm{e}^{i\left(kj\Delta x - \omega_I^\Delta n\Delta t\right)}$ to obtain the dispersion relation yields

$$\mathrm{e}^{-i\omega_I^\Delta \Delta t} = 1 - Co + Co\,\mathrm{e}^{-ik\Delta x}. \tag{B.25}$$

The term on the RHS is called the von Neumann growth factor, $G = \mathrm{e}^{-i\omega_{\mathrm{FD}}\Delta t}$, so

$$G = 1 - Co + Co\,\mathrm{e}^{-ik\Delta x} \tag{B.26}$$

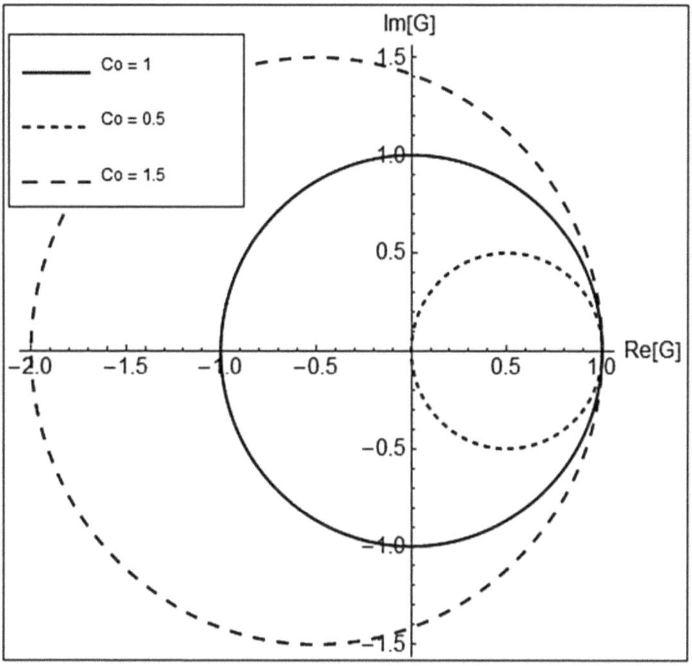

Fig. B.10 Growth rate diagram for FOU one-way wave equation

such that

$$u_j^{n+1} = G u_j^n. \tag{B.27}$$

Figure B.10 shows the growth factor plotted in the imaginary plane for $0 < k\Delta x < 2\pi$ and $Co = 0.5,\ 1,\ 1.5$. The most important feature of this plot is that $|G| \leq 1$ for $|Co| \leq 1$. It turns out that the growth factor becomes greater than 1 when $Co > 1$. This is the well-known Courant–Friedricks–Lewy stability condition. The next question is what happens when the equation is unstable.

Consider the following explicit scheme for the unstable wave equation,

$$\frac{u_j^{n+1} - u_j^n}{\Delta t} + c\frac{u_j^n - u_{j-1}^n}{\Delta x} = a u_j^n. \tag{B.28}$$

The growth factor for this equation turns out to be

$$G = 1 + a\Delta t - Co + Co\,\mathrm{e}^{-ik\Delta x} \tag{B.29}$$

which is the same as the growth factor for the stable wave with the addition of $a\Delta t$. Therefore, for an unstable equation the von Neumann criteria for stability require that the magnitude of the growth factor for a first-order scheme satisfies the inequality

$$|G| \leq 1 + a\Delta t. \tag{B.30}$$

However, to compare the stability of the numerical method with the stability of the differential model it is more convenient to use the growth rate than the growth factor. The relation between the growth factor and the growth rate, ω_{FD}, is given by

$$G = \mathrm{e}^{-i\omega_{\mathrm{I}}^{\Delta}\Delta t} \rightarrow \omega_{\mathrm{I}}^{\Delta} = -i\frac{\ln[G]}{\Delta t}. \tag{B.31}$$

Figure B.11 shows the growth rate compared to the dispersion relation with $Co = 0.5, a = 2$, for mesh sizes $\Delta x = 0.001,\ \ 0.01\,\mathrm{m}$. It can be seen that the upwind scheme damps the growth rate of the short wavelengths, but the growth rates tend to the right asymptotic value for long wavelengths. This effect is called numerical viscosity.

Figure B.12 shows the comparison of the wave speed, ω/k compared with the one obtained by the dispersion relation (i.e., $c_{\mathrm{wave}} = 1\,\mathrm{m/s}$). Here, the wave speed approaches the exact value very quickly beyond the minimum resolved wavelength, $2\Delta x$, so the wave dispersion is minimal.

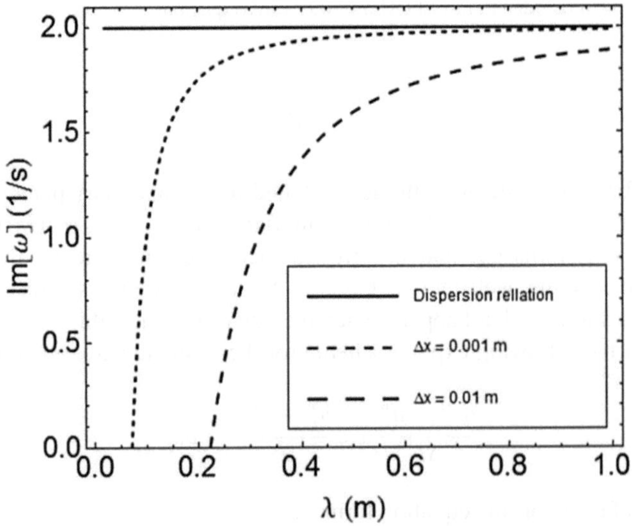

Fig. B.11 Von Neumann analysis and dispersion relation for FOU one-way wave equation

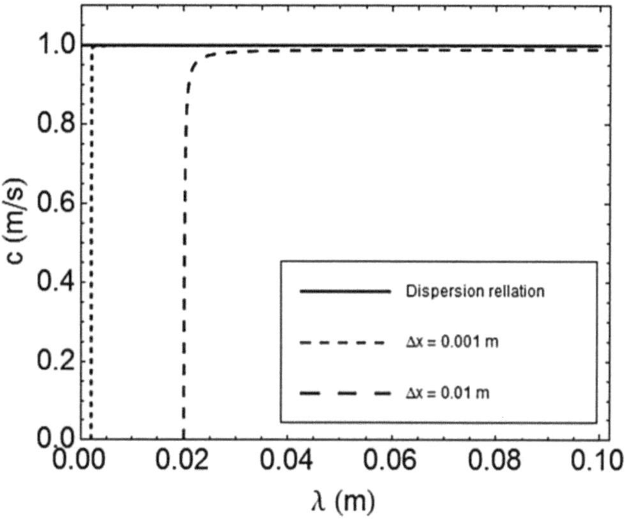

Fig. B.12 Wave speed for FOU one-way wave equation

The next step is to consider the effect of diffusivity,

$$\frac{u_j^{n+1} - u_j^n}{\Delta t} + c\frac{u_j^n - u_{j-1}^n}{\Delta x} = au_j^n + v\frac{u_j^{n+1} - 2u_j^{n+1} + u_{j-1}^{n+1}}{\Delta x^2}, \tag{B.32}$$

where the second derivative on the RHS is implicit to forego stability constraints. This is standard practice in 1D TFM codes. Defining the diffusion number $N_D = \frac{\nu \Delta t}{\Delta x^2}$, the dispersion relation is

$$G = \frac{1 + a\Delta t - Co + Co\,e^{-ik\Delta x}}{1 + 2N_D[1 - \cos(k\Delta x)]}. \tag{B.33}$$

Figure B.13 is the comparison of the von Neumann growth rates with the dispersion analysis of the model without diffusivity. The result is similar to the previous case except that the diffusivity is added to the numerical viscosity already present.

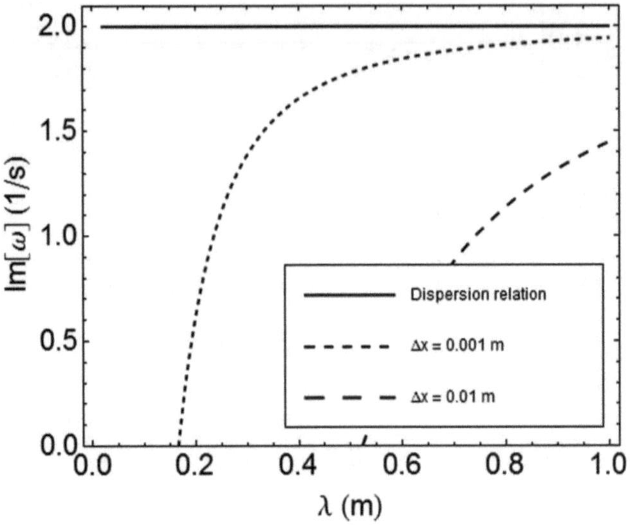

Fig. B.13 Von Neumann analysis and dispersion relation for FOU one-way wave equation with viscosity

B.4.2 First-Order Implicit Upwind Scheme

The one-way wave equation may also be solved numerically with the implicit upwind scheme

$$\frac{u_j^{n+1} - u_j^n}{\Delta t} + c\frac{u_j^{n+1} - u_{j-1}^{n+1}}{\Delta x} = 0 \tag{B.34}$$

which may be written as

$$(1 + Co)u_j^{n+1} - Co\, u_{j-1}^{n+1} = u_j^n. \tag{B.35}$$

The von Neumann growth factor now is

$$G = \frac{1}{1 + Co - Co\,\mathrm{e}^{-ik\Delta x}}. \tag{B.36}$$

It is clear that the magnitude of G is always less than one. Figure B.14 shows the growth factor plotted in the imaginary plane for $0 < k\Delta x < 2\pi$ and $Co = \frac{1}{2}, 1,\ 5$. The most important feature of this plot is that all the circles lie within the unit circle so $|G| \leq 1$ unconditionally.

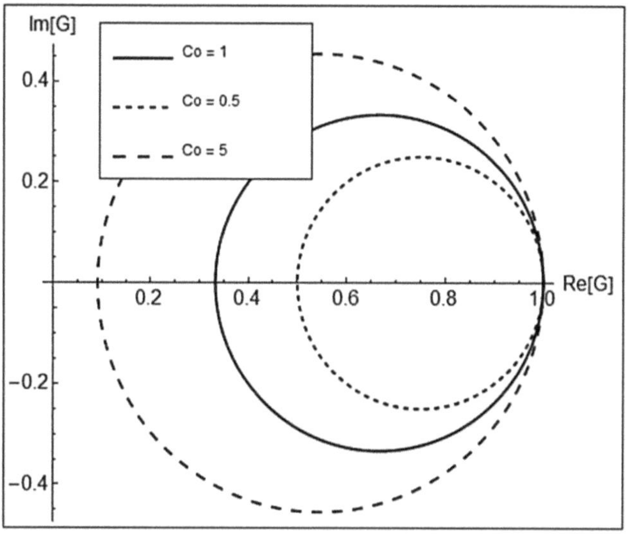

Fig. B.14 Growth rate diagram for implicit FOU one-way wave equation

Consider now the implicit scheme for the unstable wave equation

$$\frac{u_j^{n+1} - u_j^n}{\Delta t} + c\frac{u_j^{n+1} - u_{j-1}^{n+1}}{\Delta x} = au_j^{n+1}. \tag{B.37}$$

The growth factor for this equation turns out to be:

$$G = \frac{1}{1 - a\Delta t + Co - Co\,\mathrm{e}^{-ik\Delta x}} \tag{B.38}$$

Now for an unstable equation the von Neumann criteria for stability require that the magnitude of the growth factor for a first-order scheme satisfies the inequality

$$|G| \leq 1 + a\Delta t \approx \frac{1}{1 - a\Delta t} \tag{B.39}$$

which is always satisfied unlike to the explicit result.

Figure B.15 shows the growth rate of the finite difference model and the differential model for $a = 2$, mesh sizes $\Delta x = 0.001,\ 0.01\,\mathrm{m}$, and $Co = \frac{1}{2},\ 5$. For $Co = \frac{1}{2}$. It can be seen that for small values of Co the upwind scheme damps the growth rate of the short wavelengths in a similar fashion as the explicit scheme but for $Co = 5$ the numerical damping effect is much greater.

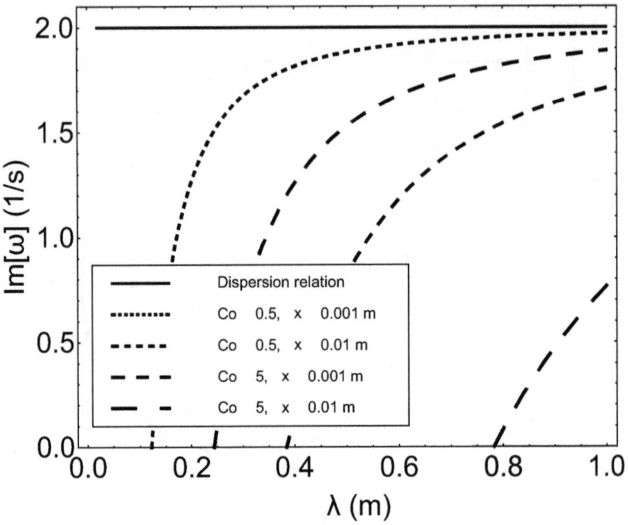

Fig. B.15 Von Neumann analysis and dispersion relation for implicit FOU one-way wave equation

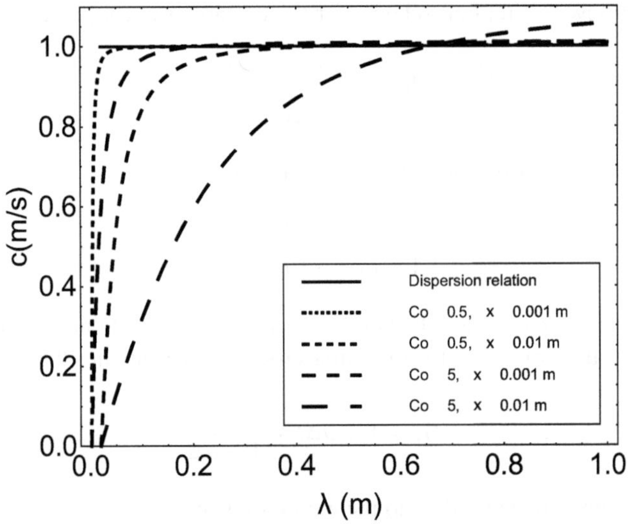

Fig. B.16 Wave speed for implicit FOU one-way wave equation

Figure B.16 shows the wave speed from the dispersion relation. The implicit scheme adds more diffusion to the solution than the explicit scheme and the effect for $Co = 5$ is large again. So the cost of numerical stability for $Co > 1$ is the loss of fidelity in wave propagation predictions. To illustrate this point Fig. B.17 shows the simulations of a wave with the explicit and the implicit schemes with $Co = 0.9$. The wave travels to the right and numerical diffusion spreads it as it propagates. It is clear that the fidelity of the implicit solution is poorer.

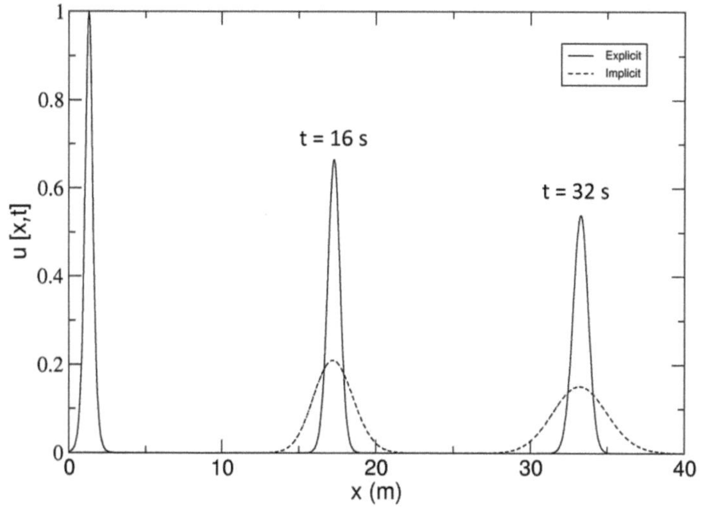

Fig. B.17 First-order schemes, $Co = 0.9$

B.4.3 Second-Order Explicit Scheme

The one-way wave equation may be solved numerically with the well-known explicit second-order Lax–Wendroff scheme (Strang 2007),

$$\frac{u_j^{n+1} - u_j^n}{\Delta t} + c\frac{u_{j+1}^n - u_{j-1}^n}{2\Delta x} = c^2\frac{\Delta t}{2}\frac{u_j^n - 2u_j^n + u_{j-1}^n}{\Delta x^2} \tag{B.40}$$

An interesting aspect of this scheme is that the convection term is center differenced and without the diffusion term on the RHS the scheme would be unconditionally unstable. The use of numerical diffusion terms, somewhat similar to this, is a common practice to stabilize the ill-posed TFM in industrial codes, although the magnitude of the numerical diffusion is much greater. Equation (B.41) may be written as

$$u_j^{n+1} = u_j^n - \frac{Co}{2}\left(u_{j+1}^n - u_{j-1}^n\right) + \frac{Co^2}{2}\left(u_{j+1}^n - 2u_j^n + u_{j-1}^n\right) \tag{B.41}$$

and the equation for the growth factor as

$$G = 1 - Co^2 + \frac{1}{2}\left(Co^2 - Co\right)e^{ik\Delta x} + \frac{1}{2}\left(Co^2 + Co\right)e^{-ik\Delta x} = 0. \tag{B.42}$$

Figure B.18 shows the growth factor plotted in the real–imaginary plane for $0 < k\Delta x < 2\pi$ and $Co = 0.5$, 1.0, 1.5. The stability condition traces an ellipse at the point of contact with the unit circle and this reflects the second-order accuracy of the scheme (Richtmeyer and Morton 1967). Figure B.19 shows the growth rate for $Co = 0.5$ and $\Delta x = 0.01$, 0.001 which should be compared with Fig. B.11 to appreciate the improvement in accuracy between the first- and second-order numerical schemes.

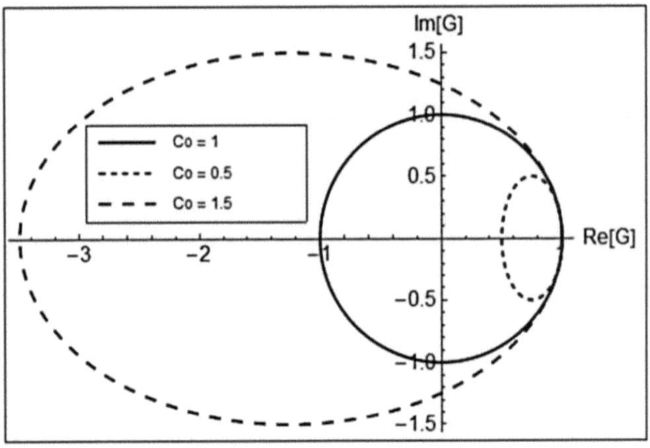

Fig. B.18 Growth rate diagram for Lax–Wendroff one-way wave equation

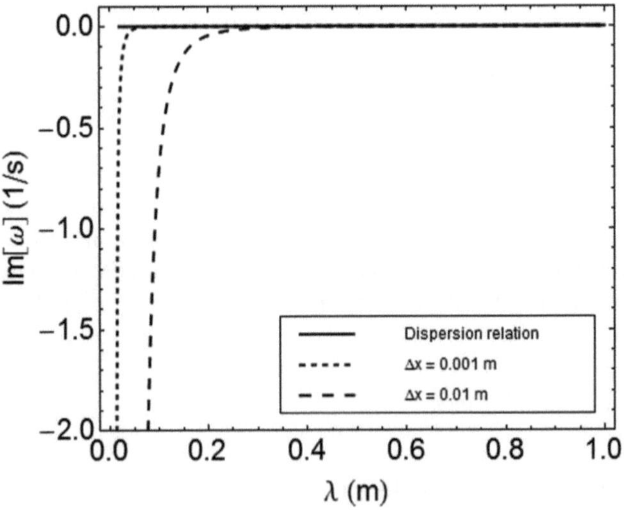

Fig. B.19 Von Neumann analysis and dispersion relation for Lax–Wendroff one-way wave equation

B.4.4 Second-Order Implicit Scheme

The well-known second-order implicit Crank–Nicholson or trapezoidal scheme,

$$u_j^{n+1} = u_j^n - \frac{Co}{4}\left(u_{j+1}^{n+1} + u_{j+1}^n - u_{j-1}^{n+1} - u_{j-1}^n\right), \tag{B.43}$$

is used for the calculations shown in Fig. B.20 with $Co = 0.9$. It is instructive to compare the Crank–Nicholson scheme with the Lax–Wendroff scheme. While the numerical diffusion is significantly reduced compared to first-order schemes for both explicit and implicit schemes, the implicit scheme introduces a significant amount of numerical dispersion to the solution, which in a second-order scheme becomes a distinct oscillatory feature.

Figure B.21 shows the solution for $Co = 5$. The numerical dispersion has become the dominant part of the solution and the result is of poor fidelity. The conclusion is that explicit or semi-implicit schemes are better for problems of wave propagation. On the other hand, implicit schemes are more convenient for quasi-steady problems when long time steps are desirable and wave propagation phenomena are not important.

While second-order schemes converge much faster than first-order schemes they are seldom used in industrial reactor safety codes because the TFMs used in the codes are ill-posed. The engineering solution has been to use first-order upwind schemes which provide significant numerical viscosity to dampen the short wavelength component the solution, but the necessary levels of numerical viscosity are only attained with coarse grids.

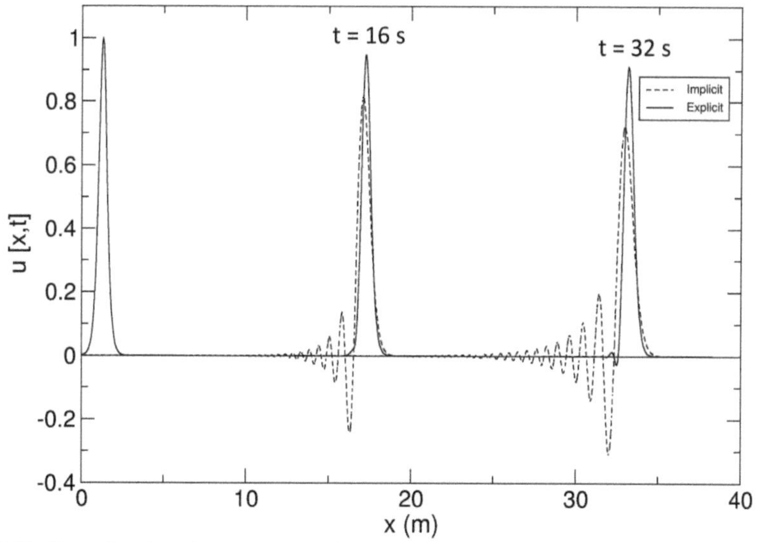

Fig. B.20 Second-order schemes, $Co = 0.9$

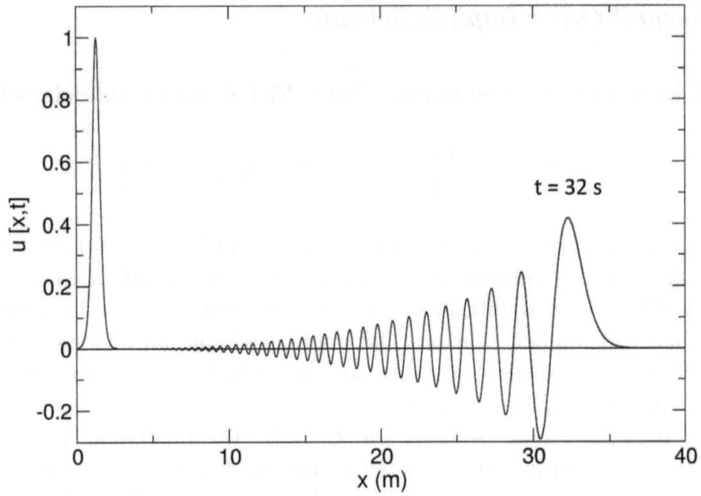

Fig. B.21 Second-order implicit scheme, $Co = 5$

B.5 Shallow Water Theory

SWT or the river flow equations is a 1D two-equation model of great significance because it turns out to be the foundation of the TFM material wave stability analyses presented in Ch. 2. Surface waves are well understood and they are described in extensive detail by Whitham (1974) who used nonlinear analysis to obtain the shocks and fans already presented for Burgers' equation.

B.5.1 Dispersion Relation

The shallow water theory equations are (Wallis 1969; Whitham 1974)

$$\frac{\partial \alpha}{\partial t} + u\frac{\partial \alpha}{\partial x} + \alpha\frac{\partial u}{\partial x} = 0, \tag{B.44}$$

$$\frac{\partial u}{\partial t} + u\frac{\partial u}{\partial x} - C\frac{\partial \alpha}{\partial x} = F, \tag{B.45}$$

where C and F are unspecified functions of the independent variables. As we shall see in Chap. 2, C is related to the Kelvin–Helmholtz stability condition and F includes interfacial and wall drag forces as well as gravity. Also of importance is the definition of the kinematic wave speed, or continuity wave speed (Wallis 1969) obtained from the kinematic condition, i.e., $F = 0$,

$$v_w = u + \alpha \left(\frac{\partial u}{\partial \alpha} \right)_F . \tag{B.46}$$

The linear stability of Eqs. (B.44) and (B.45) will now be considered. Firs the equations are written in matrix notation,

$$\boldsymbol{A} \frac{d}{dt} \underline{\phi} + \boldsymbol{B} \frac{d}{dx} \underline{\phi} + \underline{F} = 0, \tag{B.47}$$

where

$$\underline{\phi} = \begin{bmatrix} \alpha \\ u \end{bmatrix}, \quad \boldsymbol{A} = \boldsymbol{I}, \quad \boldsymbol{B} = \begin{bmatrix} u & \alpha \\ -C & u \end{bmatrix}, \quad \underline{F} = \begin{bmatrix} 0 \\ F \end{bmatrix}. \tag{B.48}$$

The dispersion relation for these equations is obtained from

$$\mathrm{Det}[-i\omega \boldsymbol{A} + ik\boldsymbol{B} + \boldsymbol{F}'] = 0, \tag{B.49}$$

where

$$\boldsymbol{F}' = \frac{\partial F}{\partial \underline{\phi}} = \begin{bmatrix} 0 & 0 \\ \dfrac{\partial F}{\partial \alpha} & \dfrac{\partial F}{\partial u} \end{bmatrix} \tag{B.50}$$

that results in

$$\omega = uk + i\frac{1}{2}\frac{\partial F}{\partial u} \pm \sqrt{\left(i\frac{1}{2}\frac{\partial F}{\partial u} \right)^2 - i\frac{\partial F}{\partial u} v_w k - (ck)^2} \tag{B.51}$$

where $v_w = u + \alpha \left(\frac{\partial u}{\partial \alpha} \right)_F = u - \alpha \frac{\partial F / \partial \alpha}{\partial F / \partial u}$ and $c = u \pm \sqrt{-C\alpha}$ is the dynamic wave speed. The dispersion relation may be rewritten as

$$\omega = uk + i\frac{1}{2}\frac{\partial F}{\partial u} \pm \sqrt{\left(ck + i\frac{1}{2}\frac{\partial F}{\partial u} \right)^2 + i\frac{\partial F}{\partial u}(c - v_w)k}. \tag{B.52}$$

If $v_w = c$ the last term in the square root cancels and the two roots become

$$\omega_1 = (u + c)k, \quad \omega_2 = (u - c)k + i\frac{\partial F}{\partial u}. \tag{B.53}$$

The first root has a zero imaginary component and the second root has a dissipative imaginary component. It can be shown that this is the marginal stability condition and that for $v_w > c$ the first root, ω_1, develops a positive imaginary component so the kinematic waves will grow.

B.5.2 Roll Waves on Spillways

Now to consider the physics that leads to a simple surface kinematic wave we add a wall friction force coefficient and a gravity force to the momentum equation, i.e.,

$$C = \frac{g_y H}{\alpha}, \quad F = -\frac{1}{H\alpha}\frac{f}{2}|u|u + g_x, \tag{B.54}$$

where f is the wall friction coefficient and g_x is the component of the gravitational force in the flow direction. Then

$$\boldsymbol{F}' = \begin{bmatrix} 0 & 0 \\ \frac{f|u|u}{2H\alpha^2} & -\frac{fu}{H\alpha} \end{bmatrix}. \tag{B.55}$$

It follows that $v_w = u + \frac{u}{2}, c = u + \sqrt{gH\alpha}$ and

$$\omega = uk + i\frac{fu}{\alpha} \pm \sqrt{\left(ck + i\frac{fu}{H\alpha}\right)^2 + i\frac{fu}{H\alpha}(c - v_w)k}. \tag{B.56}$$

This theory leads to roll waves when $\frac{u}{2} > \sqrt{gH\alpha}$ and it was first investigated by Jeffreys (1925). The dispersion relation shown in Fig. B.22 for the unstable system with $u = 2\,\mathrm{m/s}$, $g_x = 0.8\,\mathrm{m/s^2}$, $\alpha = 0.5$, $H = 0.05$ m, $f = 0.01$, and $C = -1$ indicates that the problem is well-posed, i.e., that the growth rate is not infinite at zero wavelength even though it is the maximum. Thus, the kinematic instability is well-posed, but it is missing some short wavelength mechanisms such as surface tension to make it more physical.

In Sect. B.5.4 it will be shown, using nonlinear analysis, that kinematic waves develop into shocks and fans like Burgers' equation. The behavior of these shocks will be considered next.

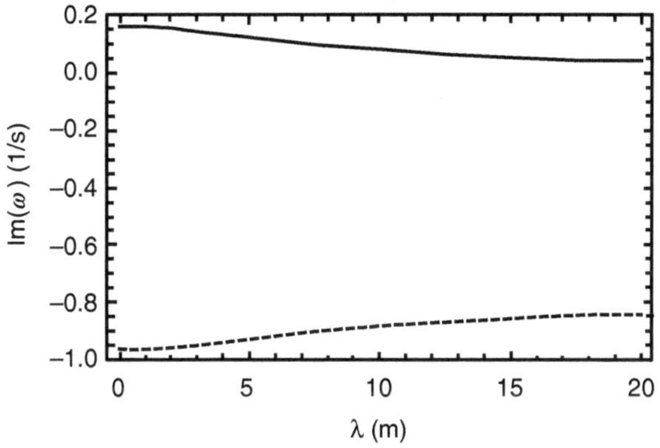

Fig. B.22 SWT dispersion relation under kinematic instability

B.5.3 Kinematic Shocks

To obtain the analytic expression of a kinematic shock we integrate the continuity equation in a finite domain, $x_1 < x < x_2$, that includes the shock, as was done in Sect. B.3.1 with Burgers' equation, to obtain

$$\frac{\partial}{\partial t}\int_{x_1}^{x_2} \alpha\, dx + \alpha_2 u_2 - \alpha_1 u_1 = 0 \tag{B.57}$$

which results in

$$\frac{dX}{dt}(\alpha_1 - \alpha_2) + \alpha_2 u_2 - \alpha_1 u_1 = 0 \tag{B.58}$$

or

$$\frac{dX}{dt} = \frac{\alpha_2 u_2 - \alpha_1 u_1}{\alpha_2 - \alpha_1}. \tag{B.59}$$

In the limit of small waves the equation reduces to

$$\frac{dX}{dt} = \frac{\partial \alpha u}{\partial \alpha} = u + \alpha \frac{\partial u}{\partial \alpha} \tag{B.60}$$

or

$$\frac{dX}{dt} = v_{\mathrm{w}} \tag{B.61}$$

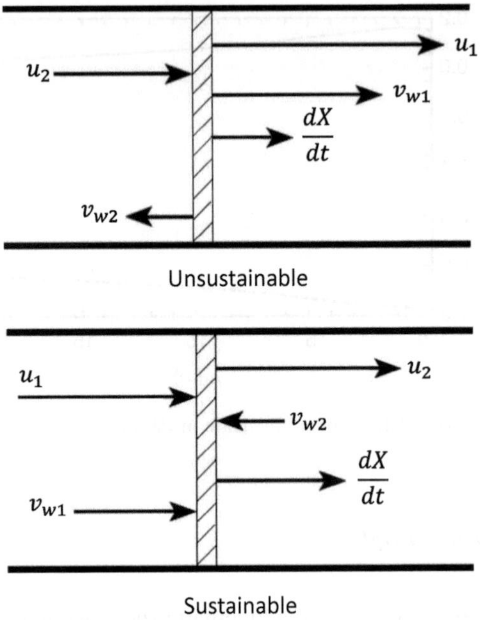

Fig. B.23 Shock wave schematic (Wallis 1969)

The sustainability of the shock depicted in Fig. B.23 depends on whether the kinematic waves on both sides flow toward it, otherwise the shock fans out.

It is convenient to rewrite the momentum equation in conservative form. This is accomplished by multiplying the continuity equation by u, adding it to the momentum equation, neglecting the forces, setting $C = -1$ and regrouping terms to obtain

$$\frac{\partial}{\partial t}\alpha u + \frac{\partial}{\partial x}\left(\alpha u^2 + \frac{\alpha^2}{2}\right) = 0, \tag{B.62}$$

which may be integrated to obtain

$$\frac{dX}{dt}(\alpha_1 u_1 - \alpha_2 u_2) + \alpha_2 u_2^2 - \alpha_1 u_1^2 + \frac{\alpha_2^2}{2} - \frac{\alpha_1^2}{2} = 0. \tag{B.63}$$

Equations (B.59) and (B.63) apply to bores in rivers and channels, where the shock is moving, and are also known as hydraulic jump conditions when the shock is stationary. Furthermore, they describe steep wave fronts which, as we shall see, are important in the nonlinear analysis of the unstable TFM.

From these two equations in three unknowns, α_2, u_2, and $\frac{dX}{dt}$, it is easy to obtain expressions for the shock speed and the liquid speed at the upstream side:

$$\frac{dX}{dt} = \sqrt{\frac{\alpha_2(\alpha_1 + \alpha_2)}{2\,\alpha_1}}, \tag{B.64}$$

$$u_2 = \frac{\alpha_2 - \alpha_1}{\alpha_2}\sqrt{\frac{\alpha_2(\alpha_1 + \alpha_2)}{2\,\alpha_1}}. \tag{B.65}$$

B.5.4 Nonlinear Shallow Water Theory

Consider a simplified form of Shallow Water Theory (SWT), given by Eqs. (B.44) and (B.45) with $F = 0$ and $C < 0$. SWT predicts breaking waves and fans with solutions similar to Burgers' equation, e.g., in Peregrine's (1967) prediction of a long wave breaking on a sloping beach the formation of the shock in 1D leading to the wave breaking is the outstanding nonlinear feature. The analysis of the formation of shocks using nonlinear analysis has been performed by Whitham (1974) expanding the solution around the steady flow α_0, u_0 as power series of $\xi = x - ct$,

$$\alpha = \alpha_0 + \xi\alpha_1(t) + \frac{1}{2}\xi^2\alpha_2(t) + O(\xi^3), \tag{B.66}$$

$$u = u_0 + \xi u_1(t) + \frac{1}{2}\xi^2 u_2(t) + O(\xi^3). \tag{B.67}$$

The zeroth order equations become

$$(u_0 - c)\alpha_1 + \alpha_0 u_1 = 0, \tag{B.68}$$

$$-C\alpha_1 + (u_0 - c)u_1 = 0, \tag{B.69}$$

which can be solved for the wave propagation speed

$$(u_0 - c)^2 + C\alpha_0 = 0 \;\rightarrow c = u_0 \pm \sqrt{-C\alpha_0} \tag{B.70}$$

and also for

$$u_1 = \left(\frac{c - u_0}{\alpha_0}\right)\alpha_1. \tag{B.71}$$

The first-order equations are

$$(u_0 - c)\alpha_2 + \alpha_0 u_2 + \frac{d\alpha_1}{dt} + 2\alpha_1 u_1 = 0, \tag{B.72}$$

$$-C\alpha_2 + (u_0 - c)u_2 + \frac{du_1}{dt} + 2u_1^2 = 0. \tag{B.73}$$

Multiplying Eq. (B.72) by C, Eq. (B.73) by $(u_0 - c)$, and adding them together eliminates the variables α_2, u_2 with the aid of Eq. (B.70) to obtain

$$C\frac{d\alpha_1}{dt} + 2C\alpha_1 u_1 + (u_0 - c)\left(\frac{du_1}{dt} + 2u_1^2\right) = 0. \tag{B.74}$$

Inserting Eq. (B.71) into it to replace u_1 and combining with Eq. (B.70) one finally obtains

$$\frac{d\alpha_1}{dt} + \frac{3}{2}\left(\frac{c - u_0}{\alpha_0}\right)\alpha_1^2 = 0 \tag{B.75}$$

which has the solution

$$\alpha_1(t) = \frac{1}{\frac{1}{\alpha_1(0)} + \frac{3}{2}\left(\frac{c - u_0}{\alpha_0}\right)t}. \tag{B.76}$$

The solution is singular for $\alpha_1(0) > 0$ if $c - u_0 < 0$ and vice versa. Then $\frac{d\alpha}{dx}$ is singular and a shock occurs. Otherwise a fan occurs. This important result is the only nonlinear analysis performed in this book; numerical simulations are performed instead, partly because the TFM is more complicated and partly because it is possible to go further with simulations, although they lack the generality of mathematical analysis.

B.5.5 The Water Faucet Problem

The SWT equations have been solved analytically for cases with discontinuities or shocks. For instance, it is well known that the model results in one-dimensional shocks when simulating "breaking" waves on beaches. In fact the model predicts breaking quite accurately in some instances even though the long wavelength approximation that the model is based upon is not strictly valid. These kinematic shocks were described by Wallis (1969) and by Whitham (1974) and they resemble the waves of Burgers equation. The prototypical case is the Riemann or dam break problem. The water faucet problem of Ransom (1984) is a variation on the Riemann problem with $C = 0$ that consists of an initially uniform liquid flow with a fixed inflow of water entering a vertical tube from the top. The boundary conditions at the inlet are held constant at α_0 and u_0 which are the initial conditions of the column. The initiation of the transient is due to a discontinuous change in the momentum source, i.e., gravity is 'turned on.' Therefore, the transient solution will also be discontinuous. With the acceleration of the liquid column under gravity, a narrowing of the jet develops and a kinematic discontinuity propagates downward as shown in Fig. B.25. Eventually, a steady-state solution is attained which describes the acceleration of the column due to gravity and the associated narrowing cross-section of the liquid profile. The problem can equivalently be described as a transition from one steady state solution to another, i.e. the initial condition to the final condition. This problem is used to verify nuclear reactor TFM codes. It is a severe convergence test because it is a contact discontinuity which means that the characteristics are parallel in the region around it so the numerical errors accumulate, unlike a shock where the errors are swept into it as the characteristics merge, i.e., shocks are a sort of numerical "garbage collectors."

The problem is solved by the method of characteristics. The momentum Eq. (B.46) is simply

$$\frac{Du}{Dt} = g \tag{B.77}$$

so the solution for $t_0 \geq 0$ is

$$u = u_0 + g(t - t_0). \tag{B.78}$$

The characteristic speed obtained from Eq. (B.51) is

$$\frac{dx}{dt} = u. \tag{B.79}$$

Therefore, the characteristics are $x = u_0(t - t_0) + \frac{1}{2}g(t - t_0)^2$. In particular, the characteristic with $t_0 = 0$, corresponding to the discontinuity, divides the characteristic plane, Fig. B.24, into two regions. In region 1 the wave has not arrived so $\alpha(x,t) = \alpha_0$ and $u(x,t) = u_0 + gt$. In region 2 the wave has passed so the solution depends only on the position, i.e., $\frac{\partial \alpha}{\partial t} = 0$. Therefore, the continuity Eq. (B.44) in region 2 reduces to

$$\frac{\partial \alpha u}{\partial x} = 0 \tag{B.80}$$

whose solution is $\alpha(x,t) = \frac{\alpha_0 u_0}{u(x,t)}$. Finally, combining Eqs. (B.79) and (B.80) and integrating one obtains $u(x,t) = \sqrt{u_0^2 + 2gx}$ so that the solution to the water faucet problem is

$$x < u_0 t + \frac{1}{2}gt^2 \quad \rightarrow \quad \alpha(x,t) = \frac{\alpha_0 u_0}{\sqrt{u_0^2 + 2gx}} \tag{B.81}$$

and otherwise

$$\alpha(x,t) = \alpha_0 \tag{B.82}$$

Figure B.25 shows $\alpha(x, t)$ from Eqs. (B.81) and (B.82) for $L = 12\,\text{m}$, $t_{max} = 0.6\,\text{s}$, $\alpha_0 = 0.8$ and $u_0 = 10\,\text{m/s}$.

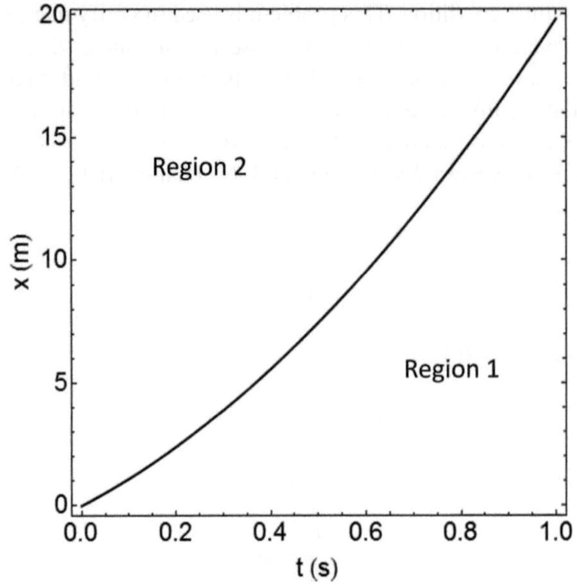

Fig. B.24 Dividing characteristic for water faucet problem

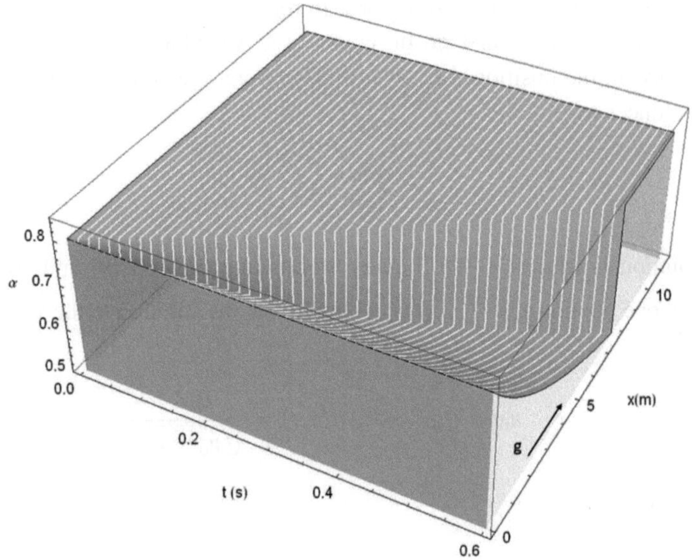

Fig. B.25 Water faucet problem wave propagation up to $t = 0.6$ s

B.5.6 Kinematic Instability

We now seek a solution to the SWT equations with viscous terms

$$\frac{\partial \alpha}{\partial t} + u\frac{\partial \alpha}{\partial x} + \alpha\frac{\partial u}{\partial x} = \varepsilon\frac{\partial^2 \alpha}{\partial x^2} \tag{B.83}$$

$$\frac{\partial u}{\partial t} + u\frac{\partial u}{\partial x} - C\frac{\partial \alpha}{\partial x} = \nu\frac{\partial^2 u}{\partial x^2} + F \tag{B.84}$$

with $u = 2\,\mathrm{m/s}$, $g_x = 0.8\,\mathrm{m/s}^2$, $\alpha = 0.5$, $H = 0.05\,\mathrm{m}$, $f_1 = 0.01$ and $\varepsilon = \nu = 0.02\,\mathrm{m}^2/\mathrm{s}$. The artificial viscosity is added to Eqs. (B.44) and (B.45) for numerical stability. $C = -1$ so the system is at the same kinematically unstable conditions of the analysis in Sect. B.5.2. The advantage of this simulation over the linear analysis is that the nonlinear effects may now be appreciated. The initial perturbation of the void fraction is $\alpha(x, 0) = 0.5 + 0.1\frac{1}{\sqrt{2\pi}\,\sigma}\mathrm{e}^{-(x-x_0)^2/2\delta^2}$, with $x_0 = 5\,\mathrm{m}$ and $\delta = 1\,\mathrm{m}$. The result is shown in Fig. B.26. In accordance with Sect. B.5.2 there are two waves, the slowest forward moving wave is dissipated so fast that it cannot be seen in the figure, as may be deduced from the growth rates shown in Fig. B.22. Meanwhile the fastest forward moving wave initially grows exponentially as predicted by linear stability theory, i.e., Eq. (B.56). Eventually the wave front develops into a shock while the rear of the wave fans out, as discussed in Sect. B.5.4. The formation of the kinematic shock is similar to Burger's equation. Furthermore, a trailing wave develops and as time goes on a wave train results. Roll waves such as this are typically seen in spillways and rivers. The waves in the figure stop growing due to the nonlinear viscous dissipation at the wave fronts.

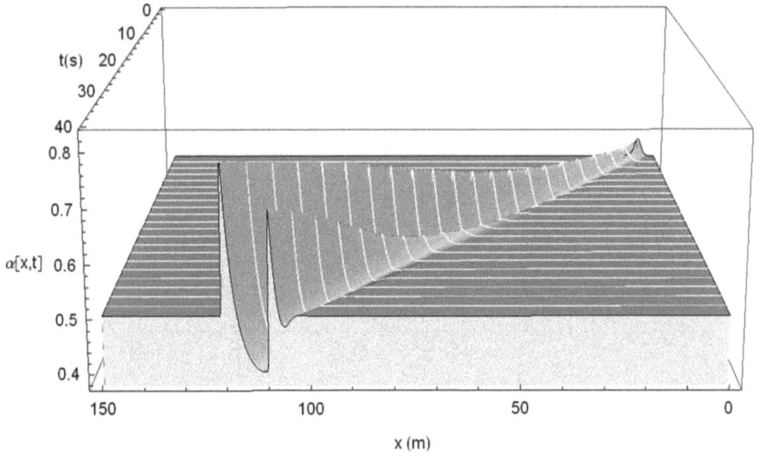

Fig. B.26 SWT wave under kinematic instability

B.6 Nonlinear Dynamics and Chaos

B.6.1 Harmonic Oscillator

The simplest equation of an oscillator is the linear harmonic oscillator

$$m\ddot{x} + \beta\dot{x} + kx = 0 \tag{B.85}$$

which models the dynamics of a mass m attached to a wall through a spring, where k and β are the elastic and damping coefficient, respectively. The solution $x(t)$ tends to zero if all the constants are positive. A convenient way to visualize geometrically this behavior is to plot the instantaneous values of x and $\dot{x}$ in a plane, called a phase space (see Fig. B.27).

If we force the oscillator with a periodic excitation, we get the following equation:

$$m\ddot{x} + \beta\dot{x} + kx = \sin \omega t. \tag{B.86}$$

In this case, the oscillator tends to a sustained regime ending up oscillating with the frequency of the excitation, ω. The amplitude of the final oscillation depends on ω, and furthermore it is independent of the initial conditions. In the phase plane (Fig. B.28) the final oscillation appears as an ellipse, and it is called a 'limit cycle,' because it tends to it in the asymptotic limit $t \to \infty$.

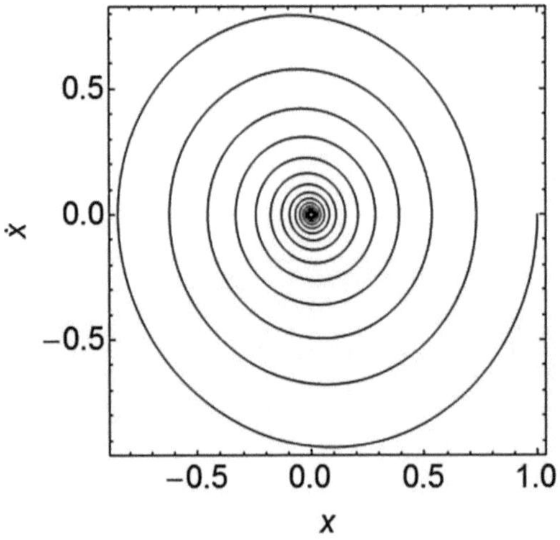

Fig. B.27 Phase-plane trajectory of a damped harmonic oscillator ($m = 1,\ \beta = 0.1,\ k = 1$)

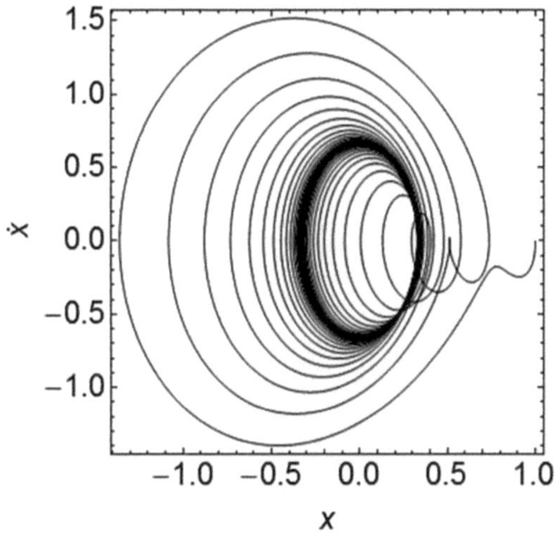

Fig. B.28 Phase-space trajectory of a forced harmonic oscillator ($m = 1,\ \beta = 0.1,\ k = 1,\ \omega = 2$). The *dark ellipse* is the asymptotic sustained regime, called 'limit cycle'

B.6.2 Van der Pol Oscillator

Consider now an unstable harmonic oscillator

$$\ddot{x} - a\dot{x} + x = 0 \tag{B.87}$$

with $a > 0$. The corresponding solution is

$$x = A\,e^{at}\sin(t - t_0). \tag{B.88}$$

In the phase plane Eq. (B.88) corresponds to a trajectory spiraling out from the origin (Fig. B.29).

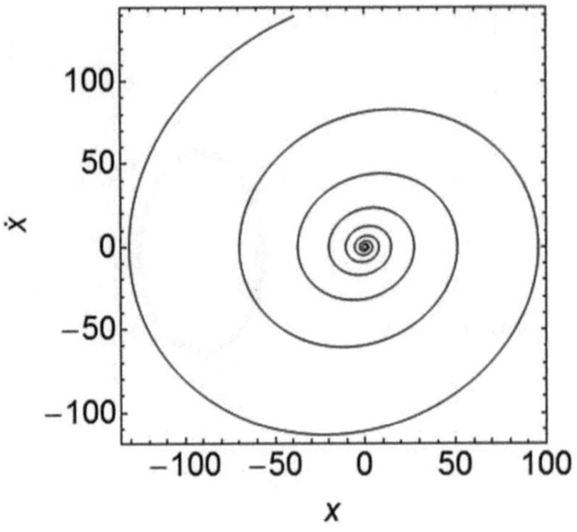

Fig. B.29 Phase-space trajectory of the unstable harmonic oscillator given by Eq. (B.88) $(a = 0.1)$

One way to bound the divergence of the trajectory is to control the term proportional to $\dot{x}$, forcing it to be positive when x increases beyond a certain value. An example that meets that requirement is the Van der Pol oscillator, given by:

$$\ddot{x} + \left(x^2 - a\right)\dot{x} + x = 0 \tag{B.89}$$

In this case there is a limit cycle (Fig. B.30), but it is produced autonomously by the oscillator state x itself, without external intervention, i.e., there is no term explicitly dependent on the time. The limit cycle is independent of the initial conditions. Hence, it is said that it "attracts" any trajectory no matter where it starts. For this reason it is called an 'attractor.' Actually, the point (0, 0) in the damped harmonic oscillator is also an attractor, called a 'point attractor'; whereas limit cycles are called 'periodic attractors.'

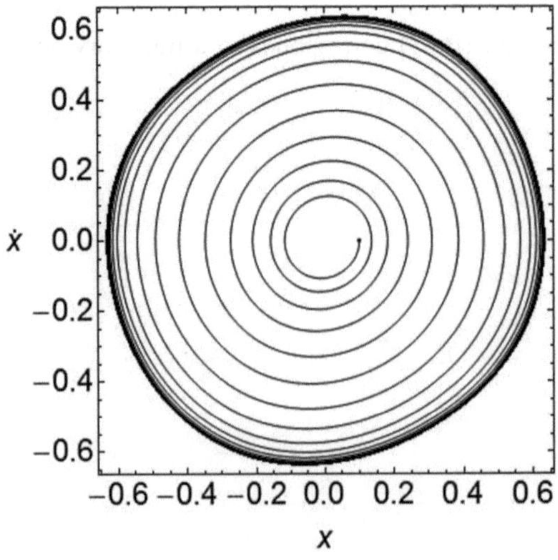

Fig. B.30 Phase-space trajectory of the Van der Pol oscillator $(a = 0.1)$

From a mathematical point of view, an attractor is a set of points that does not vary if any point of the set is taken as an initial condition of the corresponding differential equation. Accordingly, they are called 'invariants' of the differential equation. Figure B.31 shows different evolutions toward the limit cycle of the Van der Pol oscillator in the temporal domain and in the phase-plane, for $a = 0.2$, 1, and 5. It can be seen that the shape of the attractor is similar to an ellipse for lower values of a, and but changes significantly as a increases beyond unity.

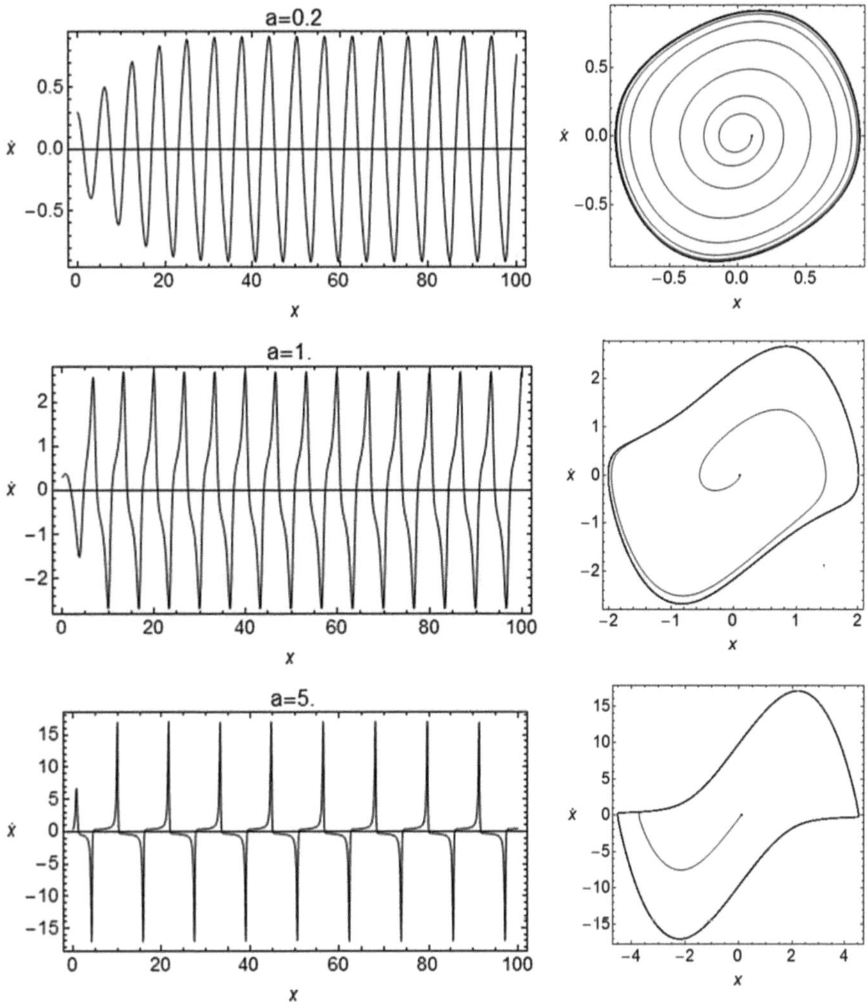

Fig. B.31 Evolution of the Van der Pol oscillator toward the limit cycle, in the temporal domain (*left*) and in the phase-plane (*right*), for $a = 0.2$, 1, and 5.

B.6.3 Rössler Oscillator

An alternative way to control the unstable harmonic oscillator is by means of an additional variable. We can write the differential equation of the unstable harmonic oscillator

$$\ddot{x} - a\dot{x} + x = 0 \tag{B.90}$$

as a set of two first-order differential equations, for example

$$\dot{x} = y, \tag{B.91}$$

$$\dot{y} = -x + ay. \tag{B.92}$$

From Eq. (B.91) it can be seen that x grows when y turns positive. Then, we may try to control the growth of x by introducing a variable z that opposes to its growth when certain value is exceeded. For example

$$\dot{x} = y - z, \tag{B.93}$$

$$\dot{y} = -x + ay, \tag{B.94}$$

$$\dot{z} = 2 + z(x - 4). \tag{B.95}$$

Thus, whenever x becomes close to 4 (see. Eq. B.95) $\dot{z}$ becomes positive, and then z grows, which hopefully would stop the growth of x. Figure B.32 shows what actually happens. For $a = 0.3$ the system effectively is bound, and it tends to a limit cycle, as in the Van der Pol oscillator. However, as a is increased further, the system shows more complicated dynamics, with a succession of period doublings, chaos, and even windows of periodic behavior in between chaotic regions.

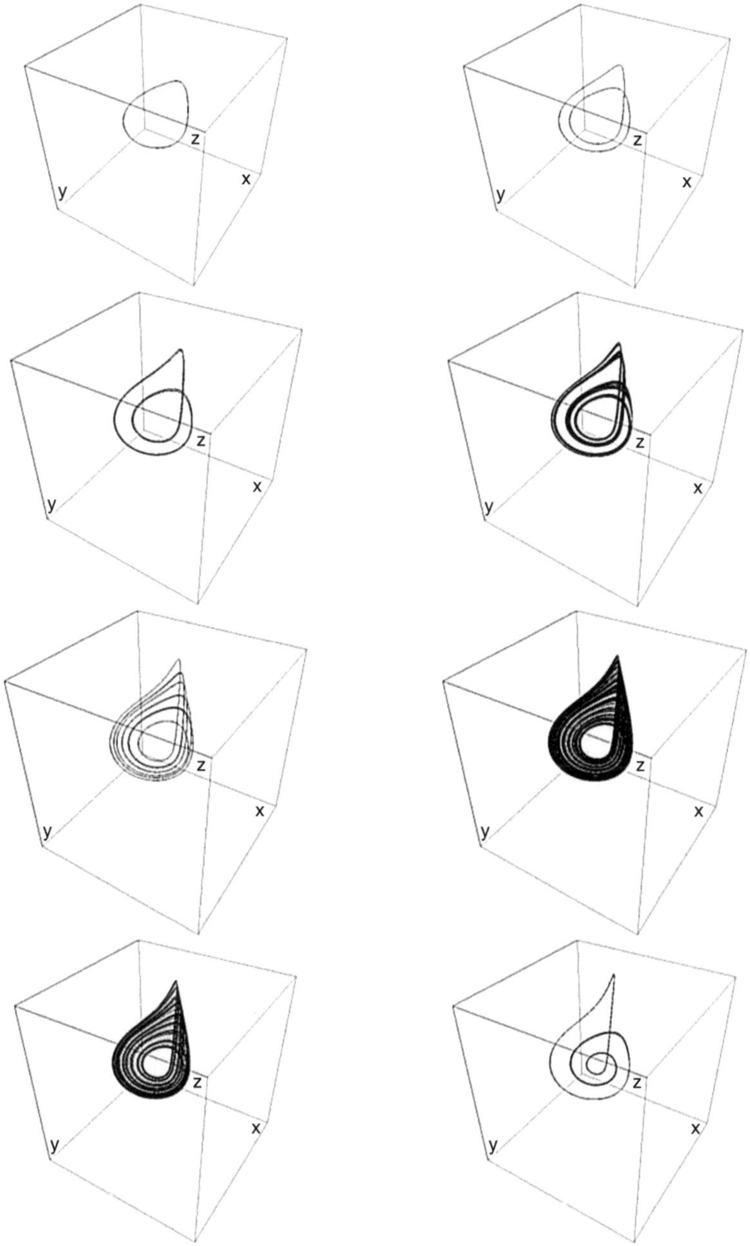

Fig. B.32 Phase-space trajectories of the Rössler attractor. First line from left to right: $a = 0.3$, 0.35; second line: $a = 0.375$, 0.386; third line: $a = 0.3909$, 0.398; fourth line: $a = 0.4$, 0.411

B.6.4 Poincaré Maps

Poincaré maps, also called return maps, are useful tools to explore the nature of the chaotic dynamics. The basic general concept is to identify a class of events in a temporal evolution. For example, for the Rössler system, represented by a trajectory in a three-dimensional phase space given by the coordinates $x(t)$, $y(t)$, and $z(t)$, it is typical to use the intersection of the trajectory with a given plane of the phase space (Fig. B.33 left) that define a discrete succession of states—for example, the successive values y_n of each intersection point n—which then can be displayed in a return map (y_n, y_{n+1}) shown in Fig. B.33(right). Thus, it is a way to explore if a given intersection point determines, or at least has a relation with, the next intersection point. If there is a relation, it should reflect in some pattern in the return map.

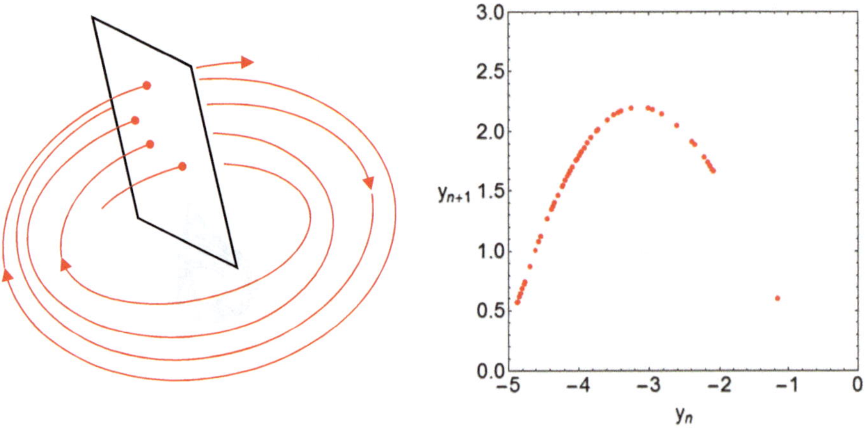

Fig. B.33 (*Left*) The Poincaré map is the set of points that intersect the phase-space trajectory of the attractor, (*right*) Return map of the Rössler attractor with $a = 0.398$ showing the y-coordinate of successive pairs of Poincaré points cutting the plane $z = 0$ ($x > 0$)

In the Rössler system for $a = 0.398$ the trajectory is a chaotic band (see Fig. B.32). However, as can be seen in Fig. B.33 (right), the corresponding return map results in a curve with a maximum (actually a zoom would show that the curve is formed by a tight bundle). This suggests that in some cases it should be possible to produce a similar dynamics using a discrete iteration map of the type:

$$y_{n+1} = f(y_n) \tag{B.96}$$

where $f(\cdot)$ is some curve with a maximum, like a parabola. As it turns out, this is true not only for a parabolic function but also for many other functions presenting at least one maximum, the most famous being the one known as the logistic map.

B.6.5 Logistic Map

The usual form of the logistic map is (Strogatz 1994)

$$x_{n+1} = \mu x_n(1 - x_n) \tag{B.97}$$

where μ is a constant control parameter. The iteration can be displayed in the plane (x_{n+1}, x_n), where the 45° line (i.e., $x_{n+1} = x_n$) is used to transport $x_{n+1} \rightarrow x_n$ after each iteration (see Fig. B.34). In this way, the map can be viewed as a successive bouncing horizontally from the parabola to the 45° line and vertically from the 45° line to the parabola. Starting with a small value of μ the map first converges to a fixed point, then to a limit cycle (period 2), then the period the cycle is doubled (period 4), and eventually to chaos (Fig. B.35).

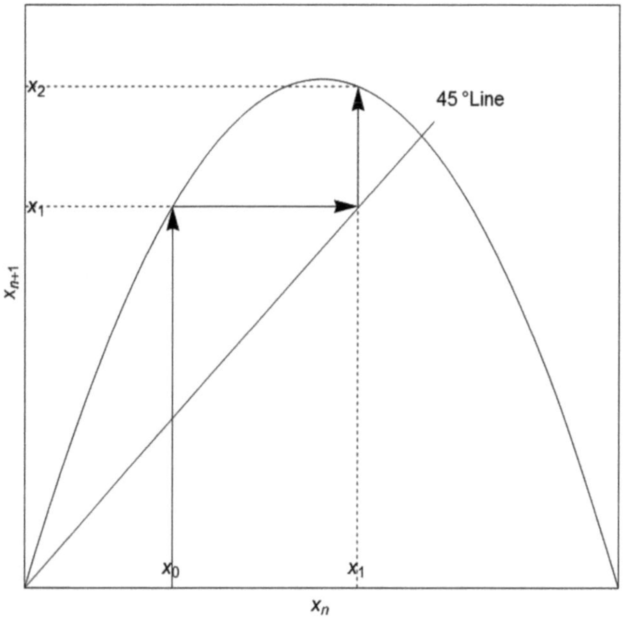

Fig. B.34 Visualization of the logistic map in the (x_{n+1}, x_n) plane

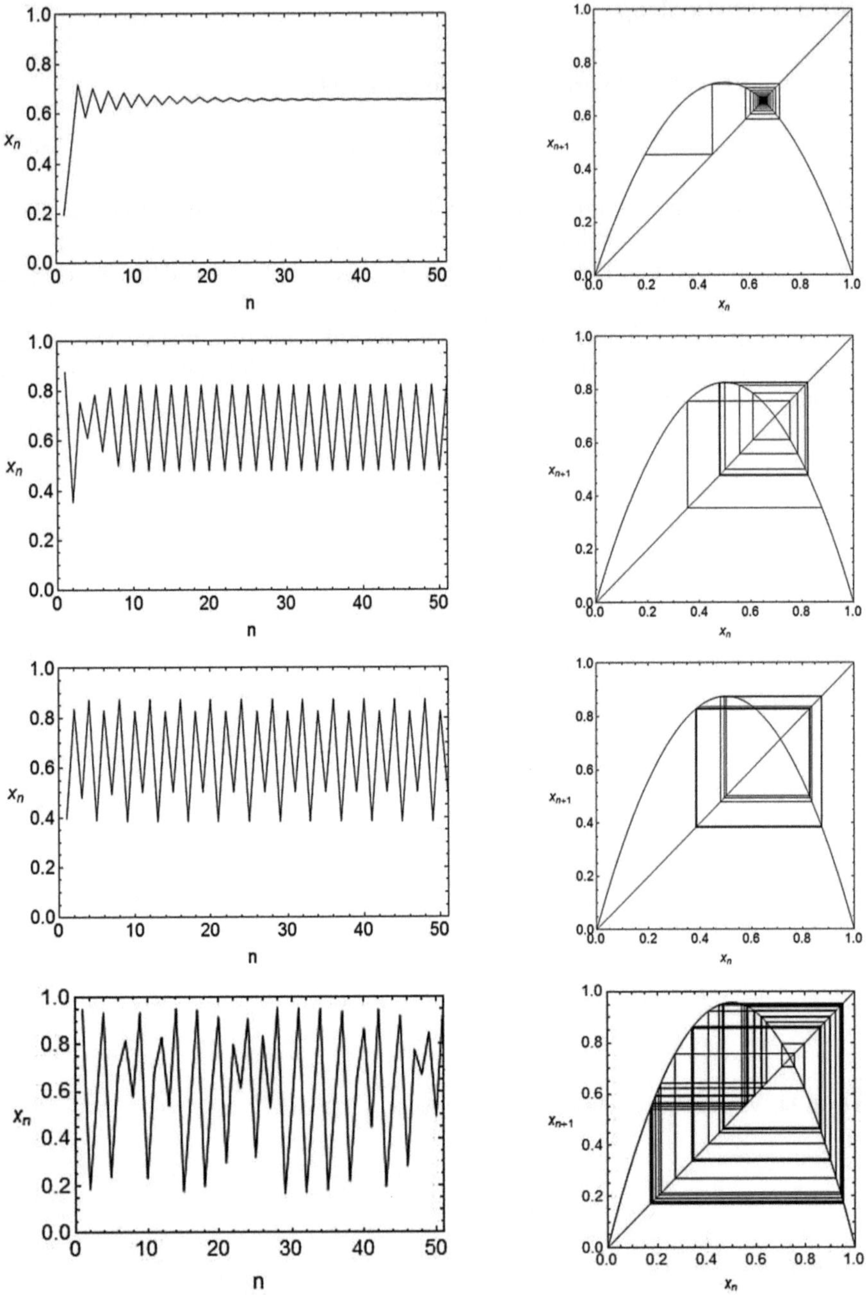

Fig. B.35 Logistic map, evolution of x with the iteration step n (*left*), geometrical visualization in the (x_{n+1}, x_n) plane (*right*). Values of the control parameter from *top* to *bottom*: $\mu = 2.9$, 3.3, 3.5, 3.83

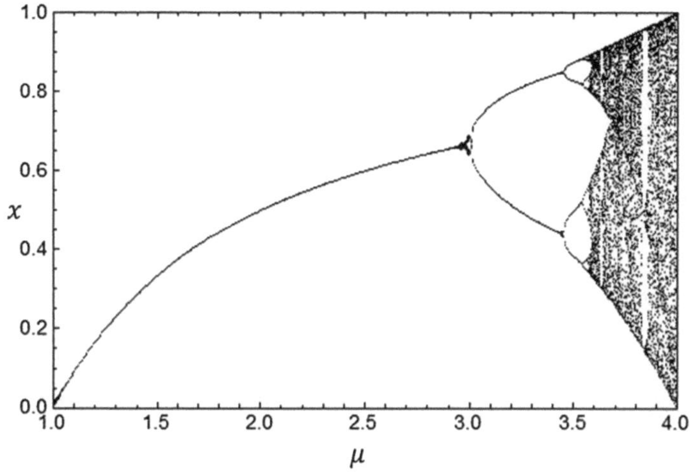

Fig. B.36 Bifurcation diagram of the logistic map

To examine the effect of the parameter μ on the behavior of the logistic map it is useful to plot the asymptotic attractor, reached after several iterations, against the control parameter μ. This is called a bifurcation diagram (Fig. B.36). For values of μ between 1 and 3 the solution is attracted to fixed points, i.e., a period 1 cycle. These period-1 fixed points increase in value as μ is increased. At $\mu = 3$ the solution bifurcates (called a 'pitchfork bifurcation' due to the form of the diagram). For values of μ between 3 and 3.44949 a period 2 attractor exists, and between 3.44949 and 3.54409 the period is 4, followed in sequence by attractors of period 8, period 16, period 32, and so on. This sequence carries on, doubling in period each time, until an infinite period is reached at a critical value of $\mu = 3.569945$. Thereon the behavior becomes chaotic and a strange attractor exists. This sequence is known as the period-doubling route to chaos.

B.6.6 The Lyapunov Exponent

The most important indicator that gives a signature of chaos is the exponential sensitivity of the dynamic evolution to the initial conditions. Consider the absolute difference of two sequences defined by the logistic equation with initial conditions 0.5 and $0.5 + 10^{-4}$. The absolute distance between sequential points of those succcssions, defined by:

$$|\Delta x| = |x_n(0.5) - x_n(0.5 + 10^{-4})| \tag{B.98}$$

goes for low n as:

$$|\Delta x| \cong e^{\omega_L n}. \tag{B.99}$$

Figure B.37 shows Eq. (B.98) for three values of μ. For $\mu = 3.56$ the attractor is a period-8 cycle, and it can be seen that $|\Delta x|$ decreases with n, meaning that ω_L is negative. On the other hand, for $\mu = 3.59$ the attractor is chaotic, and $|\Delta x|$ increases exponentially with n, meaning that ω_L is positive. At approximately iteration $n = 80$, $|\Delta x|$ levels off, meaning that the solution is bounded. This bounding is called Lyapunov stability. The onset of chaos occurs at $\mu \sim 3.57$ where $\omega_L \cong 0$. The parameter ω_L is called the Lyapunov exponent. This characteristic is informally known as the "butterfly effect," coined by Edward Lorenz after finding a chaotic attractor in a model of the atmosphere, suggesting that a hurricane's formation would be dependent on whether or not a distant butterfly had flapped its wings several weeks earlier (Fig. B.38).

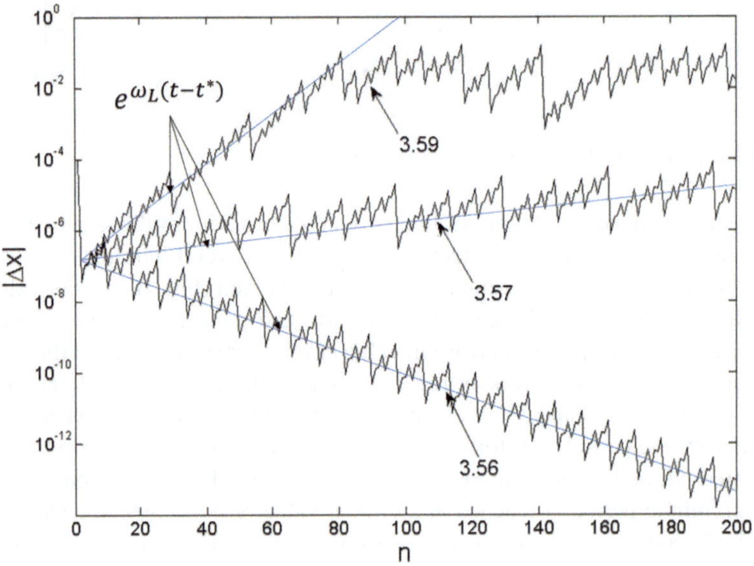

Fig. B.37 Evolution of the absolute difference between two iteration of the logistic equation starting at conditions 0.5 and $0.5 + 10^{-4}$

Fig. B.38 With permission of the author, J. L. Westover

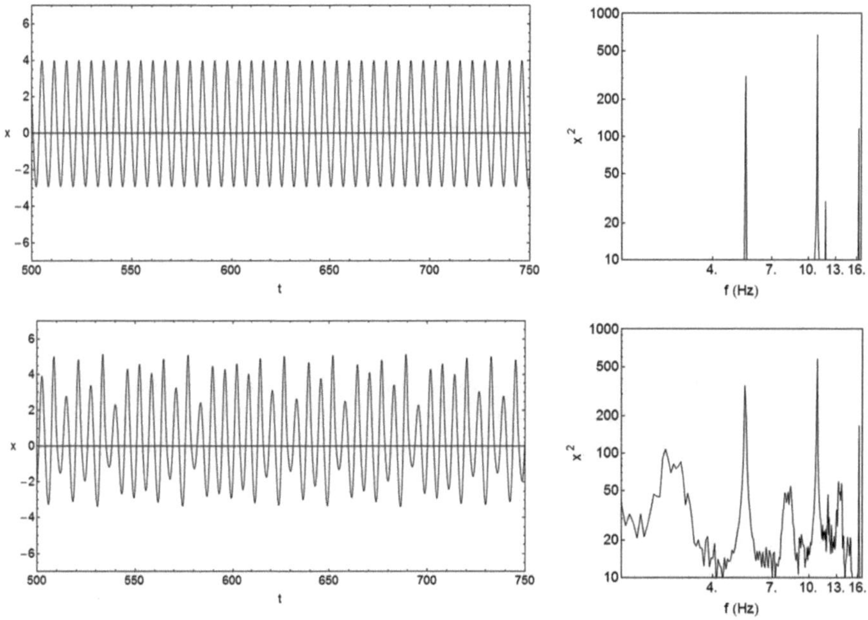

Fig. B.39 Fourier Spectrum: $a = 0.35$ (*top*: limit cycle) and $a = 0.4$ (*bottom*: chaos)

In general a system of N variables and N equations will have N Lyapunov exponents, which are a sort of generalization of the concept of eigenvalues of linear systems, i.e., the Lyapunov exponent is the nonlinear equivalent of the growth rate obtained from the linear dispersion relation. A positive Lyapunov exponent indicates expansion of the attractor in one direction. What characterizes chaotic attractors is the existence of at least one positive Lyapunov exponent, and all the others negative or null providing the bounding. On the other hand, a limit cycle has one null exponent and the others negative, hence all trajectories outside the cycle are attracted to it. There are systems that have several null Lyapunov exponents and all the other negatives. In such cases, the system is generally attracted to a trajectory confined in a torus, and are called quasi-periodic attractors. The Fourier spectrum of limit cycles and quasi periodic trajectories is in general discrete, whereas the Fourier spectrum of chaotic systems covers a continuum range of frequencies. This can be appreciated in Fig. B.39 where the Fourier spectrum of the Rössler system is shown for a limit cycle case and a chaotic case.

For high-dimensional systems of equations, Lyapunov spectra with many positive exponents can be found. However, only one positive Lyapunov exponent is required for chaos and hence it is sufficient to check whether one is positive. If this is the case, then the system is chaotic. In contrast, a purely random trajectory has $\omega_{\mathrm{L}} = \infty$, for no correlation exists between two points of the trajectory, no matter how close they are, i.e. the divergence is instantaneous.

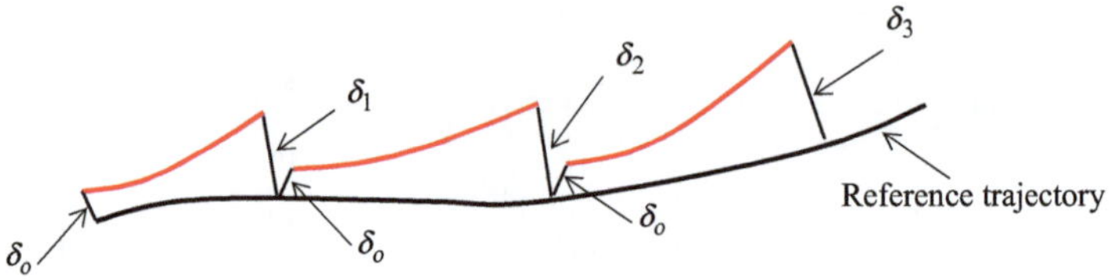

Fig. B.40 Schematic of the procedure to calculate the maximum Lyapunov exponent. The *red trajectories* corresponds to initial conditions close to each reference point

The maximum Lyapunov exponent can be assessed with the following procedure (Fig. B.40):

1. Starting with any initial conditions within the basin of attraction, calculate a significant time of the evolution.
2. Identify the moment when it is safe to say that the trajectory has reached the attractor.
3. On that point, produce a small perturbation δ_0 and then calculate the new trajectory until the divergence from the reference trajectory exceeds a certain value. Once the separation becomes too large, select another nearby trajectory, and so on (Fig. B.40).
4. The Lyapunov exponent is calculated at location k from which the perturbation started as:

$$\omega_{\mathrm{L}k} = \frac{1}{\Delta t_k} \ln \frac{\delta_k}{\delta_0} \tag{B.100}$$

where Δt_k is the duration of the perturbed calculation, and δ_k and δ_0 are the final and initial separation between trajectories. The estimation of the maximum Lyapunov exponent is then calculated as the average over all k.

B.6.7 Fractal Dimension

The fractal dimension is another magnitude that helps characterizing chaotic attractors. The fractal dimension is a generalization of the notion of Euclidean dimension of a given geometric object, i.e., a point has dimension 0, a curve has dimension 1, a surface has dimension 2, etc. One way to derive the Euclidean dimension of an object is by counting how many boxes are needed to entirely cover the object (Fig. B.41). Of course, that depends on the size of the boxes. The smaller the size, the higher the number of boxes required, and the relation between the latter and the former is in general:

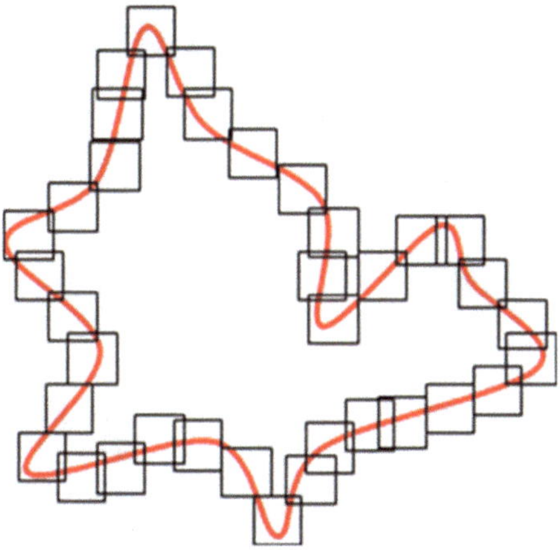

Fig. B.41 Box counting. The minimum number of boxes required to cover the object increases with the size of the unit box according to Eq. (B.102)

$$C(r) \propto r^{-D} \tag{B.101}$$

where r is the characteristic length of the box and D is called the capacity dimension, and is defined as:

$$D = \lim_{r \to 0} \frac{\log N}{\log \frac{1}{r}} \tag{B.102}$$

Now, consider the geometrical object defined as follows. Take a line segment and divide it in three equal parts. Remove the center part. Then proceed in the same way with each of the remaining segments, in infinite sequence. That is called a Cantor set (Fig. B.42).

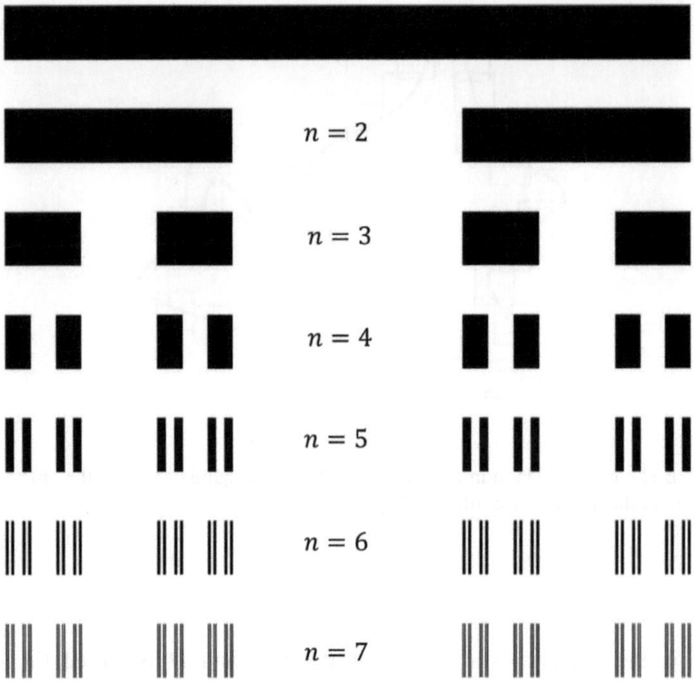

Fig. B.42 Cantor set

We can calculate the capacity dimension of the Cantor set considering the number of boxes needed to cover each set of the construction. In step 1, after removing the first third, we need 2 boxes of length 1/3 (i.e., 2^1 boxes of length 3^{-1}). In the second step, we need $4 = 2^2$ boxes of length $1/9 = 3^{-2}$. Then, in the nth step we will need 2^n boxes of length 3^{-n}. The capacity dimension is

$$D = \lim_{n \to 0} \frac{\log 2^n}{\log 3^n} = \frac{\log 2}{\log 3} = 0.63. \tag{B.103}$$

The dimension is fractional, meaning that the object is neither a finite set of points (that would have dimension 0) nor a line (which would have dimension 1).

Fractals are strange objects that have the property of self-similarity, that is, making zooms into the fractal we will find the same or similar structure at any scale. The attractor sets formed by the trajectories of a dynamical system plotted in the phase space can be viewed as geometrical objects too. A stable point attractor like the final stage of the damped harmonic oscillator has dimension 0. A limit cycle has dimension 1. And chaotic attractors have fractal dimension. The assessment of this dimension can be used to characterize the apparently random behavior, in a similar way as in thermodynamics we used some collective magnitudes, like the temperature, to characterize the chaotic movement of particles.

There is a more practical definition of the fractal dimension that is more convenient to characterize the geometry of a chaotic attractor produced from the numerical solution of differential equations. If we have a system of N ordinary differential equations, the solution defines a cloud of points in the corresponding N-dimensional phase space. If we put an N dimensional ball centered on one of the points of the cloud, the number of points contained in the ball will depends on the ball's radius r as

$$C(r) \propto r^d. \tag{B.104}$$

The exponent d is called the correlation dimension, and coincides with the Euclidean dimension in the case of a point or a limit cycle. But in the case of chaotic attractors, it will become fractional, and its value is similar to the capacity dimension in most cases.

The amount of points contained in a N-dimensional ball of radius r is calculated as the number of points satisfying the criterion

$$\sum_{i=1}^{N} (x_i - x_{oi})^2 < r^2 \tag{B.105}$$

where x_i and x_{oi} are the i coordinates of the point and the central point, respectively.

B.6.8 Embedding Dimension

In fluid mechanics and other engineering problems one usually deals with partial differential equations. In those cases, the state variables describing the system are fields, i.e., scalar or vector functions of space and time (e.g., density, temperature, velocity). Suppose that we have a chaotic solution of the fields' temporal evolution. Since the physical space is a continuum, the number of coordinates of the phase space in principle is infinite. However, we can still calculate the fractal dimensions provided that the attractor is dominated by a finite number of predominant modes. Modes can be viewed as spatiotemporal functions that persist and prevail over other functions, like sustained waves for example. If a finite number of modes are predominant and persistent, then we should be able in principle to describe the dynamics with a finite number of variables.

Following this way of thinking, we can proceed calculating the fractal dimension in steps, analogously as what we did with the Cantor set. In the first step we start choosing the evolution of two variables constructed from the evolution of the fields. This selection is guided in general by some physical insight, like for example the velocity magnitude at two given locations (see Fig. B.43 left). With these variables we construct a trajectory in the phase space, and assess the correlation dimension. Then we introduce an additional variable, and again calculate the correlation dimension, and so on (see Fig. B.43 right). What we will find is that the fractal dimension increases with the number of variables chosen for the trajectory. However, the most interesting feature is that if the dynamics were dominated by few modes, the fractal dimension, given by the slope, will saturate at certain value; if we keep adding more variables the fractal dimension will not vary any more. The number of variables at which the fractal dimension saturates is called the embedded dimension, and it is very useful, for it indicates the minimum number of ordinary differential equations that in principle is sufficient to model that dynamical mode.

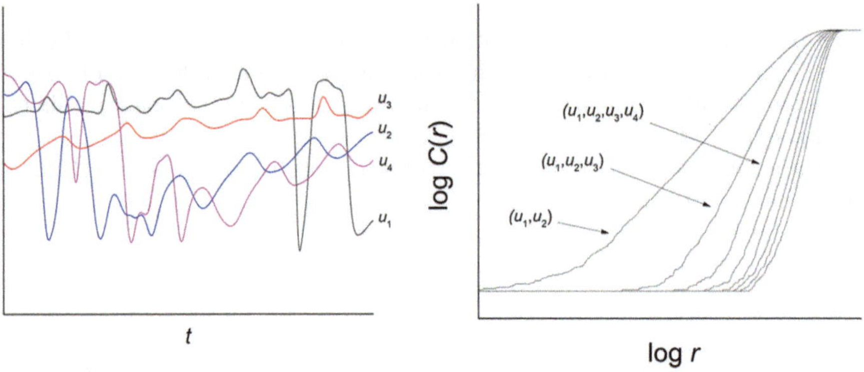

Fig. B.43 (*Left*) Velocity signals u_i at different points x_i of a channel, (*right*) number of points contained in an r-radius ball of the phase-space constructed with an increasing number of u_i. The slope of the mid part of each curve in the right figure is the correlation dimension.

References

Andritsos, N., & Hanratty, T. J. (1987). Influence of interfacial waves in stratified gas-liquid flows. *AIChE Journal, 33*(3), 444–454.
Biberg, D. (2007). A mathematical model for two-phase stratified turbulent duct flow. *Multiphase Science and Technology, 19*(1), 1–48.
Drew, D. A., & Passman, S. L. (1998). *Theory of multicomponent fluids* (Applied Mathematical Sciences). Berlin: Springer.
Fullmer, W., Lopez De Bertodano, M., & Ransom, V. H. (2011). The Kelvin-Helmholtz instability: Comparisons of one and two-dimensional simulations. In *The 14th International Topical Meeting on Nuclear Reactor Thermal Hydraulics (NURETH-14), Toronto, Ontario, Canada, September 25–29.*
Hurlburt, E. T., & Hanratty, T. J. (2002). Prediction of the transition from stratified to slug and plug flow for long pipes. *International Journal of Multiphase Flow, 28*, 707–729.
Ishii, M., & Hibiki, T. (2006). *Thermo-fluid dynamics of two-phase flow* (1st ed.). New York: Springer.
Jeffreys, H. (1925). The flow of water in an inclined channel of rectangular bottom. *Philosophical Magazine, 49*, 793.
Kocamustafaogullari, G. (1985). Two-fluid modeling in analyzing the interfacial stability of liquid film flows. *International Journal of Multiphase Flow, 11*(1), 63–89.
Kowalski, J. E. (1987). Wall and interfacial shear stress in stratified flow in a horizontal pipe. *AIChE Journal, 33*(2), 274–281.
Lopez de Bertodano, M. A., Fullmer W., Vaidheeswaran, A. (2013) One-Dimensional Two-Equation Two-Fluid Model Stability, Multiphase Science and Technology, 25(2):133–167.
Peregrine, D. H. (1967). Long waves on beach. *Journal of Fluid Mechanics, 27*(4), 815–827.
Picchi, D., Correra, S., & Poesio, P. (2014). Flow pattern transition, pressure gradient, hold-up predictions in gas/non-Newtonian power-law fluid stratified flow. *International Journal of Multiphase Flow, 63*, 105–115.
Ransom, V. H. (1984). Benchmark numerical tests. In G. F. Hewitt, J. M. Delhay, & N. Zuber (Eds.), *Multiphase science and technology*. Washington, DC: Hemisphere.
Ransom, V. H., & Hicks, D. L. (1984). Hyperbolic two-pressure models for two-phase flow. *Journal of Computational Physics, 53*, 124–151.
Richtmeyer, R. D., & Morton, K. W. (1967). *Difference methods for initial-value problems* (2nd ed.). New York: Interscience.
Strang, G. (2007). *Computational science and engineering*. Wellesley, MA: Wellesley-Cambridge Press.

M.L. de Bertodano et al., *Two-Fluid Model Stability, Simulation and Chaos*,
DOI 10.1007/978-3-319-44968-5

Strogatz, S. H. (1994). *Nonlinear dynamics and chaos: With applications to physics, biology, chemistry, and engineering* (Studies in Nonlinearity). Boulder, CO: Westview Press.

Thorpe, J. A. (1969). Experiments on the instability of stratified shear flow: Immiscible fluids. *Journal of Fluid Mechanics, 39*, 25–48.

Ullmann, A., & Brauner, N. (2006). Closure relations for two-fluid models for two-phase stratified smooth and stratified wavy flows. *International Journal of Multiphase Flow, 32*, 82–105.

Wallis, G. B. (1969). *One-dimensional two-phase flow*. New York: McGraw-Hill.

Whitham, G. B. (1974). *Linear and nonlinear waves*. New York: Wiley.

Zuber, N., & Findlay, J. (1965). Average volumetric concentrations in two-phase flow systems. *Journal of Heat Transfer, 87*, 453.

Index

A
Achard's analysis, 196
Achard's equations, 165
Adiabatic riser, boiling channel, 209, 210
  experimental validation, quasi-periodic oscillations, 218–221
  low power oscillations at low *Fr* numbers in heated channel, 213–218
  MNM equations, channel-riser system, 211–213
Air–water stratified flow, 35
Angular frequency, 238
Artificial devices, FOU scheme, 226
Artificial regularization, 66, 72
Artificial viscosity concept, 234
Artificial viscosity model, 238, 240–242
Autonomous chaotic oscillations, 209
Auxiliary algebraic equations, 212
Auxiliary variables, 205
Averaged power spectra, 121

B
Bernier's experiment, 172, 230–232
Bernoulli's principle, 249
Blasius equation, 96
Boiling boundary, 214, 215
Boiling channels, 165, 173, 174, 180, 182–184, 188, 195
  adiabatic riser, 209, 210
    experimental validation, quasi-periodic oscillations, 218–221
    low power oscillations at low *Fr* numbers in heated channel, 213–218
    MNM equations, channel-riser system, 211–213
    stability map, 214
  driven by natural circulation, 208
  effect of drift flux, nonlinear dynamics, 200, 207
  nonlinear behaviour, 196, 197
  nonlinear mapping, 197–201
  stability map, 207
  temporal evolution, inlet velocity, 199
  with moving nodes, 202–209
Boundary layer, 247, 248, 271, 287, 288
Boussinesq relation, 254
Brown boundary, 129
Bubble-center-averaged TFM, 273
Bubble-induced (BI) turbulence, 258
Bubble-induced (BI) eddy viscosity, 262
Bubble-induced (BI) turbulence, 260, 263, 264, 280
Bubble-induced (BI) turbulent kinetic energy, 259
Bubbly flow, 141–145, 148, 149, 151, 153, 154, 158, 160, 161, 226, 227, 229, 231, 235, 247–253, 256, 258, 261, 262, 267, 274, 276–278, 280, 287
  churn-turbulent, 231
Bubbly jet, 267–270
Buoyancy effects, 271
Burgers' equation, 7, 156, 173
Burgers' waves, 38

M.L. de Bertodano et al., *Two-Fluid Model Stability, Simulation and Chaos*,
DOI 10.1007/978-3-319-44968-5

C

CATHARE code, 66
CFD
  1D linear stability analysis to, 282
  TFM
    advantage, 281
    ill-posed, 281
    near-wall two-fluid model, turbulent bubbly boundary layer, 278
CFD-VOF simulation, 95
Chain rule, 20
Channel-riser system
  characteristic parameters, 214
  MNM equations, 211–213
Chaos
  FFM
    extension of Thorpe experiment, 122–123
    fractal dimension, 134
    fully developed laminar flow in pipe, 124–128
    Kelvin–Helmholtz instability, 128–130
    Lyapunov exponent, 132–133
    nonlinear simulations, 122, 130–132
    numerical convergence, 133
  Fourier transform, 134
  and Kreiss and Yström equations
    fractal dimension, 114–116
    Lyapunov exponent, 112–114
    nonlinear simulations, 109–111
    numerical convergence, 120–122
    route to chaos, 116–119
    sensitivity to initial conditions, 111
Chaos theory, 7
Characteristics, 14–17, 134, 143, 151, 154
  acoustic, 66
  analysis, 49–50, 61, 148–149
  linear stability, 68–69
  of RELAP5/MOD3.3, 243
  PB, 230
  polynomial *vs.* wave speed, 15
  TFM, 69, 228–230
Chimney effect, 262
Churn-turbulent bubbly flow, 231
Code verification, 49
Collision force, 149–151
Compressible model equations, 142
Compressible TFM
  characteristics, 14–17
  compressible model equations, 142
  one-dimensional model equations, 13–14
  virtual mass force, 143–144
Conventional wall function approach, 271
Convergence, 27, 53, 55–59, 61, 83, 92–93, 100–103, 120–122, 133, 136, 159, 226, 234, 239, 240, 244, 248, 271
  near-wall grid, 247
  numerical, 288
  study, 283
  test, 278
Correlation dimension, 115, 116
Coupled system of boiling channel, 210
Courant criterion, 43
Courant–Friedrichs–Lewy (CFL) criterion, 39
Courant limit, 87
Covariance terms, 125
Cutoff wavelength, 241, 242

D

Darcy friction factor, 96
Darcy-type model, 96
1D Drift-flux model (DFM), 1
Degree of instability, 49, 50
Delay DFM, 179–184
Density waves (DW), 195
  instability, 1, 164, 165, 183
    DFM, 192
    HEM, 187–188
    map, 191
    transfer function, 189–190
  non-linear behaviour, 196
  oscillations, 196, 197
    in boiling channel, 221
    stability map of, 206, 208
1D Euler equations, 71
Diffusion force model, 265
Diffusivity matrix, 70
Dimensionless density, 175
Dimensionless mixture density, 178
Direct numerical simulation analysis, 259
DISCON code, 74
Dispersed phase void fraction, 228
Dispersion, 235
Dispersion analysis, 70–72, 79, 129
  growth rates of, 235
  TFM, 234–236
Dispersion relation, 51–53, 70, 112, 116, 129, 130, 151–153
  KH instability, 23–28
  SWT instability, 28–33
Distorted bubbles, 153–155, 173
Distribution parameter, 166
Divergence of numerical method, 55
1D model, 248
1D Navier–Stokes equations, 26

Dominant mechanism, 196
Donored values, 237
Double wave problem, 60
D-partition method, 165
Drag force, 249–250, 281
  laminar regime (Stokes flow), 152–153
  turbulent regime (distorted bubbles), 153–155
Drainage, 170–172
1D Reynolds stress, 95
Drift flux, 196, 202, 216, 221
  effect on nonlinear dynamics, boiling channel, 200, 207
Drift-flux model (DFM), 23, 196, 221, 230, 233
  delay, 179–184, 197, 201
  density wave instability, 192
  dimensionless, 202
  dynamic
    integral momentum equation, 176–179
    mixture momentum equation, 173–176
  flow excursion, 186–187
  non-linear dynamics of, 195
  void propagation equation, 165–167
    applications, 167–173
Drift-flux velocity, 173
Drift-flux void propagation equation, 165, 167
Drift velocity, 166, 167, 231, 232
1D TFM, 3–6, 16, 67, 72, 86, 233, 236, 240, 241, 243
  derivation, 7
  simplifications, 6
1D two-fluid model, 97
Dynamic and kinematic wave growth rates, 33
Dynamic DFM
  integral momentum equation, 176–179
  mixture momentum equation, 173–176

E

Eigenvalues, 238
Embedded dimension, 115, 116, 135
Enskog equation, 149
Enthalpy nodes, 202
Euler 1D TFM, 2
Eulerian turbulent transport models, 265
Euler TFM, 71, 83, 239
Explicit diffusion, 44

F

Fanning friction factor, 227
Favre-averaged turbulent diffusion force, 265
FFT spectra, convergence of, 285
Filter, discrete and differential growth rates, 241
Finite difference equations (FDEs), 236
Finite-difference functions, 46
Finite-difference scheme, 196
Finite-volumes scheme, 196
First-order fully implicit scheme (with viscous terms), 44–45
First-order semi-implicit scheme, 77–78
  inviscid, 35–41
  with viscous terms, 42–43
First-order upwinding (FOU), 46, 66, 233
  donoring scheme, 237
  implicit scheme stability, 45
  schemes, 66, 74, 86, 93, 226, 233, 238
  solutions, 88
  stability, 40, 41, 43
Fixed flux assumption, 150
Fixed-flux model (FFM), 3, 12, 49–59, 65
  chaos
    extension of Thorpe experiment, 122–123
    fractal dimension, 134
    fully developed laminar flow in pipe, 124–128
    Kelvin–Helmholtz instability, 128–130
    Lyapunov exponent, 132–133
    nonlinear simulations, 122, 130–132
    numerical convergence, 133
  derivation, 19–23, 145–147
  time series of experiment, 133
  verification
    characteristic analysis, 49–50
    dispersion relation, 51–53
    Kreiss–Yström equations, 49
    method of manufactured solutions, 53–57
    water faucet problem, 57–59
  waveform of experiment, 130
Flow dynamics, 209
Flow excursion
  DFM, 186–187
  HEM, 184–186
  inlet restriction on, 186
  stability map, 185
Flux limiter, 46, 47, 61
FORTRAN Livermore Solver, 213
Fourier approach, 100
Fourier spectrum of limit cycle, 160
Fourier transform of chaos, 134
Fractal dimension, 108, 116
  chaos and Kreiss and Yström equations, 114–116
  FFM chaos, 134
Free-slip boundary conditions, 282
Froude numbers, 207, 213, 215
Fully developed laminar flow in pipe, FFM, 124–128

G
Gas drift velocity, 230
Gas–liquid stratified flow, 128
Gaussian initial condition, 156
Gedanken experiment, 282
General-piecewise limiter (GPL), 47, 85
Geometric method, 273
Geometric modeling
  approach, 274
  vs. data, 275
  schematic, 274
Gradient ratio, 47
Grid-dependent viscosity, 35
Grid generated turbulence, decay of
  single time-constant model, 257–258
  two time-constant model, 258–261

H
HAWAC experiment, 34
Higher resolution pseudo-spectral method, 110
Hoft bifurcation methods, 197
Homogeneous decay of turbulence, 255
Homogeneous equilibrium model (HEM), 6, 165, 184–188
Homogeneous model, 196
Homogeneous two-phase flow, 202
Homogenous decay, turbulence, 257, 261
Hopf bifurcation, 196, 217
Hydraulic diameters, 96
Hydrodynamic force, 251

I
Ill-posed, 11–13, 16, 17, 20, 23, 24, 27, 28, 37, 40, 43, 49, 50, 52, 59–61, 67, 69, 71, 72, 83, 89, 90, 103, 141, 143, 157, 230, 233, 235, 236, 239, 240, 243, 244, 287
Implicit diffusion, 43
Implicit pressure Poisson equation, 78–79
Inclined channel flow, 18
Incompressible multidimensional TFM
  drag force, 249–250
  interfacial momentum transfer, 249
  laminar pipe flow, 252
  lift force, 250–251
  model equations, 248–249
  wall force, 251
Incompressible TFM, 17–23, 66–68
  FFM derivation, 19–23, 145–147
  interfacial pressure, 145
  one-dimensional model equations, 17–18
Initial and boundary conditions (IBCs), 90
Initial and boundary value problem (IBVP), 49
Instantaneous void fraction contours, 283
Integral momentum equation, 176–179
Interfacial collision force model, 149
Interfacial Fanning friction factors, 68
Interfacial momentum transfer, 249
Interfacial pressure, 141, 145, 148
Interfacial shear, 97, 125
Intermittency, 118, 119, 135
Interphase momentum transfer, 251

K
Kelvin–Helmholtz (KH) instability, 2, 3, 5, 7, 21, 23–28, 52, 59–60, 71–73, 98, 128–130, 151, 242, 243
Kelvin–Helmholtz (KH) stability, 69
Kinematic and dynamic instabilities, 12, 30
Kinematic condition, 164, 165, 167, 173, 176, 192
Kinematic instability, 13, 26, 28–30, 32, 33, 37, 61, 122, 151–152, 154, 161
Kinematic viscosity, 129
Kinematic waves, 232
Kinematically unstable waves in Guinness, 158–161
Kinetic energy transport equation, 256
Kolmogorov turbulence spectrum, 136
Kreiss–Yström equations, 49, 109, 110
  chaos
    fractal dimension, 114–116
    Lyapunov exponent, 112–114
    nonlinear simulations, 109–111
    numerical convergence, 120–122
    route to chaos, 116–119
    sensitivity to initial conditions, 111
Kuramoto–Sivashinsky (KS) equation, 121
k–ε model, 247, 253, 266
  diffusivity of momentum in, 267
  single-phase, 253–256
  single time constant extension of, 258
  TFM averaged, 253
  TFM with, 262
  time constant, 261
  two-phase, 256–257
k–ε turbulence model, 271

L
Lagrangian approach, 265
Lagrangian turbulence time constant, 266
Lagrangian void propagation, 165

Laminar pipe flow, 252
  revisited, 275–276
Laminar regime (Stokes flow), 152–153
Laplace transform, 3, 165, 189
Large Eddy Simulation (LES) model, 280
Largest Lyapunov exponent, 4
Lax's equivalence theorem, 83, 239
Lax–Wendroff second-order numerical scheme, 46
Ledinegg boundary, 208
Ledinegg instability, 3, 183, 184, 200
Ledinegg stability, 184, 187
Leibnitz's rule, 203
Level swell, 167–170
Lift force, 250–251, 281
Limit cycle, 207, 215
  amplitude, 199, 200
  with different periods, 217
  by inlet velocity of boiling channel, 200
  obtained with delay drift-flux model, 201
  obtained with MNS model, 206
  stable, 201, 221
  unstable, 201, 202
Linear perturbation analysis, 192
Linear stability, 23–35, 68–73, 148–155, 183, 190
  analysis, 4, 236
  characteristics, 68, 69, 148–149
  collision force, 149–151
  dispersion analysis, 70–71
  dispersion relation, 151–152
    KH instability, 23–28
    SWT instability, 28–33
  drag force
    laminar regime (Stokes flow), 152–153
    turbulent regime (distorted bubbles), 153–155
  KH instability, 71–73
  sheltering effect, 33–35
  theory, 141
Linear superposition model, 262
Liquid flux, 230–232
Liquid flux condition, 231
Liquid momentum equation, 144
Liquid phase bubble-induced turbulence, 149
Liquid phase subscript, 256
Liquid phase turbulence, 262
Logarithmic law
  dimensionless velocity profiles, 278
  single-phase, 271, 272, 276, 278, 279
  two-phase, 271, 273
Logarithmic wall function approach, 271
Lower dimensional subspace, 217
Low-pass filter method, 243
Low pass filter regularization, TFM
  artificial viscosity model, 240–242
  dispersion analysis, 234–236
  numerical viscosity, 236–240
  water faucet problem, 242–243
LSODE, 213
Lyapunov analysis, 136
Lyapunov coefficient, 108
Lyapunov exponent
  chaos and Kreiss and Yström equations, 112–114
  FFM chaos, 132–133
Lyapunov stability, 72, 108, 120, 133, 136, 287

M

Manufactured solutions, method of, 53–57
Marie et al. (1997)
  convergence, volume fraction and velocity distributions, 277
  double-step function approximation of, 272
  geometric model *vs.* data, 275
  modified logarithmic law of wall, 271–273
Markov law, 266
Mass transfer rate, 227
Mass transfer source, 174
Material shocks, propagation, 172–173
Material wave speeds, 25, 148, 154
Mean amplitude spectrum of simulation, 101
Method of manufactured equations, 54
Microscale bubbles, 149
Minmod (SSP2-MM), 87
Mixing length theory of Prandtl, 271
Mixture momentum equation, 164, 173–176, 192
Mixture velocity, 178
Modified differential equations (MDEs), 238
Modified equation, 82
Modified virtual mass model, 228
Modified water faucet problem, 91, 92
Moving nodes model (MNM), 196, 202, 205, 206, 209, 221
  equations, channel-riser system, 211–213
Multidimensional CFD simulation, 128
Multidimensional TFM, 248
MUSCL scheme, 47, 86

N

Natural circulation, 207, 208, 213, 221, 222
Navier–Stokes equations, 2, 6, 26, 28, 66, 120, 136, 253

Navier–Stokes theory of turbulence, 120
2*N*-dimensional Euclidean norm, 113
2*N*-dimensional system of ODE, 112
Near-wall averaging
convergence of volume fraction and liquid velocity profiles, 276
method, 278
technique, 275
TFM, 273–276
Near-wall grid convergence, 247
Near-wall two-fluid model, 271–280
laminar pipe flow revisited, 275–276
near-wall averaging, 273–275
turbulent bubbly boundary layer, 276–278
turbulent pipe flow revisited, 278–280
two-phase logarithmic wall, 271–273
wall boundary conditions, 271
Neutronic kinetics, 196, 202
Nobel theory of superfluidity of Helium-4, 1
Nonequilibrium thermodynamics, 3
Nonlinear, 195–201, 205, 213, 214, 222
algebraic-differential equations, 205
analysis, 142
coupling, 259
evolution of Gaussian wave, 98
flux limiter, 47
ordinary differential equations, 213
oscillations, geometric structure of, 217
stability analysis, 2, 225
stabilization mechanisms, local, 233
Nonlinear simulations
chaos and Kreiss and Yström equations, 109–111
convergence, 100–103
FFM chaos, 122, 130–132
interfacial shear, 97
kinematically unstable waves in Guinness, 158–161
single nonlinear wave, 97–98
stable wave evolution, 155–158
Thorpe experiment, 93–94
Thorpe experiment validation, 98–100
viscous stresses, 95–96
wall shear, 96
Normalized variable (NV) SMART scheme, 86
Numerical convergence, 248
chaos and Kreiss and Yström equations, 120–122
FFM chaos, 133
Numerical discretization, 236
Numerical method, 236
Numerical regularization, 82–84
Numerical scheme, 196
Numerical solutions, 99
Numerical stability
first-order fully implicit scheme (with viscous terms), 44–45
first-order semi-implicit scheme, 77–78
first-order semi-implicit scheme (inviscid), 35–41
first-order semi-implicit scheme (with viscous terms), 42–43
implicit pressure Poisson equation, 78–79
numerical regularization, 82–84
second-order semi-implicit scheme, 46–48, 84–87
staggered cell structure, 74–77
TFIT TFM, 74
von Neumann analysis, 79–82
well-posed numerical model, 35
Numerical time-domain simulations, 196
Numerical viscosity, TFM, 234, 236–240

O

One-dimensional Burgers' waves, 38
One-dimensional characteristics, 12
One-dimensional drift-flux algebraic model, 166
One-dimensional models, 12, 95, 166
compressible TFM, 13–14
incompressible TFM, 17–18
One-dimensional TFM, 144
One-way wave equations, 167
Ordinary differential equations, 203, 205, 211
FORTRAN Livermore Solver, 213
non-linear, 213
Orr–Sommerfeld analysis, 5, 34, 94, 128
Orr–Sommerfeld equations, 1, 12, 28, 128
Orr–Sommerfeld stability
analysis, 123
boundary, 129
Oscillation
measurement, sustained, 219
phase-plane trajectory, 220

P

Parallel plate channel dimension, 129
Particle pressure, 149, 161
Pauchon and Banerjee (PB) *vs*. RELAP5/MOD3.3 model, 228–230
Peclet number, 43
Periodic boundary conditions, 131
Perturbation analysis, 196
Perturbations, 53, 60, 111, 113, 132

Phase diagram, stable and unstable cases, 160
Phase-plane invariants, parameters, 215, 216
Phase-plane trajectory, oscillation, 220
Phase-space and gray scale map, 117–119
Phase-space dimension, 115
Phase transfer rate, 166
Pipe flow, 247, 251
  laminar, 252, 275–276
  turbulent, 261–265, 278–280
Pipe two-phase flow, 126
Plane bubble plume, 281–287
Poincaré maps, 217, 218
Pointcaré maps, 222
Potential flow theory, 143
Prandtl mixing length model, 253
Prandtl's mixing length theory, 255
Preliminary analysis, 240
Pressure drop, 181, 185, 210
Pressure implicit scheme, 76
Pressure Poisson equation, 84
Pressure–velocity coupling, 76, 77
Probability density function (PDF), 265
Pseudo-turbulence, 95, 260, 261

Q
Quasi-periodic oscillation, 215, 218–221
Quasi-periodic torus, 217

R
Radial distribution function (RDF), 149, 150
RANS equations, 253
RANS TFM, 247
RANS two-fluid model
  decay of grid generated turbulence
    single time-constant model, 257–258
    two time-constant model, 258–261
  Reynolds stress stabilization, 253
  single-phase k–ε model, 253–256
  turbulent pipe flow, 261–265
    bubbly jet, 267–270
    turbulent diffusion force, 265–267
  two-phase k–ε model, 256–257
Rayleigh instability, 28
Reference enthalpy unit, 202
Regularization, 12, 66, 82, 226, 233–243
  artificial, 66, 72
  numerical, 74, 82–84
RELAP5, 226
RELAP5/MOD3.3 code, 66
RELAP5/MOD3.3 model, 228–232
  vs. Bernier data, 231–233
  vs. PB model, 229
  TFM equations, 226
RELAP5 TFM
  material waves
    adiabatic TFM, 226–228
    Bernier's experiment, 230–232
    characteristics, 228–230
Relative velocity, 230
Reynolds-averaged Navier–Stokes (RANS) model, 247, 253
  two-fluid model (*see* RANS two-fluid model)
Reynolds number, 94, 96, 120, 123
Reynolds stress stabilization, 253
Reynolds stress tensor, 254
Reynolds stresses, 74, 95, 128, 247, 261, 267, 281
Route to chaos, 116–119
Runge–Kutta method, 48

S
Sato's model, 265
Second-order center differencing, 46, 47
Second-order finite-difference numerical method, 135
Second-order semi-implicit scheme, 46–48, 84–87
Second-order strong-stability preserving (SSP2) scheme, 86
Second-order temporal scheme, 282
Semi-implicit FOU numerical method, 77
Semi-implicit scheme, 74, 76
Sensitivity to initial conditions, 111
Shallow water theory (SWT), 3, 12, 19, 147
Shear-induced (SI) component, 262
Shear-induced (SI) turbulence, 258
Shear-induced (SI) eddy viscosity, 262
Shear-induced (SI) turbulence, 260, 261, 280
Shear-induced (SI) vortices, 259
Sheltering effect, 33–35
Simulation, 65, 87, 89, 92, 196, 197, 219
  3D, 67
  mean amplitude spectrum, 101
  nonlinear, 93–103
  numerical, 73
  water faucet problem, 89
Sine wave, 87
Single nonlinear wave, 97–98
Single-phase axisymmetric jets, 268
Single-phase 1D Euler equations, 2
Single-phase flow, 71
Single-phase k–ε model, 253–256

Single-phase logarithmic law, 271, 272, 279
Single time-constant model, 257–258
SMART (SSP2-SMART), 87
SMART flux limiter, 58
SMART scheme, 47
Sound speed model, 16
Sparger, 232
SSP2-MM scheme, 58, 90
SSP2-MM solutions, 88, 90
SSP2-SMART scheme, 58, 91, 92, 98
SSP2-SMART solutions, 88
Stability, 196, 198–201, 209, 215, 219
  condition, 37
  of lift force, 251
  map of boiling channel, 207
  stabilization of TFM, 233
  URANS TFM, 280
Stable model, 39, 40
Stable wave evolution, 155–158
Staggered cell structure, 74–77
Staggered continuity cells, 36
State variables, boiling channel, 205
Static Ledinegg instability, 164
Steam Tables, 226
Stiffness matrix, 70
Stokes bubble, 148
Stokes flow, 152–153
Stokes regime, 153
Stratified flow, 12–15, 17, 28, 30, 31, 34, 35, 59, 61, 66, 67, 73, 94, 103, 123, 126, 128, 134
Stroboscopic visualization, 217
Strong stability preserving (SSP) third-order method, 48
Subcooling number, 197, 201
Subcritical bifurcations, 201, 221
Supercritical bifurcations, 201
Superposition model, 259
Surface tension, 72, 74, 82, 129
  stabilization, 130
  tensor, 21
Sustained oscillation
  measurement, 219
  inlet velocity, 220
  phase-plane trajectory, 221
SWT instability, 28–33, 108

T

Taylor expansion, 32
Taylor length scale of turbulence, 270
Taylor series, 21
Taylor series expansion, 237
Taylor wavenumber, 25
TFIT, 242
  code, 75
  TFM, 74
Third order, three stage (3-3) SSP Runge–Kutta method, 48
Thorpe experiment, 66, 93–94, 98
  extension, 122–123
  schematic, 94
  validation, 98–100
Thorpe's water–kerosene experiment, 71
Time constant, 260
  Lagrangian turbulence, 266
  ratio, 151
  of small bubbles, 266
Time delay, 165, 177, 189
Tollmien–Schlichting instability, 94
Total-variation diminishing (TVD), 48, 86
TRAC code numerical scheme, 74
TRACE V5, 227
TRACE V5 TFM, 229
Transfer function, zeroes, 189–190
Transient air–water bubbly flows, 143
Truncation error, 238
Turbulence, 1
  anisotropy tensor, 255
  bubble-induced, 260, 263, 264, 280
  damping effect, 259
  decay of grid generated, 257–261
  dissipation rate, 254
  homogenous decay, 255, 257, 261
  kinetic energy transport equation, 254
  liquid phase, 262
  plume, void fraction contours, 284
  shear-induced, 260, 261, 280
  Taylor length scale, 270
Turbulent bubbly boundary layer, 276–278
Turbulent diffusion force, 265–267
Turbulent flow, 154
Turbulent kinetic energy, 254
Turbulent pipe flow, 261–265
  revisited, 278–280
Turbulent regime (distorted bubbles), 153–155
Turbulent transport term, 254
Two-fluid model (TFM)
  applications, 1, 2
  behavior near wall, 273
  codes, 225
  compressible
    characteristics, 14–17
    compressible model equations, 142

one-dimensional model equations, 13–14
virtual mass force, 143–144
convergence of volume fraction and liquid velocity profiles, 253
Euler, 239
flow chart, 6
fundamental problem, 2
incompressible, 66–68
FFM derivation, 19–23, 145–147
interfacial pressure, 145
one-dimensional model equations, 17–18
incompressible multidimensional
drag force, 249–250
interfacial momentum transfer, 249
laminar pipe flow, 252
lift force, 250–251
model equations, 248–249
wall force, 251
with k–ε model, 262
less arbitrary treatment, 251
linear stability
characteristics, 68, 69
dispersion analysis, 70–71
KH instability, 71–73
low pass filter regularization
artificial viscosity model, 240–242
dispersion analysis, 234–236
numerical viscosity, 236–240
water faucet problem, 242–243
modeling, 225
multidimensional, 248
nonlinear simulations
convergence, 100–103
interfacial shear, 97
single nonlinear wave, 97–98
Thorpe experiment, 93–94
Thorpe experiment validation, 98–100
viscous stresses, 95–96
wall shear, 96
numerical stability
first-order semi-implicit scheme, 77–78
implicit pressure Poisson equation, 78–79
numerical regularization, 82–84
second-order semi-implicit scheme, 84–87
staggered cell structure, 74–77
TFIT TFM, 74
von Neumann analysis, 79–82
RANS, 247
RELAP5 (*see* RELAP5 TFM)
rigorous derivations, 2
stabilization, 233
verification
convergence, 92–93
modified water faucet problem, 91
sine wave, 87
water faucet problem, 88–90
Two-phase flow, 12, 13, 22, 23, 28, 67, 73, 74, 95, 96, 125, 126, 163, 164, 167, 176, 192, 248, 251, 252, 258, 260, 262, 271, 272, 275, 276, 278, 282, 283
Two-phase Kelvin–Helmholtz instability, 23
Two-phase k–ε model, 256–257
Two-phase logarithmic law of wall, 271–273
Two-phase sound speed predictions, 144
Two-phase turbulence models *vs.* grid generated turbulence bubbly flow, 258
Two time-constant model, 258–261

U
Uniform artificial viscosity, 240
Unstable model, 39, 41, 45
Upwind differencing, 74
Upwind method, 237
URANS TFM, 287
ill-posed, 282
with and without collision, 286
URANS two-fluid model
constitutive relations, 281
plane bubble plume, 281–287
stability, 280

V
Virtual mass coefficient, 227
Virtual mass force, 143–144
Virtual volume coefficient, 259
Viscous Kelvin–Helmholtz instability, 28
Viscous stresses, 95–96
Void fractions, 14, 18, 159
Void propagation equation, 164–167
applications
drainage, 170–172
level swell, 167–170
propagation of material shocks, 172–173
discontinuity experiment of Bernier, 172
Void wave simulation, 156
Volumetric flux, 19, 20, 146, 166, 167, 172
Von Neumann analysis, 3, 7, 38, 39, 42, 43, 79–83, 238, 239, 241
Von Neumann stability condition, 13

W

Wall-bounded vertical bubbly flows, 252
Wall force, 251
  method, 273
  model, 274
Wall lubrication force, 251
Wall shear, 96
Water faucet problem, 7, 22, 57–59, 61, 242–243
  modified, 91, 92
  simulation, 89
  solution, 242
  SSP2-MM solutions, 90
Water–gasoline experiment, 122
Water–kerosene experiment, 93
Wave evolution, 124
Wavy-stratified flow, 34
Well-posed 1D TFM, 108
Well-posed FFM, 157
Well-posed numerical model, 35
Well-posed TFM, 103
Well-posedness, 28

MIX
Papier aus verantwortungsvollen Quellen
Paper from responsible sources
FSC® C105338

If you have any concerns about our products,
you can contact us on
ProductSafety@springernature.com

In case Publisher is established outside the EU,
the EU authorized representative is:
Springer Nature Customer Service Center GmbH
Europaplatz 3, 69115 Heidelberg, Germany

Printed by Libri Plureos GmbH
in Hamburg, Germany